项目资助：
国家自然科学基金项目（32072518）
教育部新农科研究与改革实践项目（2-160）
山东省本科教学改革研究重点项目（Z2020055）

现代生态循环农业实践教程

Modern Ecological Circular Agriculture Practice Tutorial

（中英双语版）

范晶晶　季洪亮　曹　慧　主编　　陈志章　孔雪华　王超然　译

·北京·

图书在版编目（CIP）数据

现代生态循环农业实践教程 = Modern Ecological Circular Agriculture Practice Tutorial：汉文、英文 / 范晶晶，季洪亮，曹慧主编；陈志章，孔雪华，王超然译. —北京：科学技术文献出版社，2024.1
ISBN 978-7-5189-9983-5

Ⅰ.①现… Ⅱ.①范… ②季… ③曹… ④陈… ⑤孔… ⑥王… Ⅲ.①生态农业—教材—汉、英 Ⅳ.① S-0

中国版本图书馆 CIP 数据核字（2022）第 247059 号

现代生态循环农业实践教程（中英双语版）

| 策划编辑：魏宗梅 | 责任编辑：王 培 | 责任校对：王瑞瑞 | 责任出版：张志平 |

出 版 者	科学技术文献出版社
地　　址	北京市复兴路15号　邮编　100038
出 版 部	（010）58882941，58882087（传真）
发 行 部	（010）58882868，58882870（传真）
官方网址	www.stdp.com.cn
发 行 者	科学技术文献出版社发行　全国各地新华书店经销
印 刷 者	北京虎彩文化传播有限公司
版　　次	2024 年 1 月第 1 版　2024 年 1 月第 1 次印刷
开　　本	787×1092　1/16
字　　数	585千
印　　张	25.5
书　　号	ISBN 978-7-5189-9983-5
定　　价	89.00元

版权所有　违法必究

购买本社图书，凡字迹不清、缺页、倒页、脱页者，本社发行部负责调换

编写委员会

主　编　范晶晶　季洪亮　曹　慧

副主编　秦　旭　韩瑞东　王超然　高文胜　王爱国

参　编　高学清　薛　丽　孙福锦　张宏宇　张国伟
　　　　　王可峰　陈福俊　李建林　张新峰　韩　霞
　　　　　刘国栋　李　田　李雪峰

**　　译**　陈志章　孔雪华　王超然

Committee for Writing the Book

Editor–in–chief	Fan Jingjing Ji Hongliang Cao Hui
Deputy Editor	Qin Xu Han Ruidong Wang Chaoran
	Gao Wensheng Wang Aiguo
Coeditor	Gao Xueqing Xue Li Sun Fujin Zhang Hongyu
	Zhang Guowei Wang Kefeng Chen Fujun Li Jianlin
	Zhang Xinfeng Han Xia Liu Guodong Li Tian
	Li Xuefeng
Translator	Chen Zhizhang Kong Xuehua Wang Chaoran

前　言

2015年，"加强生态文明建设，促进经济社会可持续发展，建设美丽中国"被写入"十三五"发展规划。《"十四五"全国农业绿色发展规划》中提出全面贯彻新发展理念，落实构建新发展格局的要求。2022年，中央"一号文件"中，首次将农业农村绿色发展放到了"聚焦产业，促进乡村发展"的议题中，以上政策的实施体现了我国农业农村绿色发展实现了从污染治理到产业发展的阶段性转变。

中国是农业文明古国，也是农业生产大国。但农业发展却面临严峻的资源和环境压力。运用生态循环经济理念指导农业生产、创新发展模式、转变增长方式、实现人与自然的和谐共处，是当前一项重要而紧迫的任务。生态循环农业追求产量、质量、经济效益的高度统一，能够推动整个农业生产进入可持续发展的良性循环轨道，发展生态循环农业是我国农业发展的必然选择。

本书以生态循环农业理论为基础，结合生态循环农业资源利用模式与技术，实现其在现代生态循环农业园区规划设计中的应用，通过"理论+技术+实践"的方式，全面阐述了现代生态循环理论在现代农业中的应用。

本书一共有七章内容，第一章为生态农业发展概述；第二章为生态循环农业理论；第三章主要介绍农业废弃物循环利用；第四章主要介绍生态循环农业资源利用模式；第五章主要介绍现代生态循环农业发展模式与技术；第六章为现代生态循环农业园区规划基本理论；第七章为现代生态循环农业园区规划设计案例。

参与本书编写的人员均为国内从事生态循环农业科研、教学和农业园区规划工作的一线专家、专业技术人员，理论知识深厚，实践经验丰富，能够很好地把握我国生态循环农业的发展趋势。潍坊学院种子与设施农业工程学院为本书的编写提供了大力支持。在此，谨向所有为本书策划和编写提供帮助的单位和个人致以最诚挚的谢意！

由于编者水平有限，书中难免存在错漏之处，欢迎提出建设性建议，以便我们不断地改进。

编者

2023年6月

Preface

In 2015, "strengthening ecological civilization construction, promoting sustainable economic and social development, and building a beautiful China" was written into the 13th Five-Year Development Plan. The National Plan for Green Agricultural Development in the 14th Five-Year Plan puts forward the comprehensive implementation of new development concepts and the requirements for building a new development pattern. In 2022, the No. 1 central document for the first time put the green development of agriculture and rural areas in the "focus on industry to promote rural development" issue, reflecting the green development of agriculture and rural areas in China has achieved a stage change from pollution control to industrial development.

China is an ancient country of agricultural civilization and a big country of agricultural production. However, China's agricultural development is faced with severe pressure of resources and environment. Therefore, it is an important and urgent task to use the concept of ecological circular economy to guide agricultural production, innovate the development mode, transform the growth mode, and realize the harmonious coexistence between man and nature. Ecological circular agriculture not only pays attention to output and economic benefits, but also pursues the high unity of high yield, high quality and high economic benefits, which can promote the whole agricultural production into a sustainable development with virtuous cycle orbit, so developing ecological circular agriculture is the inevitable choice of agricultural development in our country.

This course is based on the theory of ecological circular agriculture, combined with the utilization mode and technology of ecological circular agriculture resources, to realize the application in the planning and design of modern ecological circular agriculture park. Through the way of "theory + technology + practice", the application of modern ecological circular theory in modern agriculture is comprehensively elaborated. There are seven chapters in this book. The first chapter mainly introduces the overview of the development of ecological agriculture; the second chapter is mainly about the theory of ecological cycle agriculture; the third chapter mainly shows the recycling of agricultural waste resources; the fourth chapter mainly introduces the utilization mode of ecological circular agricultural resources; the fifth chapter mainly introduces the development model and technology of modern ecological circular agriculture; the sixth chapter mainly introduces the basic theory of modern ecological circular agricultural park planning; the chapter seven mainly introduces the planning and design of modern ecological circular agriculture park and some cases.

The people participating in this book are domestic universities and scientific research experts engaged in the scientific research, teaching, professional and technical personnel, agricultural park planning, and so on. They have deep theoretical knowledge and rich practical experience, and can well grasp the development trend of ecological circulation agriculture in our country. The College of Seed and Facility Agricultural Engineering of Weifang University provided strong support for the preparation of this course. On the occasion of publication, I would like to extend my sincerest thanks to all the units and individuals who have contributed to the planning and preparation of this tutorial!

Due to the individual limits of the author, it is hard to avoid the existence of improper points and mistakes in this book, you are welcome to make some constructive suggestions for correction and further progress.

Compiler

June 2023

目 录

中文篇

第一章　生态农业发展概述 ·· 3
　　第一节　生态农业概述 ·· 3
　　第二节　生态农业发展理论 ·· 9
　　第三节　国内外生态农业发展状况及启示 ··························· 12

第二章　生态循环农业理论 ·· 19
　　第一节　循环农业 ··· 19
　　第二节　循环农业理论基础 ·· 22
　　第三节　生态循环农业发展概述 ·· 27
　　第四节　国外生态循环农业的主要模式 ······························ 29
　　第五节　我国生态循环农业的几种主要模式 ······················· 32

第三章　农业废弃物循环利用 ··· 34
　　第一节　农业废弃物的概念、来源及分类 ··························· 34
　　第二节　农作物秸秆循环利用 ··· 38
　　第三节　畜禽养殖业废弃物循环利用 ·································· 55

第四章　生态循环农业资源利用模式 ······································· 66
　　第一节　猪—沼—粮循环模式 ··· 66
　　第二节　草—兔—粪—稻循环模式 ····································· 69
　　第三节　草—羊—粪—粮循环模式 ····································· 72
　　第四节　稻—菜—蚓循环模式 ··· 75
　　第五节　菜—鱼共作模式 ·· 79

第五章 现代生态循环农业发展模式与技术 ·················· 82
- 第一节 种养空间立体结构模式与技术 ·················· 82
- 第二节 农业资源节约和物质循环利用系统模式与技术 ·················· 87
- 第三节 农家庭院现代生态循环农业模式与技术 ·················· 89
- 第四节 绿色休闲农业和农业科技（生态）产业园区模式及技术 ·················· 91

第六章 现代生态循环农业园区规划基本理论 ·················· 97
- 第一节 现代生态循环农业园区概念、类型和意义 ·················· 97
- 第二节 现代生态循环农业园区的理论及功能产业定位 ·················· 99
- 第三节 现代生态循环农业园区规划原则和内容 ·················· 103
- 第四节 现代生态循环农业园区综合评价系统 ·················· 116
- 第五节 现代生态循环农业园区规划发展中存在的主要问题及策略 ·················· 122
- 第六节 现代生态循环农业园区管理与运营 ·················· 125

第七章 现代生态循环农业园区规划设计案例 ·················· 142
- 第一节 生态循环理念在农业园区中的应用 ·················· 142
- 第二节 朝日绿源生态循环园区设计 ·················· 144
- 第三节 凤凰生态循环农业示范园方案设计 ·················· 148

参考文献 ·················· 158

英文篇

Chapter 1 Overview of the Development of Ecological Agriculture ·················· 161
- Section 1 Overview of ecological agriculture ·················· 161
- Section 2 Theory of ecological agriculture development ·················· 171
- Section 3 The development and enlightenment of ecological agriculture at home and abroad ·················· 176

Chapter 2   Theory of Ecological Circular Agriculture ·················· 186
  Section 1   Circular agriculture ·················· 186
  Section 2   The theoretical basis of circular agriculture ·················· 191
  Section 3   Overview of the development of ecological circular agriculture ······ 198
  Section 4   The main models of ecological circular agriculture in foreign countries ·················· 202
  Section 5   Several main models of ecological circular agriculture in our country ·················· 205

Chapter 3   Recycling of Agricultural Residues Resources ·················· 208
  Section 1   The concept, source and classification of agricultural residues ····· 208
  Section 2   Crop straw recycling ·················· 214
  Section 3   Recycling of residues from livestock and poultry breeding industry ·················· 239

Chapter 4   Utilization Model of Ecological Circular Agricultural Resources ·················· 254
  Section 1   Pig - biogas - grain circular model ·················· 254
  Section 2   Grass - rabbit - manure - rice circular model ·················· 258
  Section 3   Grass - sheep - manure - grain circular model ·················· 263
  Section 4   Rice - vegetable - earthworm circular model ·················· 268
  Section 5   Vegetable - fish co-cultivation model ·················· 273

Chapter 5   Development Model and Technology of Modern Eological Circular Agriculture ·················· 277
  Section 1   Three-dimensional planting model and technology of farmlands ··· 277
  Section 2   Model and technology of agricultural resources saving and recycling system ·················· 286
  Section 3   Model and technology of green low-carbon circular agriculture in courtyard ·················· 288

Section 4　Model and technology of green leisure agriculture and agricultural sci‑tech (ecological) industrial parks ······················ 292

Chapter 6　Basic Theory of Modern Ecological Circular Agricultural Park Planning ··· 301

Section 1　The concept, types and significance of modern ecological circular agricultural parks ·· 301

Section 2　Theory and functional industry orientation of modern ecological circular agricultural parks ······································ 305

Section 3　Planning principles and contents of modern ecological circular agricultural parks ··· 311

Section 4　Comprehensive evaluation system of modern ecological circular agricultural parks ··· 330

Section 5　Main problems and strategies in the planning and development of modern ecological circular agricultural parks ······················ 340

Section 6　Management and operation of modern ecological circular agricultural parks ·· 345

Chapter 7　Planning & Designing and Case of Modern Ecological Circular Agricultural Parks ··· 371

Section 1　Application of ecological circular concept in agricultural parks ··· 371

Section 2　Designing of Zhaori Lvyuan (Sunrise & Green Source) ecological circular farm ··· 374

Section 3　Designing of Phoenix eco‑circular agricultural demonstration park ··· 379

Bibliography ·· 395

中文篇

第一章　生态农业发展概述

第一节　生态农业概述

一、生态农业发展概念与特征

"生态农业"的概念最早由美国土壤学学者 Willian Albrecht 于 1970 年提出。英国农业学家 M. Worthingter 于 1981 年明确了生态农业的定义，指出生态农业是一种小型农业，经济上具有生命力和活力，生态上做到低输入、自我维持，审美、伦理及环境方面能够为社会所接受。目前我国生态农业的定义是：以保护和改善生态环境为目标，以现代科技和工程管理为基本手段，以生态学及经济学原理为理论基础，在传统农业技术经验上建立的集约经营，具有生态效益的现代农业。

2011 年，由中国产业研究报告网发布的《2011—2015 中国生态农业市场供需预测及投资前景评估报告》对生态农业的概念做了更为完整的定义，报告指出，生态农业是重要的农业发展模式，以保护和改善生态环境为前提，按照生态经济学、生态学的规律，运用现代科技和系统方法，开展集约化生产经营。从国内外生态农业定义对比可以看出，生态农业作为现代农业的主导模式，具有以下基本特征：首先，生态农业属于现代化农业，目标是实现较高的社会效益和生态效益，既做到合理利用资源，增加效益和财富，又有效改善农业生态环境；其次，生态农业可以增加农业产值、提高生产效益，通过对农村的自然—社会—经济融合，对生态系统结构进行改造和调整，同时使废弃物得到有效利用，减少农药和化肥施用量，有效降低农业生产成本。发展生态农业既能防治污染，保护生态环境，保持生态平衡，又能增强农副产品安全性，使农业从常规发展向可持续发展转变，从而使农业的发展后劲十足。

生态农业特征主要体现在以下几个方面：

第一，持续性。生态农业的基本特征就是可持续性，所谓可持续性是指人类自身发展与自然生态环境能力的相互协调，体现在生态可持续、经济可持续、技术可持续和社会发展可持续等 4 个方面。①生态农业的生态可持续，可以使农业生产在现有农业资源的基础上，能够很好地与自然环境相适应，使自然资源得到可持续利用；②生态农业的经济可持续，可以使农、林、渔、牧、加工业等产业之间得到良好协调，从而提高生产效率；③生态农业的技术可持续，可以使农业生产出更多的有机产品，对自然环境不造成危害；④生态农业的社会可持续，可以使人类粗放型的发展方式得到转变，提高食品安全，以人为本，实现人的全面发展。

第二，集约化。"集约"二字主要体现在资本集约、技术集约、劳动集约3个方面，生态农业正好具备了这3种集约。生态农业的资本集约，主要体现在产业化经营，资本向农业生产流进；生态农业的技术集约，主要表现为改变农业生产要素，完善农业生产方法，提高生产效率和产业化经营管理水平；生态农业的劳动集约，最为明显地表现在中国丰富的劳动力资源上，通过推进生态农业的生产经营，使劳动者技术水平和劳动效率得到提高。

第三，高效化。生态农业的高效化主要表现在经济效益和社会效益两个方面：一是经济效益。生态农业有着广阔的市场发展空间，有机产品具有明显的优质、高产、生态、安全的特征，随着"有机"消费者的不断增加，对有机农产品的需求也越来越大。此外，有机产品比传统农产品具有价格优势，因为其安全的特性，使得人们更多地倾向有机农产品。二是社会效益。生态农业提倡的是一种绿色技术革命理念，改善农产品供给结构，满足农产品市场出现的结构性生产过剩问题，从而实现可持续发展。

二、生态农业发展的基本原理

1. 农业生物共生原理

农作物共生是指两种不同种类的生物彼此互利地生活在一起。在农业生产中，妥善安排生物种群时空分布与互补互利关系，充分利用光、热、时、空等条件，建立多层次配置，多种生物共处，采取立体种植、立体养殖或立体种养殖相结合，发展高效持续农业。农作物共生原理主要运用在多维用地方面，即组合农业生物，从垂直、水平、时间、空间上利用土地，发挥土地承载的自然资源的生产潜力。生产模式主要有不同水层养殖、间混套种、立体种植等。长期以来，由于发展中国家片面追求农业经济的快速发展，农业生产过程中，化肥、农药的使用变得越来越广泛普遍，使得粮食、蔬菜、水果和其他农副产品中的有毒成分越来越多，严重地影响了食品安全，也使得人们对绿色生态环保食品变得越来越渴望。生态农业的发展将会从根本上改变这一局面。再者，随着农村生活垃圾产量的增加、种类的增多，不能单纯依靠环境自身的消纳能力来解决垃圾问题，由于缺少主管部门对村镇生活垃圾进行统一管理处置和监管，致使农村生活垃圾的收集、运输、处理工作处于无序状态，生活垃圾不能被及时收运处理；其次是政府资金投入低，导致环卫基础设施薄弱，缺乏必要的生活垃圾收集容器、运输机械和处理设施；再者就是乡村居民环保意识淡薄，意识不到生活垃圾的危害，不懂得怎样分类收集，更谈不上无害化处理。这些问题的存在导致随意倾倒、随意堆放的现象到处存在，不仅造成视觉污染，而且经过风吹雨淋，水流冲刷，造成土壤、水源污染，长此以往对广大群众的身体健康、生命安全带来严重威胁，特别是农村垃圾有机物含量多，农药、化肥等残留的有害性大于城市生活垃圾，部分地区出现的水源污染、食品污染将会产生严重的后果。因此尽快加强农业生态环境保护，大力推进生态农业发展必将成为现代农业可持续发展的必然要求。也可以说，推进生态农业将改善农村居民生存环境，影响和改变其生活方式，更好地提高人们的生活质量。

2. 农业多功能原理

20 世纪 80 年代末，日本政府在"稻米文化"中首次提出农业多功能性。90 年代初，《21 世纪议程》在联合国环境与发展大会上形成并通过，标志着农业多功能性提法被正式采用。1996 年的《罗马宣言和行动计划》中明确指出，承认和利用农业多功能性。20 世纪 90 年代末，欧盟在探索农业发展新路径时，提出了以农业多功能性为核心理论基础的"欧盟农业模式"。1999 年，联合国粮农组织（FAO）在召开的国际会议中及日本在《粮食·农业·农村基本法》中均强调农业具有多种功能，即除具有经济功能外，还同时具有政治、社会和生态功能。2005 年，世贸组织在香港召开的第六次部长级会议就新一轮农产品多边贸易问题进行谈判，随后，农业多功能性问题成为学术界争论的焦点。

农业多功能性就是农业产业除了具有提供农副产品的基本功能外，还能为人类提供政治、社会、环境等方面的功能，且各功能又表现为多种分功能，各功能互相联系、互相制约、互相促进，共同形成多功能有机系统。目前，改善生态环境、保护生物多样性、保障国家安全、实现农村持续稳定健康发展等非经济功能成为农业多功能性的重要内容。农业多功能性理论是在经济发展过程中，农业产业功能不断变换的背景下提出的。农业生产活动是人类生存与发展的基础，因为它提供其他产业所不能够生产的农产品。随着人类从事农业生产活动产出效率的变化，农业在国民经济发展中的地位与作用在不断地改变，社会经济发展阶段不断提高致使农业产业功能的多样性也日益增多。按照社会经济发展的阶段理论，在农业经济时代，农业产业的主要功能是为人类提供农产品；工业经济时代，则是提高农产品质量；后工业经济时代，转为改善人类生存与发展的环境。从农业多功能性理论的政策含义来看，由于商品性产出和非商品性产出存在技术上的联系，使它们容易成为联合产品。农业生产的非商品产出供给的变动是由通过农产品自由贸易量增大引起的农产品供给数量的变化而被影响，间接地对进出口农产品国家的福利水平造成一定的影响。反之，影响农产品进出口国的福利作用是截然相反的。鉴于此，农业多功能性理论是农业产业功能不断变换的产物，农业多功能性理论为农业寻求保护合理化提供理论支撑。

3. 新公共管理理论

新公共管理运动是 20 世纪 80 年代后提出的新的公共行政理论和管理模式，曾兴盛于英美等西方国家，也是近年来西方行政改革的主要指导思想之一。现代经济学、私营企业管理理论和方法是其理论基础,例如,从"理性人"的假设中获得绩效管理的依据;依据公共选择理论和交易成本理论,政府应以市场或客户为导向,提高服务效率和质量;依据成本—效益分析方法,使政府绩效目标得以界定、测量和评估;学习民营企业绩效管理、目标管理、组织发展、人力资源开发和其他的管理方法。此外,在公共选择和交易成本理论与新管理主义理论的基础上,新公共管理理论开始向不同的方向发展。弗里德曼和哈耶克提出"小政府理论",指出政府主要活动内容是提供市场无法做好的服务,即提供的公共产品和服务具有非排他性,管辖的空间范围应该缩小。哈默和钱皮发展出"流程再造"理论,主要是针对官僚体制,强调对官僚制进行重新改造和超越,以客户的需求和满意度为目标,建立一个新的过程型组织结构,使组织的成本、质量、服务等方面得到巨大改善。霍哲从政府绩效着手,将绩效评估看作改善绩效管理的方法,设计了一套具体的绩效评估流程,并强调在绩效评估过程中要使公民广泛参与,因为只有这样的评估结果和信息才会对政府政策和项目管理有更大的意义。霍哲还研究了"基于回应性政府全面质量管理"理论,主要内容是以客户为中心,持续改进,以授权和协作为基础,进行全面质量管理。奥斯本和盖布勒提出"重塑政府"理论,希望将政府塑造成起催化作用、竞争性、有使命、讲究效果、受顾客驱使、有事业心、有预见、分权、以市场为导向的政府。

三、生态农业的优势

农业是我国国民经济发展的基础。农业的发展是关系到国计民生的大事。农业生态模式是解决农村经济发展与环境问题的必经之路。相比传统农业产业化模式,生态农业具有以下优势:一是高效益的生态农业,为我国传统农业的现代化提供了一条可持续发展的途径。生态农业使人们把发展的思路拓展到关注人—地—人和谐共存的更广阔的背景之中,同时也与长期以来农民渴望脱贫致富的愿望相契合。基于"天时、地利、人和"的新型生态农业,将成为我国传统农业向高精尖、高附加值深度开发转移的农业现代化主流方向之一。二是生态农业旅游开发,这将成为国家宏观经济调整时期社会资金寻求新投资领域的必然选择,也将成为新的经济增长点。生态农业旅游凭借其开发项目的农业特色,直接受到国家投资政策的优惠。

四、生态农业发展的重要意义

从我国农业的发展历程来看,发展生态农业产业对于提高我国农村社会生产力,促进社会化大生产,优化农村产业结构,提高我国农业综合竞争力,增加农民收入水平,进一步深化我国对外开放都具有较为深远的意义。

发展生态农业产业可以进一步提高我国农村社会生产力。经过改革和开放的实践，我国农村社会生产力有了一定的发展，但同时我们应当认识到，我国农业目前这种小规模经营的格局，造成农业吸纳新技术的边际成本高，使一般农户通过采用新技术、发展高效农业来实现增加收益的动力不足；同时，由于小规模经营使土地分割过于分散，也难以推广使用大型机械和农业新技术。如何在稳定家庭承包经营的基础上，实现由传统农业向现代农业的跨越，是新形势下农业发展面临的重大课题。生态农业产业化经营作为社会化大生产的一种组织形式，以国内外市场为导向，以提高农业综合效益为中心，以龙头企业为依托，以农副产品生产基地为基础，对农业的各个产业或产品，实行区域化布局、专业化生产、企业化管理、社会化服务、一体化经营，通过组合建立专业生产联合体、合作社或大规模的农产品生产基地，就可以使过去分散的小农户通过产业化经营组织，在不改变原有家庭承包经营的方式下，探索出一条在符合中国小规模家庭经营基础上来提高农业整体规模效益的新途径。这种组织形式和经营方式，不改变农户的土地关系，改善了农业的投入机制和生产方式，可以使大量的工商资本和先进的技术、农艺措施、现代装备等有机地融入农业生产经营的各个环节，促进农业劳动生产率和农民素质的提高，加快传统农业向现代农业过渡的步伐。

发展生态农业产业可以优化农村经济结构，带动农业结构实现战略性调整。进入 21 世纪以后，我国的农业结构开始由适应性调整转为战略性调整。随着世界经济一体化进程的加快，我国农业已经融入了世界农业之中，成为全球农业结构的一个重要组成部分。因此，我国农业结构的调整，不能仅仅局限于我国经济发展状况，必须充分考虑到全球农业及世界经济的分工与合作。农业结构的调整是对农产品的品种、质量、区域布局和产后加工转化等进行全面调整的过程，也是加快农业科技进步、提高劳动者素质、转变农业增长方式、促进农业向深度进军的过程。这种结构调整，既不能用过去那种行政命令的方式，又不能单纯依靠市场的自发调节，必须通过发展农业产业化经营，对农业和农村经济结构的各个层面、各个环节进行调整和优化。通过推进生态农业产业化经营，一方面，可以加快农业结构调整步伐，在龙头企业的带动下，加快优势产业向优势区域集中，加快主导产业膨胀规模，加快品种结构优化升级；另一方面，可以加快农村经济结构调整步伐，通过大力发展农产品加工运销业，促进生产要素在城乡之间双向流动，提高第二、第三产业在农村经济中的比重，推进农村工业化进程。通过完善产业化利益联结机制，进行规模化生产经营，使广大农户能够按照市场需求，避免分散农户经营自发适应调整结构所带来的弊端，同时实现规模经济效益，实现"政府调控市场、市场引导企业、企业带动农户"的产业结构调整新机制。另外，深入推进农业产业化经营，可以促进小城镇建设，把农产品的加工、运销及相关服务业向小城镇聚集，扩大城镇规模，增强吸纳农村剩余劳动力的能力，加快农民向城镇转移的步伐，带动农村就业结构的调整。

发展生态农业有利于我国农业综合竞争力的提高。市场经济是高度竞争的经济，也是市场、产品、价格、技术和利益等各个要素的竞争。由于特殊的历史原因，我国封闭的，自给、半自给的自然经济经历了相当长的时期。直到20世纪70年代，党的十一届三中全会才制定了改革、开放、搞活的政策，我国才逐步开始向有计划的商品经济发展。此后，党的十四大又进一步明确了社会主义市场经济的发展方向。现在国际市场的竞争既有产品质量的竞争，又有贸易双方经济实力和规模的竞争。农户是我国农业生产长期不可改变的微观基础，也是我国农业参与国际竞争的微观主体。全国2.2亿多农户，经营规模小，组织化程度低，经济实力弱，难以适应加入世贸组织之后农业面临的国际国内两个市场的竞争。分散的家庭经营既不能保证产品质量，更没有与国际大财团、大商社竞争的实力。在这种形势下，发展产业化经营，更有利于应对经济全球化和加入世贸组织对我国农业的挑战，使我国农业更迅速地融入世界农业之中。通过发展农业产业化经营，形成一批有规模、有实力的市场主体，并通过他们把分散的农户组织起来，按国际市场的要求，实行标准化、规模化生产，尽快扩大我国有比较优势的农产品生产规模，提高精深加工水平，创造一批有较强竞争力的名牌产品和名牌企业。另外，实施农业国际化战略，也需要提高我国农业产业化经营水平，依靠产业化经营组织把国外资金、技术、设备、管理经验引进来，进一步发展壮大我们的实力，增强国际竞争能力。

发展生态农业产业可以提高农业比较效益，大幅度增加农民收入。当前我国国民经济发展的突出矛盾是农民收入增长缓慢，城乡收入差距越拉越大，这也是"三农"问题的关键所在。制约农民收入增长的原因很多，如农产品难卖、销售手段落后、市场化程度低，农产品生产链条短、深度加工不够、附加值低，农民收入渠道窄、对土地的依赖程度较高等。由于农业在相当程度上是社会效益高、经济效益低的弱势产业，在比较效益规律的作用下，资金、技术、人才等生产要素必然向比较效益高的非农产业流动，致使农业处于投入严重不足、发展后劲乏力的困境，这是市场经济条件下农业发展的难点所在。同时，由于城镇就业岗位受限，多余的农村劳动力不可能全部转移到城镇。深入推进农业产业化经营，不仅可以帮助农民解决农产品难卖的问题，还能保证卖个好价钱，减少农民上市销售产品的成本。另外，通过发展农产品加工运销业，将农村第一、第二、第三产业融为一体，为农民开辟新的就业领域，拉动农民向第二、第三产业转移；通过规模经营和多层次加工使农产品实现多次增值，提高农业附加值和综合效益，拓宽农民增收渠道。

第二节 生态农业发展理论

一、生态农业的产生背景与发展

在生态经济产生的大背景下，相对于石油农业发展带来的一系列问题，生态农业从产生到发展备受关注和重视。生态农业最早可追溯至20世纪30年代的欧洲，并于30—40年代在英国、瑞士、日本等国得到发展。自30年代初英国农学家A.霍华德提出有机农业概念并相应组织试验和推广以来，有机农业在英国得到了广泛发展。在美国，罗代尔（J. I. Rodale）是最早实践者，他于1942年创办了第一家有机农场，1974年在扩大农场和总结研究经验的基础上成立了罗代尔研究所，成为美国和世界上从事有机农业研究的著名研究所。但当时的生态农业多限于传统农业自我封闭式的生物循环生产模式，发展极为缓慢。

到了20世纪60年代，欧洲的许多农场转向生态耕作。70年代后，一些发达国家（尤其是美、欧、日）伴随着工业化的发展，环境污染日益严重，已危及人类的生命与健康。为确保人类生活质量和经济健康发展，掀起了以保护农业生态环境为主的各种替代农业思潮。1971年，美国密苏里大学教授阿布拉齐（W. Abrecht）提出了生态农业思想。1972年召开的联合国人类环境会议提出了社会、经济的"生态发展"战略。1974年，美国生态学家赫钦逊（G. E. Huchingson）等出版的论文集《生物圈》，提出了农业生态经济观点。1981年，英国环境科学家华盛顿（M. K. Washington）出版了《农业与环境》一书，把种植业、放牧业、饲养业、自然资源保护、家庭小型产品加工和产品销售、美化环境等视为一个农业生态经济系统，从宏观上进一步论证了发展生态农业的战略意义。这一时期，法国、德国、荷兰等西欧发达国家也相继开展了有机农业运动，并在法国成立了国际有机农业运动联盟（IFOAM）。日本生态农业的重点是减少农用热碱化，减少农业面源污染（农药、化肥），提高农产品品质安全。菲律宾是开展生态农业起步早、发展快的亚洲国家之一，最具世界影响的生态农业范例是玛雅（Maya）农场。这个时期，生态农业的发展引起了世界各国的广泛关注，无论是发达国家还是发展中国家，普遍认为生态农业是农业可持续发展的重要途径之一。

20世纪90年代后，世界各国生态农业有了较大发展。建设生态农业，走可持续发展的道路已成为世界各国农业发展的共同选择。特别是进入21世纪以来，可持续发展战略成为全球的共同行动，生态农业作为可持续发展的一种重要实践模式，进入了一个蓬勃发展的新时期，无论是在规模、速度还是在水平上都有了质的变化，由单一、分散、自发的民间活动转向政府自觉倡导的全球性生产运动。

二、生态农业的内涵与特点

1. 生态农业的内涵

为了正确认识生态农业，我们首先要对生态农业有个简要认识。关于生态农业的含义或概念有不同解释，如英国农学家 M. Worthington 对生态农业的定义是：生态上能自我维持、低输入，经济上有生命力，在环境、伦理和审美方面可接受的小型农业。

生态农业是按照生态学原理和经济学原理，运用现代科学技术成果和现代管理手段，以及传统农业的有效经验建立起来的，能获得较高的经济效益、生态效益和社会效益的现代化农业。它要求把发展粮食与多种经济作物生产，发展大田种植与林、牧、副、渔业，发展大农业与第二、第三产业结合起来，利用传统农业精华和现代科技成果，通过人工设计生态工程，协调发展与环境之间、资源利用与保护之间的矛盾，形成生态上与经济上两个良性循环，实现经济、生态、社会三大效益的统一。

生态农业是指在保护、改善农业生态环境的前提下，遵循生态学、生态经济规律，运用系统工程方法和现代科学技术，集约化经营的农业发展模式。生态农业是根据生态学与生态经济的原理，运用系统工程及现代科技方法组建起来的综合农业生产体系。它是以生态学理论为依据，在某一特定的区域内，因地制宜地规划、组织和进行农业生产。

生态农业是一个系统，是把生态系统和农业生产系统融为一体的更大复合系统，它不是各个要素的简单相加或机械堆积，而是相互联系、相互作用、相互制约、相互促进的有机整体。它以生态学理论为指导，按照农业生态系统内物种共生、物质循环、能量多层次利用的生态学原理，合理高效利用农业自然资源和保护良好的生态环境，应用农业高新技术，吸收传统农业的精华，采用现代管理手段，科学组织经营农业生产，实现生态自我维持和动态平衡、农业可持续发展，达到生态效益、经济效益、社会效益的高效统一。

2. 生态农业的特点

① 综合性。生态农业强调发挥农业生态系统的整体功能，以大农业为出发点，按照"整体、协调、循环、再生"的原则，全面规划、调整和优化农业结构，使农、林、牧、副、渔各业和农村第一、第二、第三产业综合发展，并使各产业之间互相支持，相得益彰，提高综合生产能力。

② 多样性。生态农业针对各地自然条件、资源基础、经济与社会发展水平的差异性，充分吸收传统农业精华，应用现代科学技术，以多种生态模式、生态工程和丰富多彩的技术类型装备进行农业生产，使各区域扬长避短，发挥优势，同时根据社会需求与当地实际相结合，发展各具特色的产业。

③ 高效性。生态农业通过物质循环、能量多层次综合利用和系列化深加工，实现经济增值，实行废弃物资源化利用，降低农业成本、提高效益，为农村大量剩余劳动力创造农业内部就业机会，保护农民从事农业的积极性。

④ 持续性。发展生态农业能够保护和改善生态环境、防治污染、维护生态平衡、提高农产品的安全性，使农业和农村经济的常规发展为可持续发展，把环境建设同经济发展紧密结合起来，在最大限度地满足人们对农产品日益增长的需求的同时，提高生态系统的稳定性和持续性，增强农业发展后劲。

三、生态农业的基本原理

生态农业是生态学在农业上应用的分支。研究生态农业的学科或科学是农业生态学。农业生态学的概念由3个词干组成，即来源于拉丁语的agrarius（指田地），希腊语的oikos（意为住所或家政）、logos（科学），根据上述海克尔关于生态学的定义，农业生态学可以定义为：人类为某些作物的生产所塑造的环境中生物生存条件的科学。美国科学家小米勒总结农业生态学为三定律：第一定律，我们的任何行动都不是孤立的，对自然界的任何侵犯都具有无数的效应，其中许多是不可预料的。这一定律是G. 哈定（G. Hardin）提出的，可称为多效应原理。第二定律，每一事物无不与其他事物相互联系和相互交融，此定律又称相互联系原理。第三定律，我们所生产的任何物质均不应对地球上自然的生物地球化学循环有任何干扰，此定律可称为勿干扰原理。农业生态学基本原理的应用思路是：模仿自然生态系统的生物生产能量流动、物质循环和信息传递而建立起人类社会组织，以自然能流为主，尽量减少人工附加能源，寻求以尽量小的消耗产生最大的综合效益。

生态农业按照农业生态学的基本原理进行"整体、协调、循环、再生"运行。其原理具体反映在以下几个方面：

① 生物间相互制约原理。即按食物链合理组织生产，才能挖掘资源潜力，提高效益。

② 生物与环境协同进化原理。即因地因时制宜，合理布局，合理轮作倒茬，立体种植，种养结合，取得显著效益。

③ 能量多级利用与物质循环再生原理。即进行多层次分级利用，使有机废物资源化，使光合产物实现再增殖，发挥减少污染、补充肥源的作用，同时提高综合效益。

④ 结构稳定性与功能协调性原理。遵循充分发挥生物共生优势原则，利用生物相克趋利避害的原则，利用生物相互依存的原则。

⑤ 生态效益与经济效益统一的原理。遵循资源合理配置原则、劳动资源充分利用原则、农业结构与经济结构合理化原则、专业化与社会化原则。

四、生态农业理论是绿色低碳循环农业的重要理论基础

根据上述对生态农业理论的基本认识，生态农业的基本原理是物种共生、物质循环、能量多层次利用，实现生态良性循环。实践证明，生态农业是在现有条件下兼顾经济、生态、社会效益，实现生态环境保护、资源培育和高效利用的成功模式，是解决中国农村人口过剩、经济发展与资源和环境之间矛盾的有效途径，是农业和农村经济可持续发展的必然选择。绿色生态循环农业发展同样离不开生态农业的发展原理，也要实现物质循环、能量多层次利用、生态良性循环。生态农业的发展模式与发展目标同样是生态循环农业发展的重要模式和所追求的目标，因此，生态农业的基本原理与发展思路也是生态循环农业发展的重要理论依据。

第三节 国内外生态农业发展状况及启示

一、国外生态农业主要模式

从20世纪中后期开始，以欧美等发达国家为主对世界许多地区的生态农业发展模式进行了探索实践，逐步取代高消耗高投入的农业运作方式，建立与生态环境和谐共处的农业生产模式，并形成良好的体系，有许多值得借鉴学习的地方。

1. 环保型生态农业发展模式

（1）模式的基本特点

以环境保护为主要目标的生态农业发展模式在美国的农业发展模式中得到了较为成功的体现。从20世纪70年代开始，传统的以高成本、高度专业化、高度化学集约化、高产量、高污染为特征的现代石油农业面临着前所未有挑战，在这种情况下，美国推出了生物农业、再生农业、绿色农业等运行模式，企图解决农业发展与资源环境之间日益加剧的矛盾，但都未能达到预期的目标。因此，美国推出了低投入和高效率生态农业发展模式。低投入生态农业发展的主要目标是保护农业生态环境，并通过法律法规使其予以实践，力求降低成本，尽可能减少化肥、农药等外部化学试剂的使用，围绕农业自然生产特性和农业内部资源的管理来保护和改善生态环境。这种模式的生态农业发展效果较好，但生产效率相对较低，主要包括合理轮作模式、种植业与畜牧业综合经营模式、病虫害的综合防治模式、利用农场内部有机肥土壤管理模式。随后，美国提出了高效率生态农业发展模式，注重对农业生产各个环节的科学管理，强化农业的生态原则，关键在于依靠科技进步促进农业生产效率的提高，减少化学试剂对于环境的污染，保护生态环境，创造良好的物质条件。目前用于农业生产的发达技术包括机械化技术、化学技术、生物技术、信息技术等。政府在宏观调控、政策法

规上也起了巨大的推动作用。这种模式主要是一些人多地少、自然资源非常紧缺的国家，在面临现在农业发展中的许多环境问题时所提出的。

（2）主要案例

该模式典型的成功发展模式在日本。20世纪80年代开始，日本正式提出了"绿色资源的维护与培养"原则，强调对资源的合理利用和有效保护环境，选择了环境保护型农业发展路径。在国土规划的指导和政策支持下，日本推行资源的永续利用与环保同生产率密切结合的措施，坚持农业的生态发展，主要的发展方向为生态农业和精准农业。该模式要求降低农场外部条件，如化肥、农药等的投入，以保护环境，防止对土地资源的破坏，提高耕地肥力；利用生物育种技术培育可栽植于盐碱化和荒漠化等环境不适宜地区的作物，扩大耕地面积，改善资源的不足；重视农业生产的高效率，强调农、林、牧、渔产业结构的比例与区域农业特点相结合，提高效率；对农业资源进行效益评价测算，强调森林对于物种多样性和净化空气等的作用，加强绿色资源的保护。同时，日本政府对于环境保护高度重视，从立法、政策、信贷宣传上积极配合生态农业的发展，形成了良好的发展氛围。在技术措施及产品质量安全上，日本推行大量的政策措施，在管理上切实可行地加大监管力度，并制定《生态农业发展农业法》。

2. 综合型生态农业发展模式

（1）模式的基本特点

综合型生态农业发展模式，是指能够满足人类生存需要，同时不破坏农业自然环境的生产经营方式。德国作为世界上生产和使用化肥、农药最多的国家之一，在农业生产上保持了高产，同时带来了对自然环境的严重破坏、产品质量问题及生产过剩问题。在这种情况下，德国提出了综合型生态农业发展模式，主要包括：实施综合性农业规划的同时不破坏自然环境，与生态系统的发展规律相平衡；综合植物保护和水资源保护，综合经营农业等。

（2）主要案例

该模式以荷兰为代表，荷兰是典型的人多地少的国家，因此十分重视自身资源特点优势的利用开发，发展高效益的农业。近年来，荷兰政府非常重视生态农业发展战略的推广实施，与自身的特色优势相结合，提高农业的效益。其中以花卉产业的发展为代表。荷兰的农业由种植业和畜牧业组成，其中种植业分为大田种植业和园艺业两大类。就大田种植业、园艺业和畜牧业3个层次讲，荷兰农业的结构是：畜牧业，份额超过了55%；大田种植业，所占份额为10%；其余为园艺业。荷兰的农业结构表现出了明显的专业化格局。荷兰农业分成3个主要的生产带，西部沿海是园艺生产带，中部为奶牛生产带，东部和南部主要为集约化的畜牧生产带；大部分农户实行专一化的生产模式，这大大提高了农业劳动生产率和管理水平，同时提高了农产品市场竞争力，成为农产品出口大国。

3. 集约型生态农业发展模式

（1）模式的基本特点

德国通过制定相关法律法规和政策确保综合农业的实施，如废弃物排放规定、化肥使用规定、环境污染处罚法规、再生资源开发等奖惩制度等，同时农牧结合，种植方式轮作，重视对农民、年轻农业劳动者的培训。相关法律规定开办或继承农业企业必须接受培训深造，获得相应资格。对农业从业者的教育有严格的规定，并制定了农业教育新计划，强调全面更新农民的专业知识与技能。为培养高级专门人才，政府还创办了多所农村业余大学，提高农民的素质。

（2）主要案例

法国主要采用集约化和现代化的农业发展模式，但是对环境的影响很大，负面作用日益严重，因此，法国采取了环境保护制度和农田休耕制度。以色列由于大部分国土属于干旱或半干旱地区，因此大力发展节水型生态农业发展模式，改变原始的粗放耕作模式，采用精细的耕作方式，依靠现代科技，采用高效的灌溉技术，实现了自给自足，国家政策给予了大力支持，还推行了公司与农户合作的发展模式。韩国采用了亲环境的发展模式，政府制定了极大的鼓励措施。印度也发展了低成本、高能效的发展农业模式，颁布了相关法律法规，将生态农业模式落到实处。墨西哥也高度重视农业科学技术应用于农村的综合发展。综合以上所述，在选择生态农业模式时，发达国家在拥有强大现代化生产系统的基础上，注重从本国的国情、生态、能源3方面综合考虑；而发展中国家农业重点在于增长与发展，但为避免走西方老路，在注重发展的同时也应注重保护生态环境。

二、国内生态农业主要模式

1. 契约化农业发展模式

2000年以后，重庆市的生态农业逐步发展起来。重庆市农业科学院等研究机构针对观赏蔬菜存在的问题，采取"试验+示范+推广"的方案，培育出适合重庆市的观赏蔬菜80余种，种类涵盖了盆栽、庭院栽培、鉴玩、观赏等，在开发多种观赏蔬菜品种的同时，编制以上观赏蔬菜的配套养护说明，不断提高观赏蔬菜品种的适应性。在推广过程中，重庆市针对消费者的需求差异，优化生态农业结构，提高蔬菜的生产培育技术，大大增加了农民的收入，社会经济效益明显。重庆市于2008年通过政府扶持，引进观赏果蔬76种，新建果蔬圃0.33公顷、观赏果蔬试验及种苗繁育基地近1.7公顷。为将观赏蔬菜的信息有效地利用起来，以便形成查询系统，重庆市搭建了观赏果蔬信息网络平台，健全了信息数据库，通过再次细致的筛选，推出了18类适合本地栽培的观赏蔬菜，建设2个规模化的观赏蔬菜示范基地，创造了巨大的经济效益、社会效益和生态效益。重庆市现已开发的观赏蔬菜主要是观叶类、果蔬类及赏花类。观叶类蔬菜是指观赏植物的叶形雅致、叶的色彩鲜艳明亮，比如紫甘蓝、薄荷、芦笋、香芹、玉丝菜、羽衣甘蓝、七彩菠菜、红甜菜、银丝菜、叶用辣椒、彩叶莴苣、苦苣等。果蔬类是指观赏蔬菜的果实大小、形状、色彩别致新颖，吸引顾客，例如，在蔬菜色彩上，有观赏茄子、五彩椒、鲜红辣椒等；在形状上，有飞碟状的南瓜、多角丝瓜等。赏花类主要包括食用鳞茎的百合、食用花蕾的黄花菜与锦鸡儿属植物——红花菜，以及从国外引进的彩色花椰菜和翡翠塔花椰菜。如今，重庆市观赏蔬菜的用途多为都市家居生活、城市园林绿化、乡村景观美化、节日时尚礼品、现代盆栽小景等。重庆市在引进国外品种时，不断进行品种改良，将国外观赏蔬菜与本土蔬菜进行杂交，培育自有品牌，通过大范围搜集野生资源及优质种质资源，经过长期反复试验、驯化，对其采取抗逆性和适应性监测、观赏性评价，筛选出最优质的种质资源，利用现代先进科学技术，组织培养快速繁殖，进行深入推广，其本质就是通过将种质优良的植物外源基因移植到原有的观赏蔬菜中，优化其综合适应性。重庆市发展生态农业的主要举措为：第一，加大推广力度，提高农民参与意愿，扩大产业规模；第二，通过优化产品结构，提高产品培育技术，提高农民收入；第三，政府在各个方面提供支持，促进生态农业的发展，具体如引进新品种、建造信息平台和建立示范基地等。

2. 品牌带动农业发展模式

实施农业品牌战略，打造安全品牌。例如，为了进一步提高潍县萝卜的产业效益，山东省潍坊市开发研究了贮藏保鲜技术，提高深加工水平，努力延长萝卜的供应周期，充分发挥出贮旺补淡的作用；重视采后处理工作，做好萝卜的包装，由政府出面，正确引导各个加工企业引进国内外新技术，更新设备，加强潍县萝卜的加工技术创新力度，开发出更多种类的萝卜系列保健及美容产品，如萝卜干、萝卜果脯、萝卜酱菜等广受人们喜爱的食品，提高萝卜类产品的附加价值，找出更多能够创造经济效益的发展空间；在寒亭区建立关于农产品方面的"110"式科技信息服务网络平台，以获得更多的产销信息；利用现有的信息网络技术，充分发挥营销协会的作用，设立信息中心，与全国各地大型的蔬菜批发市场进行信息联络，做好信息的收集、分析、发布等各项工作，在最短的时间内将国内外蔬菜市场的实际情况真实地反映出来；政府出面鼓励各个行业协会建立健全信息沟通设备，开发潍县萝卜产销网站，通过"网上订单"实现销售，拓宽销售渠道，建立配送服务中心，有效地缩短萝卜销售中间环节，提高潍县萝卜的经济效益。传统的潍县萝卜主要是各个农户自行种植自行销售，大多都是以零售的形式存在于市场上的，极大地限制了萝卜的生产规模。由于外地很多大的批发商家和零售商家对潍县萝卜的需求量不断增大，小规模的零散种植很难满足市场需求，因此，在寒亭区政府的引导下，按照群众自愿的原则，自主形成集生产、加工和销售于一体的萝卜种植、加工和销售组织。在比较集中的种植区域，通过建立潍县萝卜合作社，组织萝卜的种植生产，为农户开展技术培训教育，并做好各个农户种植的质量监督工作，加强产品的升级和包装，及时联系买家进行销售，解决了农户在生产过程中遇到的种种技术困难，同时提高了农户的经济效益。寒亭区的科研单位也与潍县萝卜合作社进行合作，致力于绿色萝卜的开发和培养工作，又进一步地提高了萝卜产品的品质。与此同时，寒亭区相继建立了30余家萝卜销售点以满足本地市场的需求；并且为了满足外地市场对萝卜的需求，分别在山东省济南、青岛及北京市建立了销售超市，极大地推动了潍县萝卜的持续生产。

品牌是能够帮助产品快速占有市场并稳步扩大市场份额的有效通行证。近几年，我国农产品市场的竞争不断加大，只有走品牌路线，充分发挥出潍县萝卜名优特产的优势，尽可能建立健全产业化经营模式，才能够获得更大的效益。因此，潍县萝卜还应该进一步加大宣传力度，充分发挥出其具有的品牌优势。由于潍县萝卜目前产品的质量有着比较明显的差别，寒亭区政府强调要走出自己的品牌路线，鼓励商标注册，并对绿色萝卜的认证、包装箱的质量等都做出了严格的检测规定。对于已经完成商标注册的萝卜品牌，不管是生产基地、管理设施，还是其他的包装质量等，都提出了严格的检测标准，不仅要确保萝卜具有高质量，更要发挥出其在价格上的优势。下一步，寒亭区还需要带领农户进一步开发萝卜销售市场，树立品牌形象，充分发挥出名优特产的效应，加大经营规模，提高潍县萝卜的竞争力。

三、国内外生态农业发展启示

1. 政府提供政策、法律和资金支持

作为生态农业的重要主体，政府在推动生态农业发展中起到了关键的作用，其职能主要表现在调控职能和管理职能。一是制定实施相关的政策和法律法规，重点涉及农业旅游、规划、生态、安全等，如新加坡的农业科技园规划、德国颁布的《市民农园法》等。二是对生态农业区的规划和建设提供资金支持。相对于政策延迟效应而言，资金支持则是立竿见影，大量资金的投入，在很大程度上缓解了生态农业经营者的压力，为促进产业的持续发展提供了动力。财政支持表现在多个方面，例如，德国、荷兰等欧洲国家每年都为观光农业提供大量财政补贴，还拿出专项资金进行农业科研，间接促进农业的发展；新加坡政府直接出资参与农业科技园的建设；中国重庆市政府则从多方面多角度对生态农业提供支持。

2. 非政府组织提供协调保障

在发达国家的生态农业中，政府虽然发挥着主导作用，但与此同时，行业协会和社会团体也起到不可忽视的作用。非政府组织的主要职能为第三方监督和提供服务，通过监督实现生态农业的行业自律和规范管理。特别要提的是行业协会，许多生态农业较为成熟的国家都设立了不同形式的行业协会，非政府组织中的行业协会在保障生态农业长期发展过程中功不可没。行业协会的职责是根据当地生态农业的特色，制定行业规范，管理企业行为。由于其是从产业内部为出发点，因此行业协会能够有效地促进生态农业良性发展。德国、荷兰都设立了相应的行业协会，其他欧洲国家，如法国、意大利，也有"法国农家旅社网""农业与旅游全国协会"等非政府组织。国内许多城市在促进生态农业协会方面做出了努力，借鉴国内外农业协会的组建形式和基本内容，创建了多个生态农业的协会组织，为进一步实现国际化接轨提供了契机。这些组织发挥着组织服务功能，为维护生态农业产业化的健康有序发展发挥了巨大的作用。

3. 注重观光农业产品的特色

观光农业是具有鲜明地域特色的旅游产品，但同时也是旅游市场休闲娱乐的产品，具有较大的替代性。观赏蔬菜业秉承的是可持续发展理念，必须突出自然环境的绿色生态特性。目前，大部分地区的观赏蔬菜大众化现象突出，雷同产品居多，没有体现出本地的特色。因此，要求对旅游产品进行深入挖掘，也要对游客的心理需求进行深入调查和分析。许多国家的做法是在细分消费市场的基础上，开发多种类型、多个档次的观赏蔬菜产品，产业层次涵盖了近郊及远郊，价格由高到低，层次性明显。比如，在城市近郊采取的是园区式大众文化观赏蔬菜示范园，远郊则采取个性突出、品质较高、特色鲜明的乡村型旅游产品。"乡土性"是指区别于城市文化的乡村文化，"乡土性"越浓厚，就越能够吸引生活环境差异大的游客，且差异越大，则对游客的吸引力越强。国际上许多国家在发展观光农业的同时融入当地文化特色，设计独特的文化旅游主题，这种例子不胜枚举，例如，德国的观光农业就非常注重自然特色和人文特色，具有鲜明的特点；荷兰的果蔬彩车展非常具有本国特色。这些依托本国资源开发的观光农业项目，赋予了丰富的乡村文化内涵，极大地提升了观赏蔬菜产品的品位。从这一层次上理解，观光农业不仅是促进当地经济发展的手段，同时也是本地文化宣传的平台；站在传统文化承袭角度，更是对民族文化、民俗文化的继承和保护。将地域文化作为旅游产品，对于境外游客来说有着强大的吸引力，经营者通过文化活动策划，每年可获得可观的收入。

4. 加强对观光农业的推广宣传

观光农业不同于单一的第一产业和第三产业，而是一种农村人文与自然景观相结合的体验活动。这需要农村社区居民具有共同的意识，需要农村社区居民创造一个有助于观光农业发展的社会氛围，因此需要更多的推广和宣传。要鼓励社区成员积极参加，将观光农业作为带动经济的动力，发展观光农业合作社，加强社区居民与游客的沟通交流。否则，即使具备了地域文化和乡土特色，当地观光农业也难以发展起来，更不会实现观光农业生态性和体验性的预期效果，社区居民的参与性和积极性低还将导致产业规模无法扩大，难以形成规模效应。各国的主要做法是组织当地农民组建观光农业合作社和联合会，不仅能完善产业的发展，而且为经营者提供一系列服务，以此提升经营者参与的积极性，在参与意识上保证社区居民对产业构建的认同，逐渐形成产业集群，加深经营者、农村居民、游客的日常沟通。农村社区居民对观光农业的参与及认可，有利于促进城乡一体化，带动农村社会经济发展，减少彼此之间的摩擦和冲突，整合当地农业旅游资源，营造文明和谐的观光农业社区环境。此外，有的地区为了提升参与率，促成对观光农业的认可，采取放松农地管制等做法，比如，重庆市加强对生态农业的推广，让农民了解其好处，增强了农民的参与积极性，促进了生态农业的快速发展。

第二章 生态循环农业理论

第一节 循环农业

一、循环农业的产生与发展

随着工业文明的诞生和发展，特别是随着近现代科学技术手段的不断创新，人类的活动对生物圈的影响越来越大，关系着整个生物圈的稳定和繁荣。由于人类长期以来在非科学发展观引领下的不理智行为，如乱砍滥伐、乱捕滥杀、过度放牧、环境污染及人口膨胀、社会冲突等，对包括人类在内的整个生物圈造成了直接或间接的威胁，迫使人类不得不反思自身活动的理性问题和对稀缺资源与环境的保护问题，谋求协调人与生物圈的关系，于是才有了可持续发展思想。世界各国从可持续发展的理念出发，提出了一系列变革传统经济的发展模式与战略，力图通过这些战略规划及具体措施，协调经济与环境之间的关系，以实现总体效益的最大化。循环经济就是诸多战略构想之一。作为一种可持续发展理念的新型经济发展模式，循环经济是21世纪世界经济发展与环境保护结合的必然战略选择。

循环经济的思想萌芽产生于20世纪60年代，源于美国经济学家波尔丁提出的"宇宙飞船理论"，波尔丁对传统工业经济"资源—产品—排放"的"开环"范式提出了批评。几乎同时，1962年，美国生物学家蕾切尔·卡逊出版《寂静的春天》一书，对"杀虫剂"等化学农药破坏食物链和生物链的恶果进行了控诉。此后，1972年，罗马俱乐部在《增长的极限》报告中倡导"零增长"；1992年，联合国世界首脑环发大会发表《里约宣言》和《21世纪议程》，可持续发展观深入人心；2002年，世界环发大会决定在世界范围内推行清洁生产，并制订行动计划。在上述背景下，循环经济理念应运而生，循环经济由此发展。

随着工业化、城市化的不断发展，传统农业经济的增长面临着越来越突出的资源约束（尤其是土地资源和淡水资源）问题；同时，农业生态环境的不断恶化，农业废弃物不断增加，污染面积不断扩大，农业生产过程和农产品安全性减弱，这种将资源变为产品同时也不断地变为废弃物并造成污染的农业生产方式，已经影响到农业生态安全、人民生命健康安全和农民收入水平的提高。为摆脱上述困境，伴随着循环经济产生的大背景条件，循环农业应运而生。

循环农业是可持续发展的模式之一，是把循环经济理念应用于农业系统的结果，也是国内外农业发展的新趋势。在全世界，许多国家对循环农业进行了大量卓有成效的实践，但由于国情的不同，自然资源、地理条件、气候等方面的差异，采取了不同的农业循环方式，形成了一些各具特色的循环农业模式，如物质再利用模式的日本爱东町地区循环农业、德国的绿色能源农业等，减量化模式的美国精准农业、以色列的节水农业、瑞典的轮作型生态农业模式等。

中国政府对发展循环农业十分重视，2007年12月，农业部确定河北省邯郸市、山西省晋城市、辽宁省阜新市、江西省吉安市、山东省淄博市、河南省洛阳市、湖北省恩施自治州、湖南省常德市、广西壮族自治区桂林市、甘肃省天水市10个市（自治州）开展循环农业示范市建设。邯郸市、常德市、阜新市、淄博市、吉安市已经编制循环农业发展规划。其中，邯郸市制定了"34567"循环农业发展目标；晋城市大力开展以农村沼气、秸秆气化为核心的循环农业建设；阜新市从2005年起制定了《阜新市循环经济实施方案》及《阜新市发展循环经济规划》，2007年10月印发了《阜新市发展农业循环经济管理办法》。各地形成了一些循环农业模式，如河南等地的"粮饲—猪—沼—肥"生态模式，用种植业为饲养业提供饲料，用饲养业为种植业提供肥料，充分利用生态系统中物质与能量的循环，大大提高了物质的利用率且减少了环境污染；利用自然生态系统中各生物生长特点形成生物生态群体的立体复合型生态模式，等等。

总之，循环农业在全世界如火如荼地发展，它将成为现代农业重要模式之一，成为可持续发展的重要内容，前景广阔，意义深远。

二、循环农业的内涵与特点

循环农业是一个整体的有机系统，是一项艰巨的农业系统工程，它将循环经济理论和"减量化、再利用、资源化"的原则及高新技术应用于农业生产系统过程的物质与能量的多层次循环利用，实现农业资源的可持续利用和农业生产的低消耗、少排放、高效益。其内涵包括以下几点。

1. 具有一般循环经济的3个明显特点

① 减量化，要尽量减少进入农业生产和消费过程的物质量，节约农业资源使用，减少污染物的排放，保护生态环境；② 再利用，要尽量提高农副产品和资源的利用效率，减少生产污染或达到零排放；③ 再循环，使农业物品完成使用功能后能够重新变成再生资源，进行再利用。

2. 具有农业生产的自身特点

① 农业生态食物链条循环，在农业生产物质与能量循环中的各个主体互补互动、共生共利性较强；② 清洁生产，特别强调产品的安全性，利用高新技术控制和尽量减

少化肥、农药的施用量，或者采取施有机肥和生物控制；③ 保护土地、水资源和农业生态环境，使农业资源可持续利用和不受污染，农业生态环境良性循环；④ 绿色消费，人们既满足生活对绿色产品的需要，又不浪费资源和不污染环境，当农业的主副产品消费后，对废弃物再循环利用，变废为宝，回归自然。

三、循环农业发展的思路与策略

1. 发展思路

在循环经济理论指导下，按照物质能量循环和生态系统食物链原理，充分发挥农业生态经济系统的能量转化、物质循环、价值增值和信息传递功能，促进能量和物质在食物链中不断转化循环，有效利用各种有机废弃物资源（农作物秸秆、畜禽粪便、加工下脚料等），实现最佳生产、最大效益、最适度消费和最少废弃。要有效地充分利用好植物、动物和微生物等生物资源，构建"植物生产—动物转化—微生物还原"的循环利用体系，实现人与自然的和谐发展。在农业生产过程中要把握好"四化"：① 农业生产清洁化，包括清洁投入（清洁的原料、清洁的能源）、清洁产出（不危害人体健康和生态环境的清洁农产品）和清洁生产（使用无毒无害化肥、农药等）；② 产业内部资源利用梯度化，合理安排产业内部的生产方式，优化生产空间结构，尽可能地减少水、肥、土、药等资源浪费，提高资源利用效率；③ 产业间废弃物利用资源化，合理安排农业产业的时间和空间结构，在相关产业间建立废弃物资源化循环利用的互惠互利关系，降低生产成本，提高经济效益，改善生态环境；④ 农产品消费理性化，引导消费者客观地选择自己的消费档次、产品种类和品牌，为农业生产经营者提供准确的需求信息，引导产品种类调整、组织管理方式和技术工艺改进，提高农业资源配置效率，避免结构性、泡沫性资源浪费。

2. 发展观念

发展循环农业，要树立 3 个发展观念。

① 经济效益、社会效益、生态效益的有机统一观念。发展循环农业，要改变重生产轻环境、重经济轻生态、重数量轻质量的观念。既注重在数量上满足供应，又注重在质量上保障安全；既注重生产效益提高，又注重社会效益、生态环境建设，实现经济效益、社会效益、生态效益的有机统一。

② 资源多级多层循环利用观念。传统的农业生产活动一般表现为"资源—产品—废弃物"的单程式线性增长模式，产出越多，资源消耗就越多，废弃物排放量也就越多，对生态的破坏和对环境的污染就越严重。循环农业以产业链延伸为主线，推动单程式农业增长模式向"资源—产品—废弃物"循环的综合模式转变，同时循环农业要通过对农业生态经济系统的优化设计与管理，实现农业系统光热等自然资源和可再生资源的高效利用，最大限度地减少污染物的排放，节能减排，促进低碳农业发展。

③ 依靠农业高新技术的观念。发展循环农业要依靠科技创新，加强循环农业科学技术研究，推广促进资源循环利用和生态环境保护的农业高新技术，提高农业技术含量，实现由单一注重产量增长的农业技术体系向注重农业资源循环利用与能量高效转换的循环农业技术体系转变。

3. 发展策略

① 优化农业循环结构。优化农业循环结构主要是做好以下工作：A. 产业链循环结构优化，主要是在农产品自产地到餐桌全过程中推行清洁生产、污染防控，做到源头控制、中间处理、末端循环，使污染物排放量最小。B. 农业行业内部循环结构优化，主要是农业行业内部物质、能量相互交换，废弃物排放最小化，例如，立体种植、立体养殖及农产品生态加工等都是典型的循环农业发展模式。C. 农业行业外部循环结构优化，主要是农业各行业之间相互交换废弃物，使废弃物得以资源化利用。例如，种养结合的稻田养鱼，稻田为鱼提供了良好的生长环境，鱼吃杂草、害虫，鱼粪肥田，减少了水稻化肥农药使用量，保护了生态环境，增加了经济效益，实现物质流、能量流的闭合式循环。

② 搞好循环农业产业链。循环农业产业链是在生态种植业、生态畜牧业、生态林业、生态渔业及生态加工业之间，通过废弃物交换、循环利用、要素耦合等方式形成网状的相互依存、协同作用的有机整体。各产业部门之间，在质上为相互依存、相互制约的关系，在量上按一定比例组成有机体。例如，以蔗田种植业系统、制糖加工业系统、酒精酿造业系统、造纸业系统、热电联产系统、环境综合处理系统为框架，通过盘活、优化、提升、扩张等步骤，建设制糖生态产业链，各系统内分别有产品产出，其资源得到最佳配置，废弃物得到有效利用。各系统之间通过中间产品和废弃物的相互交换来衔接，从而形成一个比较完整和闭合的生态产业网络。

③ 构建循环农业技术体系。发展循环农业要以高新技术为支撑，把经济效益、社会效益和生态效益有机结合起来，使物质和能量得到充分利用，建立适合循环农业发展的农业技术体系。循环农业技术包括清洁生产技术、低碳技术、生态技术等。

第二节　循环农业理论基础

循环农业是指运用可持续发展思想和循环经济理论与生态工程学方法，结合生态学、生态经济学、生态技术学原理及其基本规律，在保护农业生态环境和充分利用高新技术的基础上，调整和优化农业生态系统内部结构及产业结构，提高农业生态系统物质和能量的多级循环利用，严格控制外部有害物质的投入和农业废弃物的产生，最大限度地减轻环境污染，把农业生产经济活动真正纳入到农业生态系统循环中去，建立农业经济增长与生态系统环境质量改善的动态均衡机制，实现生态的良性循环和农业的可持续发展。

一、循环经济理论

循环经济（Circular Economy）是对物质闭环流动型（Closing Materials Cycle）经济的简称，一般认为美国经济学家鲍尔丁（K. E. Boulding）1962 年提出的"宇宙飞船理论"是循环经济思想的雏形；英国环境经济学家珀斯和特纳 1990 年在《自然资源和环境经济学》一书中首次正式使用了"循环经济"一词，之后，循环经济发展模式受到国际社会的广泛重视。概括地讲，循环经济是一种以资源的高效利用和循环利用为核心，以"减量化、再利用、资源化"为原则（简称 3R 原则）的经济模式，减量化（Reducing）原则要求减少进入生产和消费流程的物质量，因此又叫减物质化，这一原则有利于避免先污染、后治理的传统发展方式；再利用（Reusing）原则的目的是延长产品和服务的时间强度，减少生产和消费中废弃物的产生，这一原则可以防止物品过早地成为垃圾；资源化或再循环（Recycling）原则要求物品在完成使用功能后重新变成可以利用的资源。

循环经济的理念核心是把传统"资源—产品—污染排放""单向单环式"的线性经济，改造成"资源—产品—再生资源—产品—再生资源"的"多向多环式"与"多向循环式"相结合的反馈经济及循环经济综合模式，使传统的高消耗、高污染、高投入、低效率的粗放型经济增长模式转变为低消耗、低排放、高效率的集约型经济增长模式。在宏观层面上，循环经济要求对产业结构和布局进行调整，将循环经济理念贯穿于社会经济发展的各领域、各环节，建立和完善全社会的资源循环利用体系；在微观层面上，要求节能降耗，提高资源利用效率，实现减量化，并对生产过程中产生的废弃物进行资源化利用，同时根据资源条件和产业布局，延长和拓宽生产链条，促进产业间的共生耦合。

二、生态系统物质循环与能量流动原理

物质和能量是所有生命运动的基本动力，能流是物流的动力，物流是能流的载体，生物有机体和生态系统为了自己的生存和发展，不仅要不断地输入能量，而且还要不断地完成物质循环。进入生态系统的能量和物质并不是静止的，而是不断地被吸收、固定、转化和循环的，形成了一条"环境—生产者—消费者—分解者"的生态系统，各个组分之间的能量流动链条，维系着整个生态系统的生命。自然系统依靠食物链、食物网实现物质循环和能量流动，维持生态系统稳定；农业生态系统则要借助人工投入品及辅助能维持正常的生产功能和系统运转。在生态系统中，能流是单向流动的，并且在转化过程中逐渐衰变，有效能的数量逐级减少，最终趋向于全部转化为低

效热能,由植物所固定的日光能沿着食物链逐步被消耗并最终脱离生态系统。生态系统中某些贮存的能量,也能形成逆向的反馈能流,但能量只能被利用一次,所谓再利用是指未被利用过的部分。但这不是单向流动,而是循环往复的过程,物质由简单无机态到复杂有机态再回到简单无机态的再生过程,同时也是系统的能量由生物固定、转化到水散的过程,不是只能利用一次,而是重复利用,物质在流动的过程中只是改变形态而不会消灭,可以在系统内永恒地循环,不会成为废物。

任何生态系统的存在和发展,都是能流与物流同时作用的结果,二者有一方受阻都会危及生态系统的延续和存在。参与生态系统循环的许多物质,特别是一些生物生长所不可缺少的营养物质,既是用以维持生命活动的物质基础,又是能量的载体。以太阳能为动力合成的有机物质,沿食物链逐级转移,在每次转移过程中都有物质的丢失和能量的散逸,但所丢失的物质部分都将返回环境,最终分解成简单的无机物,然后被植物吸收、利用,而所散逸的能量则将不能被再利用。相对于生态系统而言,由于日光能为主要能源,是无限的,而物质却是有限的,分布也是很不均匀的。因此,农业生态系统如果调控合理,物质可以在系统内更新,不断地再次纳入系统循环,能量效率也得到持续提高。

三、生态位与生物互补原理

生态位(Niche)是指生物在完成其正常生活周期时所表现出来的对环境综合适应的特征,是一个生物在物种和生态系统中的功能与地位,生态位与生物对资源的利用及生物群落中的物种间竞争现象密切关联。生态位的理论表明:在同一生态环境中,不存在两个生态位完全相同的物种,不同或相似物种必须进行某种空间、时间、营养或年龄等生态位的分异和分离,才可能减少直接竞争,使物种之间趋向于相互补充。由多个物种组成的群落比单一物种的群落能更有效地利用环境资源,维持较高的生产力,并且有较高的稳定性。在农业生产中,人类从分布、形态、行为、年龄、营养、时间、空间等多方面对农业生物的物种组成进行合理的组配,以获得高的生态位效能,充分提高资源利用率和农业生态系统生产力。随着生态学概念的不断深化,已从单纯的自然生态系统转移到社会—经济—自然复合生态系统,生态位概念也进一步拓展,不再局限于单纯的种植业系统或养殖业系统,甚至拓展到整个农业经济系统。

生态系统中的多种生物种群在长期进化过程中,形成对自然环境条件特有的适应性,生物种与种之间有着相互依存和相互制约的关系,且这一关系是极其复杂的。一方面,可以利用各种生物及生态系统中的各种相生关系,组建合理高效的复合生态系统,在有限的空间、时间内容纳更多的物种,生产更多的产品,对资源充分利用及维持系统的稳定性,如我国普遍采用的立体种植、混合养殖、轮作,以及利用蜜蜂与虫媒授粉作物等。另一方面，可以利用各种生物种群的相克关系，有效控制病、虫、草害，目前正兴起的生物防治病虫害及杂草，以及生物杀虫剂、杀菌剂、生物除草剂等生物农药技术已展示出广阔的发展前景。

四、系统工程与整体效应原理

按照系统论及系统工程原理，任何一个系统都是由若干有密切联系的亚系统构成的，通过对整个系统的结构进行优化设计，利用系统各组分之间的相互作用及反馈机制进行调控，可以使系统的整体功能大于各亚系统功能之和。农业生态系统是由生物及环境组成的复杂网络系统，由许许多多不同层次的子系统构成，系统的层次间也存在密切联系，这种联系是通过物质循环、能量转换、价值转移和信息传递来实现的，合理的结构将能提高系统整体功能和效率，提高整个农业生态系统的生产力及稳定性。著名生态学家马世骏先生曾把生态学的基本原则高度浓缩概括为八个字："整体、协调、循环、再生"，其中"整体、协调"点明了生态系统合理而协调的横向关系，而"循环、再生"则蕴含着生态系统永续运转的特性。

农业生态系统的整体效应原理，就是充分考虑到系统内外的相互作用关系、系统整体运行规律及整体效应，运用系统工程方法，全面规划，合理组织农业生产，通过对系统进行生态优化设计与调控，使总体功能得到最大限度发挥，实现生态系统物种之间的协调共存、生物与环境之间的协调适应、生态系统结构与功能的协调发展及不同生态过程的协调，建立起一个良性循环机制，使系统生产力和资源环境持续保持增值与更新，满足人类社会的长远需求，达到生态与经济两个系统的良性循环。

五、农业区位及地域分异原理

19世纪初期，德国经济学家杜能（T. H. Von. Thiine）根据资本主义农业和市场的关系，探索因地价不同而引起的农业分带现象，创立了农业区位理论，它是一种从空间或地区方面定量研究自然和社会现象的理论。经济学家对这一理论又进行了发展，从自然区位向经济区位、市场区位、生态经济区位等拓展，并结合比较优势理论等，有效推进了农业区域化、规模化、专业化生产发展。农业生产受自然因素限制比较明显，农业发展必须因地制宜、扬长避短，充分发挥区位优势、经济优势、市场优势和科技优势。比较优势是区域分工的基本原则，也是进行农业结构调整的重要理论依据。

由于地形、地势、气候、土地、社会经济、人文等要素的相似与差异，区域存在着聚合与分离的现象，而农业生产是自然和人工环境与各类农业生物组成统一体，其地域分异特征显著。农业地域分异规律包括自然地理、人文地理、生物地理的差异，造成了农业生产及生态经济类型的差异性。尽管随着社会经济持续发展，农业从传统性、自给性、粗放性向现代性、商品性、集约性方向发展的规律是相同的，但农业的地域性、多样性仍将长期存在。我国幅员辽阔，自然与社会经济条件格外复杂，发展循环农业必须使物种和品种因地制宜，彼此之间结构合理，相互协调。依据地区环境，构建有特色的循环农业模式，要考虑地区全部资源的合理利用，对人力资源、土地资源、生物资源和其他自然资源等，按照自然生态规律和经济规律，进行全面规划，统筹兼顾，因地制宜，并不断优化其结构，充分提高太阳能和水的利用率，实现系统内的物质良性循环，使经济效益、生态效益和社会效益同步提高。

六、农业可持续发展原理

可持续发展的本质含义就是要当代人的发展不应危及后代人的发展能力和机会，实现资源最佳效率和公平配置，实现人与自然和谐及协同进化。自20世纪80年代可持续农业兴起以来，世界各国在理论和实践上的探索不断深入，尽管理解和做法各有不同，但总的发展目标是相同的，即保障农业的资源环境持续、经济持续和社会持续等。资源环境持续性主要指合理利用资源并使其永续利用，同时防止环境退化，尤其要保障农业非再生资源的可持续利用，包括化肥、农药、机械、水电等资源。经济持续性主要指经营农业生产的经济效益及其产品在市场上竞争能力保持良好和稳定，这直接影响到生产是否能维持和发展。尤其在以市场经济为主体的情况下，一种生产模式和某项技术措施能否推行和持久，主要看其经济效益如何，产品在国内外市场有无竞争能力，经济可行性是决定其持续性的关键因素。社会持续性指农业生产与国民经济总体发展协调，农产品能满足人民生活水平提高的需求，既要保证产品供应充足，保持农产品市场的繁荣和稳定，尤其是粮食和肉蛋产品的有效供给，又要保证产品优质、价格合理，能满足不同消费层次对优质农产品的需求，满足社会经济总体发展的需求。社会持续性直接影响着社会稳定和人民安居乐业的大局。

农业可持续的3个目标是相辅相成的，三者不可分割，即在合理利用资源和保护生态环境的基础上，努力增加产出，满足人类不断增长的物质需求，同时促进农村经济发展，提高农民收入和社会文明。偏废任何一个方面，把持续性仅理解为生态环境上的持续性是片面的和脱离实际的。

第三节 生态循环农业发展概述

一、生态循环农业的概念

生态循环农业，是遵循生态学和经济周期理论等相关原则，利用现代科学技术和现代管理工具，与传统农业中必不可少的、有效的、显著的综合经验相结合，能取得良好的经济效益、生态效益和社会效益的新型农业。它不仅局限于某种农产品的生产方式、产量和经济效益，而且是以生态为基础，通过对生产模式、发展模式、效益模式等的改造结合，形成一个可循环发展的农业生产模式，同时将经济效益、社会效益、生态效益高度统一在一起，把"环境优美、产品安全、生产科学、发展持续"的理念变为现实。

从20世纪80年代起，在世界上一部分农业生产较为发达的国家中，出现了初具雏形的生态农业，这一新的农业发展模式出现后即受到普遍重视，并快速发展。我们为了提高生态农业发展速度，加快其建设，一方面，应注意总结和推广生态农业的经验和做法，比如作物的合理轮作、有机肥（绿肥）代替化肥施用、提高田地利用率等，这些都是从事基层生产人员（如农民）非常熟悉且都愿意接受的措施；另一方面，应加紧研究和推动生态农业先进的新技术，如利用新型的光解薄膜生产生物性药品，有机肥料、秸秆合理利用，节水节肥灌溉等，只有通过这两个方面有效结合，生态农业的发展效果才可以变得更好。

自从人类发明和使用农药、化肥后，农业生产中所有的生物都或多或少受到威胁，生物圈遭受了各种有害化学物质的不断侵袭，人类自己也未能幸免，各种化学物质污染土地，形成成千上万的"死地""臭土"。为探究如何合理开发土地、生产安全的农产品，保障人类身心健康，减少污染、保护生态环境，有效保证农业可持续发展，生态循环农业应运而生。在生态循环农业模式中，种植业一般不允许像传统农业那样大量施用化肥、农药等化学物质，而是利用生产环境、人为设计、科学控制等方式生产安全、高质、高效益的产品；在养殖业方面，除了不大量使用各种添加剂等虽能促进牲畜生长获得更多产品的物质，而且还强调对生产中的废弃物（主要是粪便、尿液、生产废水等）综合利用，把种植业与养殖业中产生的废弃物综合利用，形成循环的、生态的利用模式。

二、生态循环农业的特点

1. 综合协调性

生态循环农业着重调动和发挥农业生态系统整体功能，从规模化种植业、养殖业着手，按照整体考虑、全面协调、循环利用、持续发展的原则，进行全面协调规划，最大限度地调整和优化农业产业结构，使涉及其中的农业形态与各产业综合协调发展，使各行业之间能相互支持、相互配合，从而达到提高综合生产能力的目的。

2. 模式多样性

发展生态循环农业，就要考虑到我国农业生产的实际情况。由于我国幅员辽阔，各地自然条件、资源基础、经济和社会发展水平的差异是很大的，生态循环农业充分吸收了传统农业的精华，结合现代科学技术，把多种生态模式、生态工程建设和实用技术应用到农业生产中，通过不断的、深入的分析、对比，找到不同地区的农业生产优势，改造落后的生产方式，进而整合不同的、多样的发展模式，突出发展的多样性。

3. 效益高效性

生态循环农业将物质循环、废弃物的循环利用、能量的多层次多极化利用与农产品的深加工相结合，降低农业生产成本、提高产品质量，从而提高效益，达到农业经济的增效、增值；同时，将农民从传统的"日出而作，日落而息""面朝黄土背朝天"的落后、低效、高耗的农业生产中解放出来，为他们创造新的劳动方式和就业岗位，使他们乐于接受、积极生产，从而达到稳定"三农"环节中"农民"这一发展基础。

4. 发展持续性

生态循环农业的科学发展，能够有效保护农业生态环境，防止农业生产中的化肥、农药等有害物质对土地、作物、牲畜、农产品的污染，不仅维护了生态平衡，而且提高了农产品的质量、保障了其安全性，把环境保护、农业生产和经济发展紧密结合起来，在保证农产品的供应、满足人们消费需求的同时，使我国现代农业朝更加稳定、更加和谐、更加持续的方向发展。

三、生态循环农业与传统农业的比较

传统农业经济结构是单程线性的，即在"农业生产中对土地、品种等进行资源使用，生产出初级农产品，产生大量的废弃物未加以利用"，对资源的消耗较高，对资源的利用率较低，而且生产中产生的废弃物往往排放较多，又未能加以综合利用，造成了浪费。在整个过程中，生产者（主要是农民）往往仅追求他们认为价值较高的农业生产，片面追求经济效益。生产方式、资源利用方式较粗放，无形中增加了对生态环境的破坏、造成了资源的严重浪费，而且生产出的产品价值也并不是很高的，相反有些还是"亏本"的。

而生态循环农业则强调资源的合理利用、生态环境的保护、经济效益的合理追求及农业的可持续发展，是一种循环多线程的发展，创造出了"合理利用资源，生产出安全有保障的农产品，将产生的废弃物综合利用又投入到农业生产中"的循环发展模式，把资源消耗降到最低。通过废弃物的综合利用，仅排放少量的污染物，使资源的利用率成倍提高，既保证了生态平衡，又发展了农业生产，还使农业的发展进入到可持续的"快车道"中。

生态循环农业与传统农业的区别主要表现在4个方面。

生态循环农业在理论发展上，强调把生态的、循环的、环保的思想应用到生产实际中，防止过度使用资源、坚持废弃物再利用、坚持用科学的方法减少废弃物的产生和循环利用看似废弃的物质，从内源上控制污染问题，想尽办法做到循环再利用。

生态循环农业在生产方式上不仅注重经济效益，如对一些高投入、高耗能的产业进行积极改造，对资源的利用也是用科学的方法做到利用最大化、最优化，而且追求除了经济效益以外新的技术、新的产品、新的发展模式等，使它们同样处在重要的位置。

传统农业在产业发展方面，往往局限于某一产业的固定发展，种植业无非是粮、油、棉，养殖业无非是猪、牛、鸡，这些产业之间的相互联系是很少的。生态循环农业则倡导积极调整农业产业结构，从生态循环的角度出发，既做大、做强、做精某一产业，又在此基础上将不同的产业有效结合，形成产业化，使相关产业能在产业链条内实现循环可持续发展。

传统农业注重产品的产量和经济效益，这是几千年来雷打不动的"真理"，在以前农业科技不发达、市场竞争不激烈、消费要求简单化的社会，这一"真理"确实应该奉行，但生态循环农业的出现，改变了这一格局，农业生产不能仅仅注重产品的产量和经济效益，而是要学会利用好自然资源、品种资源，搞好产业规划，链接好产业链，把产业内部的循环做好，在增产、增效的同时，降低生产成本，提高生态效益和社会效益。

第四节 国外生态循环农业的主要模式

一、物质再利用模式

1. 日本爱东町地区循环农业

爱东町地区循环农业的核心内容是发展油菜生产。一方面，油菜籽利用后遗留的油渣可以通过堆肥或饲料化处理得到优质的有机肥料或饲料；另一方面，回收废弃食用油，再加工处理成生物燃油。该地区循环农业有效促进了资源在农业经济系统中的高效再生，减少了外部资源投入和农业废弃物排放，实现了资源合理循环再利用和生态环境保护，符合生态循环农业的基本原则。

2. 德国"绿色能源"农业

通过不断的研究，德国科学家发现，可从一些农产品中提取矿物能源和化工原料替代品，实现农产品的循环再利用，这些生物质能源和原料是绿色无污染的，因此，德国政府开始重视发展此类经济作物。德国科学家对甜菜、马铃薯、油菜、玉米等进行定向选育，从中制取乙醇、甲烷，成功地研制出绿色能源；从菊芋植物中制取酒精；从羽豆中提取生物碱。油菜籽是德国目前最重要的能源作物，不仅可用作化工原料，还可提炼植物柴油，代替矿物柴油用作动力燃料。

二、减量化模式

1. 美国精准农业

精准农业追求以最少的投入获得优质的高产出和高效益，指导思想是按田间每一操作单元的具体条件，精准地管理土壤和各类作物，最大限度地优化使用农业投入（如化肥、农药、水、种子等）以获取最高产量和经济效益，减少使用化学物质，保护农业生态环境。精准农业是"减量化"的循环农业。美国是世界上实施精准农业最早的国家之一，早在1990年前后，美国就将GPS系统技术应用到农业生产领域，在明尼苏达州农场进行了精确农业技术试验，用GPS指导施肥的作物产量比传统平衡施肥作物产量提高30%左右。试验成功后，小麦、玉米、大豆等作物的生产管理都开始应用精准农业技术。

2. 以色列节水农业

为了保持区域水环境和生态的持续稳定，以色列的循环农业突出体现为完善的节水农业体系。喷灌、滴灌、微喷灌和微滴灌等技术在以色列普遍使用，80%以上的农田灌溉应用滴灌，10%为微喷，5%为移动喷灌，完全取代了传统的沟渠漫灌方式。成效最大的是农业滴灌技术：一是水可直接输送到农作物根部，比喷灌节水20%；二是在坡度较大的耕地应用滴灌不会加剧水土流失；三是经污水处理后的净化水（比淡水含盐浓度高）用于滴灌不会造成土壤盐碱化。滴灌技术比传统的灌溉方式节约用水和节省肥料30%以上，而且有利于循环利用污水。为开辟水源，以色列加大了对污水处理和循环使用的投入。以色列规划农业灌溉全部使用污水再处理后的循环水，目前，已将80%的城市污水处理循环使用，主要用于农业生产，占农业用水的20%。经处理后的污水除用于农业灌溉外，还重输回蓄水层。

三、资源化模式

1. 日本菱镇循环农业

菱镇是发展循环农业较早且较成功的地区，是将农业生产和生活中的废弃物转化为有机肥，发展废弃物资源化的循环农业模式。1988年，该镇通过了《发展自然农业条例》，规定农业生产中禁止使用农药、化肥和其他非有机肥料，生产的农产品需是无化肥、无农药添加残留、无公害的有机农产品。此后，菱镇将小规模下水道污泥、家禽粪便及企业的有机废物作为原料投入到发酵设备，产生的甲烷气体用于发电，剩余的半固体废渣进行固液分离，固态成分用于堆肥和干燥，液态成分处理后再次利用或者排放（排放时已基本对环境无害），实现了废物的高度资源化和无害化。此外，菱镇对厨房垃圾进行统一收集和处理，制成有机肥。

2. 英国"永久农业"

"永久农业"是循环经济中废物资源化的一种重要形式，特点是在节约资源和不破坏环境的基础上，通过元素的有效配置达到有利关系的最大化。种植者循环利用各种资源，节省能源，如用香烟头来收集雨水、变粪便为有机肥料、实行秸秆还田。"永久农业"寻求尽可能节约使用土地资源，强调使用多年生植物，鼓励使用自我调节系统。耕种土地时，通过多种类种植和绿色护盖等技术来保养土地，监控当地环境，构建绿色发展规划。"永久农业"不使用人造化肥和杀虫剂，通过种植多样性的植物及促使食肉动物进入生态系统来阻止害虫。例如，豆类植物苜蓿，能够释放氮气，可使害虫迷失方向。

四、生态循环产业园模式

循环农业的尺度有部门、区域、社会 3 个层次：部门层次主要指以一个企业或一个农户为循环单元；社会层次意味着"循环型农村"；区域尺度是按照生态学的原理，通过企业间的物质、能量、信息集成，形成以龙头企业为带动，园内包含若干个中小企业和农户的生态产业园。

菲律宾玛雅农场是一个成功的生态循环产业园典范。玛雅农场最初只是一个面粉厂，从 20 世纪 70 年代开始，经过 10 年建设，形成了一个农、林、牧、副、渔良性循环的生态系统。面粉厂产生大量麸皮，为了不浪费麸皮，建立了养殖场和鱼塘；为了增加收入，又建立了肉食加工和罐头制造厂，对畜产品和水产品进行深加工。到 1981 年，农场拥有 36 公顷的稻田和经济林，农场饲养有 2.5 万头猪、70 头牛和 1 万只鸭。为了控制畜禽粪肥污染、循环利用加工厂的废弃物，农场建起十几个沼气车间，每天生产的沼气能满足农场生产和家庭生活所需的能源。从产气后的沼渣中，还可回收一些牲畜饲料，其余用做有机肥料。产气后的沼液经处理后，送入水塘养鱼养鸭，最后再取塘水、塘泥用来肥田。农田生产的粮食又送到面粉厂加工，进入下一次循环。玛雅农场不用从外部购买原料、燃料、肥料，却能保持高额利润，而且没有废气、废水和废渣的污染，充分实现了物质的循环利用。

第五节　我国生态循环农业的几种主要模式

一、农田内循环模式

1. 不同作物间的套种

例如，黄豆和水稻间的套种，两种作物所需的营养成分不同，就可以在种间进行交换，水稻用不上的养分黄豆可以利用，黄豆吸收不了的养分可以吐纳给水稻，形成种间的循环。

2. 废弃物还田、留田

一般以水稻、小麦为主，进行秸秆的利用。把收割后的秸秆作为下一轮作物所需的养分留在田中，或者通过堆肥的方式施用到田中。

3. 田内种养

例如，在稻田内种植水稻，田内水域用来养鱼、养鸭和其他小型牲畜或者特种经济动物（如牛蛙等），动物可食用田内杂草、害虫，排出的粪便又可作为水稻的肥料，提供给水稻养分。

二、种养间循环模式

1. 单纯直连循环

将种植业中的作物作为养殖业中动物的饲料，例如，通过种植饲料作物玉米、籽粒苋、饲料菜、苦荬菜、菊苣、鲁梅克斯、串叶松香草、御谷、紫花苜蓿、沙打旺等，收割后饲喂牛、羊、猪、鸡、鱼、鹅、鸭等，动物粪便经过处理回田，作为作物肥料。

2. 种养结合沼气

这是目前较为常用的生态循环农业模式，例如，饲养生猪产生粪污，粪污通过沼气工程的处理产生"三沼"（沼气、沼液、沼渣），沼气作为能源物质用于生产、生活，沼液、沼渣作为肥料施用到作物田内，作物又可作为猪的饲料，如此循环。

3. 种养结合菌基

即养殖业产生的畜禽粪便混合种植业作物产生的秸秆，生产出食用菌的培养基料，培养出食用菌后，培养基料又可作为作物的肥料还田。

4. 生物处理

这一模式中最著名的案例要数悉尼、北京奥运会中利用蚯蚓处理废弃物。上述两届奥运会中所产生餐厨垃圾、马术项目中赛马的粪便，均通过蚯蚓的吞食、排泄后转变成肥性极好的生物有机肥。在农业生产中，亦可将蚯蚓的吞食、排泄物转变成肥性极好的生物有机肥。还可将蚯蚓引入堆肥的再处理、污水处理厂沉淀污泥池中的污泥再消化，产生的有机肥可用作还田。

5. 种养结合水产

早在周朝就有记载养蚕取丝的技术，生态循环农业的发展将其进一步发展成立体养殖，即种植桑树生产桑叶提供给蚕，蚕沙又可施用到鱼塘，作为水产的饵料，塘泥又可作为基肥施用到桑田。

三、产业化循环模式

这种模式不仅仅局限于各种种养品种间的循环，而是扩大到产业实体中，在从事相关的作物种植公司与养殖公司间建立起产业循环模式，如种植茶叶的实体公司向养殖肉牛、奶牛的实体公司收购有机肥等。

四、工农复合型模式

在工业尚不发达的年代，城市垃圾主要是有机物，一般都通过收集又回到农村作为肥料。随着工业的迅速发展、城市化进程不断加快，处理这些垃圾的主要手段多为填埋、焚烧，其中不少有机物无法收集返回到农业中。通过广泛的宣传和相关规定的出台，有机物被有效分离出来直接回田，或设立工农产业循环园区，工业生产中的废弃物经过处理成为有机物又回到农业生产中，农业生产又借助工业生产进一步发展。

五、农业与地球生物化学循环模式

农业的物质输入和输出本身就是全球物质循环的一部分，其中，植树造林和保护森林，对碳循环、氮循环、磷循环、硫循环等有重要作用，通过植树造林、保护现有森林，可以帮助调节大气中二氧化碳的浓度，稳定自然界原有的物质平衡。

第三章 农业废弃物循环利用

第一节 农业废弃物的概念、来源及分类

一、农业废弃物的概念

我国学术界对农业废弃物的界定并不统一,在其内涵与外延的探讨上各存己见。孙振钧是最早对农业废弃物进行界定的学者,以后的学者在他的基础上通过不同学科角度对农业废弃物进行了阐述。学术界认为,农业废弃物(Agricultural Residue)是指在整个农业生产过程中被丢弃的有机类物质,主要包括农林业生产过程中产生的植物类残余废弃物、畜牧渔业生产过程中产生的动物类残余废弃物、农业加工过程中产生的加工类残余废弃物和农村生活垃圾等。通常,农业废弃物主要指农作物秸秆和畜禽粪便,包括植物类废弃物(农林生产过程中产生的残余物)、动物类废弃物(畜牧渔业生产过程中产生的残余物)、加工类废弃物(农林牧渔业加工过程中产生的残余物)和农村生活垃圾等四大类。根据资源废弃化理论,农业废弃物是农业生产和再生产链环中资源投入与产出在物质和能量上的差额,是资源利用过程中产生的物质能量流失份额。也有学者认为,农业废弃物包括农业生产过程中的废弃物和农村居民生活废弃物。从循环经济学的角度出发,在目前技术资金和劳动力等条件允许的情况下,农业生产或农产品加工业的副产品中能作为原材料被再生利用的那部分农业废弃物是物质和能量的载体,是以特殊形态存在的资源,是农业生产和农产品加工过程中不可避免的一种副产品。

科学技术部中国农村技术开发中心对农业废弃物的定义为:"在农业生产过程中,除了目的产品外而抛弃不用的东西,是农业生产中不可避免的一种非产品产出。按其来源不同,可划分为种植业产生的各种农作物秸秆、养殖业产生的畜禽粪便及屠宰畜禽而产生的废弃物、对农副产品加工而产生的废弃物,农业生产过程中残留在土壤中的农膜也是主要的农业废弃物之一。"

目前,我国相关法律法规并没有对农业废弃物的定义进行统一界定,在不同场合和不同领域,我国的法律和官方文件对农业废弃物的内涵与外延使用不同。但是我国已经制定了部分法律、部门规章及地方性法规,对农业废弃物与农业废弃物概念进行了规定。2001年3月20日,经国家环境保护总局局务会议通过的《畜禽养殖污染防

治管理办法》规定，畜禽废渣是指畜禽养殖场的畜禽粪便、畜禽舍垫料、废饲料及散落的毛羽等固体废弃物。2013年10月，国务院第26次常务会议通过的《畜禽规模养殖污染防治条例》，对2001年的《畜禽养殖污染防治管理办法》中的"畜禽废渣"的定义进行了延伸，指出，畜禽养殖废弃物是指畜禽粪便、畜禽尸体、污水等。中华人民共和国第十一届全国人民代表大会常务委员会第四次会议于2008年8月29日通过了《中华人民共和国循环经济促进法》，其中第四章第三十四条中阐述"对农作物秸秆、畜禽粪便、农产品加工业副产品、废农用薄膜等进行综合利用，开发利用沼气等生物质能源"。虽然其中没有明确使用"农业废弃物"这一概念，但实际上已经将农业废弃物的内涵表达了出来。2010年9月，浙江省人民政府第五十六次常务会议审议通过的《浙江省农业废弃物处理与利用促进办法》是第一部对农业废弃物进行定义的地方性法规。其中规定，农业废弃物是指在种植业、畜牧业生产中产生的废弃物，包括畜禽养殖废弃物、农作物秸秆、食用菌种植废弃物、废弃农膜及县级以上人民政府确定的其他农业废弃物。2012年9月，宁夏回族自治区人民政府第一百二十三次常务会议通过的第48号文件《宁夏回族自治区农业废弃物处理与利用办法》第一章第三条中指出，农业废弃物是指在种植业、畜牧业等农业生产过程中产生的废弃物，包括畜禽养殖废弃物、农作物秸秆、废弃农膜等。结合上述对农业废弃物概念的诸多表述及我国的实际国情，农业废弃物可定义为：种植业、林业、畜牧业和水产养殖业生产过程中，以及与其产品生产、加工相关的活动中，农村居民日常生活中，或为日常生活提供服务的活动中产生的不具有原有利用价值、其所有人、使用人已经或准备或必须丢弃的部分物质或能量。农业废弃物是农业生产过程中排放的废弃物总称。

二、农业废弃物的来源

1. 种植业农业废弃物

种植业农业废弃物主要来源于农田和果园，主要指种植业、林业生产过程中或收获活动结束后，除了果实以外的、被丢弃的物质或能量，如粮食作物的秸秆、蔬菜瓜果的落果残叶、藤蔓等。

我国是一个农业大国，种植业生产中会产生大量的废弃物，它们种类繁多，数量巨大，以农作物秸秆为主。2009年，我国农作物秸秆总产量为6.87亿t，其中，稻草为2.5亿t、小麦秸秆为1.5亿t、玉米秸秆为2.65亿t、豆类秸秆为2726万t、棉花秸秆为2584万t。2018年，全国秸秆产量9亿多t。

自古以来，秸秆被作为农户的生活燃料、牲畜的粗饲料；或者还田作为肥料，用来保水抗旱，防虫除虫；还可作为建筑材料；少量用作造纸等工业原料。现在，秸秆还可以转化发电，被加工压块成燃料制取煤气，作为食用菌基料培养生产平菇、香菇等食用菌。可见，种植业农业废弃物的可再生利用价值高。据调查，2009年我国的秸秆未利用资源量为2.15亿t，2018年秸秆综合利用率超过82%。

2. 养殖业废弃物

养殖业废弃物主要来源于畜牧、渔业生产过程中产生的残余物，主要是指在畜禽养殖过程中产生的畜禽粪便、污水、废饲料、羽毛等废弃物及水产养殖过程中的含饲料残渣、农药残余的养殖塘泥等废弃物质。其中，畜禽粪便在畜禽废弃物中占据了较大的比例，而且不同畜禽品种产生的畜禽废弃物存在差异。我国是世界上的畜禽养殖业大国，每年产生大量的畜禽粪便。据统计，2011年，我国畜禽粪便的产生量为21.21亿t，其中牛粪便产生量较大，其次是羊、猪、肉鸡和蛋鸡。2017年，我国产生的畜禽粪污量约为38亿t。我国畜禽粪便2008—2011年的增加量相当于2010年我国工业废弃物产生量的50%左右。与我国每年产生的各类农作物秸秆约6.5亿t相比，我国畜禽粪便的产生量约是秸秆产生量的3.26倍。可见，养殖业废弃物占我国农业废弃物的比例最大，此类废弃物如处理不当，对环境危害最大。2010年，环境保护部、国家统计局和农业部联合发布的《全国第一次污染源普查公报》显示，畜禽养殖业的污染排放已经成为我国最重要的污染源之一。近年来，国家对规模化畜禽养殖业的污染进行了治理，并于2013年颁布《畜禽规模养殖污染防治条例》，使监管治理做到有法可依。但是，由于大部分养殖场未能对畜禽粪便进行有效的处理和利用，并且将未经处理的粪便随意堆放，导致大量的氮、磷流失，造成水体、土壤和空气污染。畜禽粪便中含有的大量病原菌及饲料中含有的重金属添加剂，不仅对环境造成污染，同时还会影响到人们的身体健康。我国自古以来就有畜禽粪便的利用传统。畜禽粪便可以当作有机肥料用于农作物耕种，不仅有利于改良土壤，还节约了肥料资源。北方少数民族有用牛粪作为燃料的传统，用于做饭和取暖。目前，可以通过先进的科学手段将畜禽粪便进行加工，作为饲料用于鱼、猪等的养殖；畜禽粪便经过集中处理还可以生产沼气，作为能源用于农民的生活当中。

3. 农产品加工类废弃物

农产品加工类废弃物是来自于农产品加工过程中产生的废弃物，包括肉食加工工业废弃物、制糖业的甘蔗和甜菜渣、罐头食品厂加工废弃物、木材加工后剩余的边角废料和各类经济林抚育管理期间育林剪枝所获得的薪材等。

这类废弃物年产量现已超过1亿t。随着粮食产量的增长，预计到2020年，这类废弃物年产量会增长到2亿t，其中大部分可以综合利用，例如，肉食加工工业的废弃物可以用于制造皮革制品、肥皂、动物胶、生物药剂、羽绒、骨粉等；农作物秸秆可作为纸、板生产的原材料，利用其木质素或纤维素可进一步加工制造化工产品。农产品加工类废弃物如果处理不当，其产生的废渣废水进入环境当中，当积累到一定程度，会对环境构成污染及威胁。

4. 农村生活垃圾

农村生活垃圾是农村居民日常生活中产生的废弃物,如人类粪便尿液、剩菜剩饭等。改革开放以后,在农村经济进一步发展的形势下,农民的生活水平进一步提高,与此同时,农村生活垃圾的数量及种类也逐渐增多。这类废弃物对环境污染的威胁程度及对人体健康的影响都在逐渐增加。农村居民环保意识不够,农村的管理体制不到位,法律法规欠缺和基础设备落后等,都逐渐成为农村生态环境的重要影响因素。

三、农业废弃物的分类

1. 按照农业废弃物的来源分类

① 种植业废弃物:主要指粮食、蔬菜、瓜果、糖料等植物性农产品生产及收获过程中产生的废弃物,如秸秆、残株、杂草、落叶、果实外壳、藤蔓树枝和其他废弃物。

② 动物养殖业废弃物:主要指畜牧、渔业生产过程中产生的残余物,包括畜禽粪便、脱落的羽毛、畜禽饲料残渣、畜禽圈舍的垫料、水产养殖塘污泥等,还包括病死的畜禽尸体、养殖过程中产生的污水等。

③ 农产品加工业废弃物:指农产品加工过程中产生的残余物,包括农、林、牧、渔业加工过程中产生的残余物。农产品在初加工过程中产生的废弃物主要包括稻壳、玉米芯、花生壳、甘蔗渣等。

④ 农村生活垃圾:主要指人类粪便、尿液及生活废弃物,主要包括塑料袋、建筑垃圾、生活垃圾等组成的混合体。

2. 按照农业废弃物的利用价值分类

① 资源性农业废弃物:指在当前的技术、资金和劳动力等条件允许的情况下,农业或农产品加工业的副产品中能作为原材料被再生利用的部分,包括资源性农业生产废弃物(植物性废弃物秸秆、食用菌栽培废料及动物性废弃物,如畜禽粪便等)和资源性农产品加工废弃物(甘蔗渣、屠宰污血和污水等)。

② 非资源性农业废弃物:指在现有技术、资金等条件下,农业生产和农产品初加工过程中产生的废弃物及农村居民生活垃圾中不能被循环利用的部分,如化肥农药、废弃农膜、杀虫剂、农业温室气体、废弃塑料袋等。

3. 按照农业废弃物的化学性质分类

① 有机农业废弃物:指农业生产和农民生活过程中产生的有机类物质的总称,具有可资源化利用的特点,主要包括农业生产或产品初加工过程中产生的植物类废弃物(如农作物秸秆、果壳、杂草落叶、果渣等)、动物类废弃物(如畜禽粪便)和农村居民的生活垃圾(如人类粪便、厨余垃圾等)。

② 无机农业废弃物：指农业生产过程中产生的残余物或农村居民的生活垃圾中无法再生利用、投入环境中无法自然降解，需要采用特殊措施进行处理的、自身物质由无机成分构成的农业废弃物，包括农膜农药残留包装、化学药剂、废旧家电和电池等。

4. 按照农业废弃物的形态分类

① 农业固体废弃物：指在整个农业生产过程中和农民生活中产生的固体类废弃物，如畜禽养殖废弃物、农作物秸秆、农用塑料残膜等。

② 农业液体废弃物：指在整个农业生产过程中或农民生活中产生的液体废弃物，主要包括畜禽等动物性农产品生产过程中产生的污水、农民生活中产生的生活废水等。

③ 农业气体废弃物：指农业生产过程中和农业产品加工过程中排放的 CO、CH_4、N_2O 等温室气体，主要来源于农村用电、农业机械总动力、农用柴油使用、化肥施用等。

5. 按照农业废弃物的性质分类

① 危险农业废弃物：指对环境或者人体健康造成有害影响、不排除具有危险特性、需要按照特殊措施进行相应处理才能降低或者消除其危害的农业废弃物，如化肥、农药包装物、生活用品中的塑料制品、农膜等。

② 一般农业废弃物：指对环境或人体健康不具有明显的危害性、可以自然降解、不需要特殊处理的农业废弃物，如农作物秸秆、畜禽粪便等。

第二节 农作物秸秆循环利用

随着我国传统农业向现代化农业转变，以及改革开放以来我国经济、社会快速发展，传统的秸秆利用途径发生了翻天覆地的变化，尤其是随着农业科学研究和农业废弃物综合利用科学研究的日渐深入，科技进步和技术创新为农作物秸秆利用开辟了新方法、新途径。人们已经认识到农作物秸秆的巨大的、潜在的经济价值，农作物秸秆可以变废为宝、多渠道综合利用，农作物秸秆可以被消解，农民从中可获得额外的经济利益，政府可取得较好的社会效益，一举多得。

一、农作物秸秆循环利用的原则

1. 资源化原则

农作物秸秆是农作物的重要副产品，也可以是工业、农业生产的重要资源，因此，处理农作物秸秆废弃物应该从资源化和能源化两方面着手。

2. 因地制宜原则

农作物秸秆废弃物循环利用应该具体问题具体分析，根据不同地区的农作物秸秆产量及当地的经济水平、科技水平等情况，选择适宜的农作物秸秆资源化利用技术。

3. 可持续化原则

农作物秸秆废弃物资源化应该以提高综合利用率为目标，重点是突出农作物秸秆利用价值，降低农业生产成本，促进农业的可持续性绿色发展。

二、农作物秸秆循环利用途径分析

目前，我国农作物秸秆循环利用途径主要体现在肥料化利用、饲料化利用、能源化利用和工业化利用等方面。

我国秸秆利用中，根据刘建胜的8省调查研究，在各类农作物秸秆综合利用途径中，秸秆用作燃料的比例占主导地位；排在第2位的是饲料化，但是棉花秸秆基本不作为饲料，因为其木质素含量高，难以消化，一般都是用作燃料，利用比例高达83%；排在第3位的是秸秆还田，用作肥料。

1. 秸秆肥料化

秸秆作为肥料的循环利用方式主要有秸秆直接还田和秸秆间接还田两种，秸秆直接还田包括秸秆覆盖还田、秸秆翻压还田、秸秆留高茬还田；秸秆间接还田包括秸秆过腹还田、秸秆过腹沼肥还田、秸秆菌糠还田、秸秆菌糠沼肥还田、秸秆堆沤还田、秸秆沼肥还田、外置式生物反应堆、草木灰还田等。因为农作物秸秆中含有的有机质和氮、磷、钾、镁等元素都是农作物生长的必需营养物质，故其是丰富的肥料资源。当农作物收割以后，农作物秸秆以适合的方式还田后，会大幅增加农业有机肥，使土壤中的氮、磷、钾等元素增加，尤其是钾元素的增加最为明显，而且土壤的活性及有机质也有一定的增加，这对于改善土壤结构有着重要作用。在实践过程中，农作物秸秆的不同利用方式大多数是相互结合、互为循环的，最终实现能量的高效梯级利用。

2. 秸秆饲料化

农作物秸秆饲料化主要是通过生物法、物理法、化学法、干贮法等方式，改变秸秆的长度、粗度、硬度等，把秸秆转化为优质的饲料，提高其适口度和可消化率，饲喂牛、马、羊等大牲畜，并将其粪便还田，即过腹还田。此过程对提高秸秆饲料的营养成分等作用显著，具有简单易行、省工省时、便于长期保存、全年均衡供应饲喂等特点，既解决了冬季牲畜饲料缺乏的问题，又节省了饲料粮，具有广阔的推广应用前景。

3. 秸秆能源化

农作物秸秆的能源化利用方式主要有直接燃烧、转化成气体燃料和转化成液体燃料3种。具体有秸秆生产沼气、秸秆固体成型燃料、秸秆气化、直接发电、秸秆生产乙醇等方式，其中秸秆气化（气化成沼气、水煤气等）目前在国内已经开始得到较大规模的推广应用。

4. 工业化

农作物秸秆的工业化利用范围非常广泛。秸秆作为原料可用于建材、造纸、生产板材、轻工、纺织和化工等领域。此外，农作物秸秆还可以转化为基础料，主要用于培养食用菌基料、育苗基料、花木基料、草坪基料等。

三、农作物秸秆污染防治技术

1. 秸秆还田技术

秸秆还田是将秸秆发酵后施用于农田当中，或者将秸秆粉碎后埋于农田中进行自然发酵。秸秆还田是改良土壤、提高土壤中有机质含量的有效措施之一。农作物秸秆作为种植业中产量最大的农业废弃物，为了防止秸秆焚烧等带来的环境污染，通过各种形式的秸秆还田可以实现"粮食生产—农作物秸秆—还田—有机肥料—粮食生产"的农业循环，做到农业可持续发展。

在我国，几十年来，由于化肥的肥效快、施用方便、可以使粮食等农产品产量大幅增加，化肥在我国农村广泛使用。目前，我国已经成为世界上使用化肥量最多的国家之一。但是我国的化肥利用率较低，大约为30%，大量的化肥流失，不但造成了农业资金投入的浪费，而且由于化肥的使用不当，对农村环境和农业生态环境，尤其是对农田土壤环境造成的不良影响及危害日益严重。此外，我国大部分地区并没有采取有效的还田措施，导致耕地连年种植，土壤有效养分得不到及时补给，土壤中的有机质含量下降，肥力逐年降低。这种种大于养、产大于投的种植方式，造成土壤板结酸化、农作物营养不良、病虫害增多等严重后果，既不利于农业生产，也不利于生态环境良性发展。

秸秆还田在我国具有悠久历史，秸秆还田可以以草养花、以草压草，实现用地、养地相结合，达到培肥地力的目的。同时，秸秆还田可以减少化肥的施用量，增加土壤中有机质含量和速效养分的含量，缓解氮、磷、钾肥比例失调的矛盾；还可以调节土壤的物理性能，改善土壤结构，形成地面覆盖，对土壤水分蒸发、贮存具有调节作用；降低病虫害发生率，从而达到有机农业生产的基本要求，改善农业生态环境质量，避免秸秆就地焚烧造成的环境污染。农田覆盖秸秆后，冬天低温时节地下温度可提高 0.5~0.7 ℃，夏天高温时节地下温度可降低 2.5~3.5 ℃，土壤的水分可以提高 3.2%~4.5%，杂草可以减少 40.6% 以上。所以说，农作物秸秆还田是农作物秸秆污染防治和推进农业循环经济的重要技术。

农作物秸秆还田的方式有很多，都是根据我国不同地域的农业生产所总结出的简便易行、成效显著、便于推广的农业还田模式。

(1) 秸秆直接还田技术

秸秆直接还田主要采用机械作业,机械化程度高,秸秆处理时间短,腐烂所需时间长。

1) 机械直接还田技术

机械秸秆还田技术以机械粉碎、破茬、深耕和耙压等机械化作业为主,将秸秆粉碎后直接还田,增加土壤有机质,提高作物产量,减少环境污染,是争抢农时的一项综合配套技术。秸秆机械粉碎还田技术是大面积实现"以田养田"、保护环境,建立高产、稳产农业的有效途径。

秸秆粉碎还田是采用机械一次作业将田间直立或者铺放的秸秆直接粉碎还田,使手工还田多项工序一次完成,这种方式能使生产效率提高 40 倍。使用秸秆粉碎根茬还田机可以使秸秆粉碎与旋耕灭茬融为一体,能够加速秸秆在土壤里腐解,从而使营养物质被土壤吸收,改善土壤的团粒结构和理化性能,增加土壤肥力,促进农作物持续增产增收。这种形式主要在我国的华北地区采用,水热条件好、土地平坦、机械化程度高的地方都可以使用秸秆粉碎还田技术。农作物秸秆粉碎的长度要小于 10 cm,均匀地撒在农田里,在使用的过程中配合施用氮肥 300~600 kg/hm^2。同时,为了夯实土壤、加速农作物秸秆的腐化,必须在整好地后浇好"蹋墒水"。

秸秆整秆还田技术主要是指小麦、水稻和玉米秸秆的整秆还田机械化,可将田间直立的农作物秸秆翻埋或者平铺于地面。

机械直接还田技术高效低耗、省时省工,容易被广大农民接受。而且这项技术可通过地表秸秆残茬覆盖保护土壤,提高土壤蓄水保墒能力,达到农业高效、高产、优质、低耗的可持续发展。但是其也有缺点,如在丘陵、山地中,由于耕地面积小,机械使用不方便,影响还田效果;而且前期投入比较大,成本较高,推广起来较难。

2) 秸秆覆盖还田技术

秸秆覆盖还田技术是将农作物秸秆粉碎后或者使整秆直接覆盖在农田里。秸秆覆盖可以使土壤饱和导水率提高,土壤蓄水能力增加,可以调控土壤供水,提高水分利用率,促进地上的作物生长。农作物秸秆在农田里腐烂以后,还能增加土壤的有机质,补充土壤中的氮、磷、钾等营养成分,改善土壤的物化性能。秸秆覆盖还可以调节土壤温度,有效缓解气温变化对农作物的伤害。

农作物秸秆覆盖还田技术在使用时要注意农作物秸秆的覆盖厚度为 3~5 cm,覆盖均匀,地表的秸秆覆盖率大于 30%,以此来保证顺利地完成播种等种植任务。整秆覆盖法比较适宜于干旱地区及北方地区的小面积人工整秆倒茬间作,而高留茬覆盖还田技术主要应用于我国麦、稻种植区。

目前,在我国农村常用的秸秆覆盖还田技术有以下几种形式:

A. 农作物秸秆直接覆盖还田。这种方式十分简单，仅是把农作物秸秆直接覆盖在农田土壤的表面，与免耕播种相结合，农田的蓄水、保水、增产效果都十分明显，而且工序少、成本低，便于就地抢农时播种。

B. 农作物秸秆高留茬覆盖还田。这种方式主要应用于小麦、水稻等秸秆，分为高茬覆盖打碎覆盖和高茬休闲覆盖两类。高茬覆盖打碎覆盖适合旋耕播种、硬茬播种、先撒籽后旋耕播种等播种方式。高茬休闲覆盖主要包括旋耕覆盖、深松覆盖、整秆覆盖，具体操作为：在收割时，留秸秆高茬 20~30 cm，农作物秸秆的还田量约为 2250 kg/hm^2，同时配合使用氮肥 150~225 kg/hm^2，然后用拖拉机犁将其翻入土中，实行秋冬灌溉及早春保墒。这种方式可以就地翻压、省时省力、还田均匀，但是这种形式，因为农作物秸秆还田量少，不足以弥补土地肥力的消耗。

C. 超高茬麦田套稻秸秆还田。这种方式是指在小麦收割前的麦田中撒播稻种，稻种发芽出苗，在小麦收割时留高茬秸秆还田，灌水后麦田直接转化为稻田。超高茬麦田套稻将轻型栽培、节水旱育、免耕及秸秆还田等栽培技术融为一体，可以培肥土壤、保护环境，省工省本（不用育秧和插秧）。

D. 带状免耕覆盖。这是一种新型的保护性耕作技术，使用带状免耕播种机，在农作物秸秆直立的状态下直接播种，实现带状免耕的农作物秸秆集垄覆盖、垄际耕作播种，具有适应性强、生产工序少、生产成本低、应用效果好等优点。

3）机械旋耕翻埋还田技术

这种形式主要适用于玉米秆，这是由于玉米秆木质化程度低，秆壁脆嫩，容易折断。当玉米收获以后，由拖拉机挂旋耕机横竖两遍旋耕，就可以将玉米秆切成 20 cm 左右长的秸秆并旋耕入土。由于玉米秆通气组织发达，遇水易软化，腐解速度快，其营养成分当季就可以被利用。如果按照每公顷秸秆还田量 30 000 kg 计算，相当于每公顷投入碳酸氢铵 345 kg、氯化钾 150 kg、过磷酸钙 975 kg，可以使每公顷水稻增产 1.2~1.65 t。

（2）秸秆间接还田技术

1）秸秆堆沤还田技术

秸秆堆沤还田技术就是使农作物秸秆充分高温腐熟以后，对其进行人为调节和控制，加入畜禽粪便和多种微量元素、生物菌，加工成生物有机肥还田。这种方法就地取材、操作简单，可以解决干旱少雨地区的农作物秸秆不易腐烂的问题，尤其适用于农户分散的小规模应用。秸秆堆沤还田是解决我国当前有机肥源短缺的主要途径，也是中低产田改良土壤、培肥地力的一项重要措施。

秸秆堆沤还田技术根据堆肥条件的差异可分为好氧堆肥和厌氧堆肥两种。

农作物秸秆好氧堆肥技术也可以称为高温堆肥技术，主要是在有氧状态下，利用好氧微生物在高温条件下对农作物秸秆进行降解，加速农作物秸秆中的木质素、纤维素和半纤维素的腐烂降解，从而形成堆肥。

农作物秸秆厌氧堆肥技术，也可以称为自然发酵堆肥技术，主要是将收集的秸秆粉碎后使其与畜禽粪尿充分混合，封闭不通风，使其自然发酵，经发酵的秸秆可加速腐殖质分解，制成质量较好的有机肥，作为基肥还田。为了缩短堆肥时间，还可以采用添加发酵菌、营养液和降解菌等措施。这种农作物秸秆利用技术是我国传统且普遍使用的方法，深受农民欢迎。

根据资料显示，500 kg 腐熟堆肥的肥效相当于 15.2 kg 尿素、24 kg 磷肥。如果长期使用堆腐肥，不仅能够减少环境污染，还可以提高土壤中的有机质含量，同时减少化肥的使用量，提高农产品的质量。

2）秸秆过腹还田技术

农作物秸秆过腹还田在我国有着悠久的历史，是一种效益很高的秸秆利用方式。秸秆过腹还田技术就是对农作物秸秆进行青贮、微贮、氨化处理后，喂给畜禽食用，经过畜禽的肠胃消化再把畜禽粪便当作肥料还田。这种方法既可以达到畜牧业增值增收的目的，同时实现了有机肥还田，形成了"粮食—秸秆—饲料—畜禽—有机肥—粮食"的良性循环，真正形成了节粮型畜牧业结构。实践证明，农作物秸秆过腹还田一举多得，一物多用，既可以增加畜牧业产量，促进农业生产，缓解粮食和饲料供需矛盾，同时还可以提高农作物秸秆循环利用率，减少农作物秸秆对环境的污染。

3）草木灰还田技术

用户用农作物秸秆烧制草木灰，利用农作物秸秆中含有的钾元素制取钾肥。这种技术并不可取。虽说农作物秸秆中的钾含量最高，约为 0.9%，露天焚烧农作物秸秆可以得到一定量的天然钾肥，但是秸秆中其他有机物和氮肥等都会白白浪费掉。钾肥的主要作用是单一的，对于农作物生长来讲，仅对其茎秆坚韧、预防农作物生长期的倒伏而造成减产有一定作用。而农作物生长所需的营养成分是多方面的，不是只有钾肥就可以的。

最重要的一点是，露天焚烧秸秆、烧制草木灰还田会产生大量的浓烟，造成空气污染，影响交通安全等，是农作物秸秆污染环境的罪魁祸首，因此不提倡这种形式，甚至国家和各地政府出台政策将秸秆焚烧作为环境污染的防治对象。

4）沼渣还田技术

沼渣还田技术即"秸秆气化、废渣还田"，是一种生物质热能气化技术。秸秆汽化后，其生成的可燃性气体（沼气）可作为农村生活能源集中供气，汽化后形成的废渣经处理后可作为肥料还田。生产实践证明，沼渣还田技术是一种效益较好的农作物还田方式。秸秆发酵后产生的沼渣、沼液是优质的有机肥料，无毒无害、优质高效，营养成分丰富，腐殖酸含量高，肥效缓速兼备，是生产无公害农产品、有机食品的良好选择。

5）秸秆菌糠还田技术

这种方式是利用农作物秸秆培育食用菌，再经过菌糠进行还田，这种技术可谓经济效益、社会效益、生态效益三者兼得，既可以节省成本，又可以减少化肥污染，保护农田的生态环境。

2. 秸秆饲料化技术

我国农村自古就有利用农作物秸秆作为畜禽饲料的传统，但是直接使用农作物秸秆作为饲料，不是最好的选择。因为农作物秸秆的粗纤维含量较高，粗蛋白、可溶性糖类、胡萝卜素和各种矿物质元素的含量比较低，而且硅酸盐的含量也较高，这就使秸秆作为饲料的适口性差，动物的采食量低、消化率低。如果直接将秸秆作为饲料喂畜禽，而不经过加工处理，不仅会造成畜禽的食用量大增，而且也不能满足畜禽的生理要求。因此，要提高农作物秸秆的营养价值，必须对其进行合理的加工处理，用物理、化学方法使秸秆中的木质素等降解，转化为含有丰富菌体蛋白、维生素等成分的生物蛋白饲料，提高其消化率，使秸秆的饲料利用率得以提高。

（1）农作物秸秆饲料加工改进方法

1）物理方法

物理方法改进农作物秸秆的使用价值主要是通过机械加工、辐射处理、蒸汽处理等方式进行的，在实施过程中主要根据秸秆在口粮中的比例、饲养畜禽的种类及经济条件等因素采取不同方法。

A. 机械加工。主要是通过机械加工使农作物秸秆的长度变短、颗粒变小，使畜禽对秸秆的采食量、消化率及代谢能的利用率都发生改变。这种方法的优势是便于饲喂，防止畜禽挑食，减少浪费。

B. 辐射处理。利用X射线对小麦秸秆、大麦秸秆等农作物秸秆进行照射，以提高秸秆饲料的体外或者体内消化率。

C. 蒸汽处理。通过高温水蒸气，对秸秆化学键进行水解，以达到提高秸秆饲料消化率的目的。但是蒸汽处理耗能太多，推广较难。

2）化学方法

主要是利用化学制剂作用于农作物秸秆，使秸秆内部结构发生变化，从而有利于瘤胃微生物的分解，达到改善秸秆营养价值、提高其消化率的目的。

用于农作物秸秆处理的化学制剂很多，包括甲酸、乙酸等酸性制剂，NH_4HCO_3等盐类制剂，$NaOH$、NH_3等碱性制剂及其他品种。化学方法中，应用最多的是$NaOH$处理和氨化处理。

A. $NaOH$处理。最早源于20世纪初期，有"湿法"和"干法"两种。"湿法"是将$NaOH$配比为秸秆体积10倍的溶液，用其浸泡秸秆一定时间后，用水洗净余碱，剩下的秸秆用来喂畜禽。"干法"是20世纪60年代后期被提出的，主要是将高浓度生物$NaOH$溶液喷洒在秸秆上，通过充分混合，使溶液渗透到秸秆里，不用水洗直接饲喂畜禽。这两种方法都会污染环境，而且畜禽长期大量采食这种方法处理的秸秆饲料，会引起体内矿物质的失衡，影响畜禽健康。因此，从20世纪70年代后，这种方式逐渐被淘汰。

B. 氨化处理。主要是利用秸秆中低含量的氮与氨相遇后发生氨解反应，可以破坏木质素并与多糖链间的酯键结合形成铵盐，作为反刍动物瘤胃微生物的氮源，这是强化饲料消化作用的关键。

3）生物处理

利用生物学方法处理农作物秸秆、进行饲料加工，具有能耗低、成本低、效果佳的优点。生物方法处理秸秆主要有两类：一是用秸秆作为基质进行单细胞培养，可以直接在秸秆上培养能够分解纤维的单细胞生物，也可以用化学或酶的作用来水解秸秆的多聚糖变为单糖，然后培养酵母生产高质量饲料；二是主要分解木质素，破坏纤维素—木质素—半纤维素的复合结构，以此来提高秸秆消化率。现在应用广泛的生物处理法是青贮技术。

4）复合处理

在实际生产中，单一的处理方法并不能达到很好的效果，往往需要通过各种方法结合使用。例如，青贮过程中添加精料是物理处理与生物处理的结合，以碱化和微生物发酵同时进行农作物秸秆处理则是化学处理与生物处理的结合。

目前，热喷技术与碱化—发酵处理是较为理想的农作物秸秆饲料加工技术，适于在经济较为发达的农村进行推广。以上几种方法各有所长，需要根据当地具体条件因地制宜地选择合适的方法。

（2）农作物秸秆青贮技术

青贮技术属于生物处理技术，是利用乳酸菌等微生物的生命活动，通过发酵作用，将青贮原料中的糖类等碳水化合物变成乳酸等有机酸，增加青贮料的酸度，以厌氧的青贮环境抑制霉菌的活动，来保证青贮料的长期保存。

适用于青贮技术的主要是含水率在60%左右，具有一定含糖量的秸秆，如玉米秆、高粱秆、花生蔓等。

青贮的设备种类很多，主要有青贮窖、青贮袋、青贮塔等，最为常见的是青贮窖。青贮设备的要求是密闭、抗压、承重及装卸料方便，设备原材料采用混凝土、塑料制品等。以青贮窖为例，其建筑结构以砖砌混凝土为主，一般采用长方形的半地下式，一端留斜坡，便于运输；青贮窖宽度一般为 2.5~3 m，深度不超过 3 m，长度根据养殖规模而定。青贮窖应选择在干燥、排水好、地势高、土质坚硬、避风向阳、没有粪场、距离畜舍较远的地方。

青贮饲料的调制方法主要有地上青贮法、水泥池青贮法、窖内青贮法和土窖青贮法。青贮技术对原料的含水率要求为70%左右，因此，在操作过程中对含水过高的原料应该适当晾晒，或是混入含水较少的原料进行调配；原料水分偏低的时候，应该均匀地喷洒清水或者是混入含水较多的饲料。青贮的原料应适时收割，以免影响原料产量或者青贮质量，或者导致青贮失败。由于农作物秸秆乳酸菌的生长繁殖要求湿润、厌氧和有一定数量糖类的环境，因此，在青贮时要将农作物秸秆原料切成2～3cm长的段，装填时要层层压紧、压实，排出空气，营造厌氧环境，防止发酵失败。青贮饲料在30～50天就可以取用喂畜禽。取饲料时，一般要从阴面一端开始，逐层逐段取料，每次取料后，对剩余料要立即用塑料薄膜盖好，且严实压紧，防止空气进入使青贮料发霉变质。

青贮饲料添加剂有氨水、尿素、甲酸、丙酸、稀硫酸、盐酸、甲醛、食盐、糖蜜和活干菌等。

影响青贮饲料质量的因素有农作物的品种、生长期、土壤和肥力、气候与地形等；决定青贮发酵程度和方式的主要因素有干物质含量（青贮成败的关键）、水溶性碳水化合物、缓冲能力、硝酸盐含量、氧气和青贮保存能力等。总之，要提高青贮质量，需要对农作物秸秆收获前和收获后进行良好管理，适期收割，且原料水分适当。已经收获的农作物秸秆要尽快切短、装窖、压实、密封，开窖后要加强管理。收获制作过程越快，青贮质量越高。

青贮饲料的品质鉴定主要采取感官鉴定，通过色泽、气味、味道等来评判青贮饲料的质量。优质的青贮饲料为黄绿色或者绿色，有水果的弱酸味或者酒糟味，质地柔软，叶脉明显，茎叶保持原状。

（3）农作物秸秆微贮处理技术

农作物秸秆微贮处理是借助以乳酸菌为主的微生物作用，使秸秆在厌氧的状态下发酵，既可以抑制或者杀死各种微生物，又可以降解秸秆中的可溶性碳水化合物而产生醇香味，提高饲料的适口性。

微贮处理技术需要添加微生物添加剂，如秸秆发酵活干菌、酵母菌等，它们可以将农作物秸秆中的纤维素、半纤维素及木质素等有机碳水化合物转化为糖类、乳酸和其他一些挥发性脂肪酸，来提高秸秆的利用率。

农作物秸秆微贮处理技术在实施过程中应该注意：农作物秸秆原料的糖分要高；农作物秸秆原料的含水率要较高，为55%～65%；要具有密封的厌氧环境条件。

（4）农作物秸秆氨化处理技术

农作物秸秆氨化处理是在密闭的条件下用液氨或者尿素等对秸秆进行处理。从20世纪80年代开始，我国就开始了秸秆氨化的研究和试验，到了20世纪90年代已经在全国普及，氨化秸秆总量达到1000万t，位居世界第一。农作物秸秆氨化处理技术成本低、方法简单、对环境无害，因此使用较为广泛。

农作物秸秆氨化处理技术使秸秆的消化率提高15%～30%，含氮量增加1.5～2倍，且处理后的秸秆营养价值大大提高，适口性好，畜禽采食量增加。

1）农作物秸秆氨化的主要氨源

我国的秸秆氨化的主要氨源有尿素、液氨、碳铵和氨水4种。

A. 尿素 [CO(NH$_2$)$_2$]。尿素含氮量为 46.67%，在秸秆氨化过程中吸收水分，在适宜的温度和脲酶的作用下分解成氨和二氧化碳。尿素在运输、使用过程中不需要专用设备，比较适合农民使用，但是由于秸秆内脲酶含量低，尿素分解不完全，需要在氨化过程中加入 1% 左右的豆饼粉作为脲酶来源，并提高氨化温度。

B. 液氨。液氨分解效果比尿素好，成本低。但是液氨需要用高压容器来贮存，需要当地政府建立专用设备站、提供贮存液氨的氨瓶等。

C. 碳铵（NH$_4$HCO$_3$）。碳铵含氮量在 16% 左右，可以在一定温度下分解成氨、水和二氧化碳。

D. 氨水。氨水浓度为 18%～20%，也是常用氨源，比较适用于运输条件方便、距离氨水厂近的地方。

2）农作物秸秆氨化的主要方法

A. 氨化池氨化法。氨化池为长方体或圆柱体，选址在向阳、避风、地势高、土质硬、方便管理和运输的地方。氨化原料需要是粉碎的或是 1.5～2 cm 的小段农作物秸秆。将秸秆质量 3%～5% 的尿素用温水溶解制成溶液，均匀喷洒在秸秆上，并搅拌均匀，保证每层秸秆都被尿素水溶液喷洒到 1 次，并对其踩实，最后用塑料薄膜盖好池口，四周用土壤覆盖密封。

B. 塑料袋氨化法。塑料袋大小依个人方便为宜，选取结实多层的塑料袋，将切断的秸秆用配置好的液氨溶液均匀喷洒，装满后封严袋口，放在阳光充足、避风、干燥的地方即可。需要注意的是，为保证氨化过程，塑料袋要密封良好。

C. 窖贮氨化法。这种方式选取的设备形式有窖、池等。窖的大小视具体而定，窖可建设为地下式或半地下式。选址也要在向阳、避风、地势高、土质硬、方便管理和运输、距离畜禽舍近、方便取拿的地方。此法要求窖不能漏气、不能漏水、窖壁光滑平整。氨化窖建好后，所需要的氨化原料为长 1.2～2 cm 的农作物秸秆段，用尿素溶液均匀喷洒，装满后需要在原料上盖 5～20 cm 厚的秸秆，再覆盖上 20～30 cm 厚的土并踩实。封窖时，原料要高于地面 50～60 cm，防止雨水渗入。氨化过程中需要时常观察窖的密封性。

以上 3 种秸秆氨化形式是我国比较普遍使用的方法，适用于个体农户的小规模生产。无论是窖还是池都可以多用，既可以氨化，又可以青贮，利用率高，建造成本低，使用时间长，而且由于池和窖都有固定尺寸，便于测量农作物秸秆的量。

D. 堆垛氨化法。这种秸秆氨化法需要选择的场地为地势干燥、无鼠害的平地。准备工作是在地上覆上一层厚约 0.2 mm 的塑料薄膜，长宽需要根据秸秆堆垛的大小而定，一般来讲堆高不要高于 2.5 m。塑料薄膜上堆垛 15～20 kg 的秸秆，秸秆应捆成垛、加上氨水并调整含水率，塑料薄膜的周边留出大约 70 cm 的宽度，再在垛上盖塑料薄膜，将上下薄膜的边缘包卷起来，埋土或用重物压住密封。

E. 氨化炉法。这种方法是将加氨秸秆在氨化炉内加温到 70~90 ℃，保温 10~15 小时，然后停止加热，保持密闭状态 7~12 小时，开炉后让余氨散发 1 天，就可以用于饲喂了。氨化炉的结构可以为砖砌混凝土结构或钢铁结构，需要安装温度自动控制装置、轨道，制作专门的草车，沿轨道运输。氨化炉一次性投入大、成本较高，优势是经久耐用，一天一炉，氨化处理时间短，不受季节、天气等因素影响，生产效率高。

F. 真空氨化法。这种方法主要是在澳大利亚等发达国家使用，将秸秆装入容器后，先用真空泵抽出一部分空气，然后用氨泵注入液氨。

3）农作物秸秆氨化的注意事项

首先，原料的品质对于秸秆氨化后的改进幅度有很大影响。其次，氨化的秸秆要求为含水率为 30% 左右的干秸秆，含水率过高不便于运输操作，且具有霉变的可能性。再次，氨的经济用量关乎秸秆的消化率，因此用液氮、尿素等处理秸秆时要根据各自的含氮量进行计算，要求是在秸秆干物质质量 2.5%~3.5% 的范围内。另外，环境变化及氨化时间也是需要注意的因素，一般环境温度与氨化时间成反比。用时随用随取，取料后一定要注意密封，秸秆氨化后的饲料不能直接饲喂畜禽，需要晾晒 1~2 天再饲喂。

（5）农作物秸秆颗粒化处理技术

农作物秸秆颗粒化处理需要借助机械，将农作物秸秆粉碎、揉搓成一定长度之后，再按照配方把各种原料搭配并且混合一定时间后，用特定型号的颗粒机制成颗粒饲料。

这种方式的优点是易将维生素、添加剂等成分强化进入颗粒饲料中，提高饲料的营养价值，实现营养均衡，改善适口性。这种技术操作简单、实用性强、饲喂效果明显，而且投资不多，比较适合农民使用，能够很好地解决秸秆本身存在的对规模发展畜牧业的制约，避免了大部分秸秆利用技术就地处理使用的弊端。其生产模式主要有自动化高效生产模式、半自动化生产模式和户用型生产模式。

（6）农作物秸秆热喷处理技术

农作物秸秆的热喷处理是将饲料原料（秸秆、鸡粪、饼粕等）装入饲料热喷机内，向机内通入热饱和蒸汽，经过一定时间，使物料受到高压热力的处理，然后对物料突然降压，迫使物料从机内喷爆于大气中，从而改变其结构和某些化学成分，并经过消毒、除臭过程，使物料变成更有营养价值的饲料。这是一个压力和热力加工过程，需要特殊的热喷装置及独特的工艺流程来完成。

农作物秸秆经过热喷加工处理后，消化率可以提高 50%，全株采食率可提高到 90% 以上，还可以对菜粕、棉粕等进行脱毒，使其利用率提高 2~3 倍。此方法还可对鸡鸭粪便、牛粪进行除臭、灭菌等处理，使其成为正常的蛋白质饲料。

（7）农作物秸秆膨化处理技术

农作物秸秆的膨化处理技术现已被广泛应用于秸秆饲料的生产加工过程中。在膨化机内部高温高压的作用下，农作物秸秆被膨化，最终达到熟化、膨化的效果。其工作原理是秸秆通过螺杆挤压方式被送入膨化机,在这个过程中,螺杆螺旋推动物料向着轴线方向流动,在螺杆螺旋、机筒与物料的摩擦作用下,物料被强烈挤压、搅拌和剪切,最终达到被细化、均化的目的。机器内部的压力和温度会随之升高,使得其内部已经被粉碎的秸秆物料熟化,由粉末状变成糊状,从膨化机的模孔喷出。在物料被喷出膨化机的瞬间,物料周围的温度和压力迅速降低,物料在这种压力差的作用下膨化、失水,形成疏松、多孔的膨化饲料。

3.秸秆能源化技术

生物质能是太阳能以化学能的形式贮存在生物质中的能量,可以直接或间接来源于绿色植物的光合作用,是唯一可存储和运输的可再生能源。农作物秸秆作为生物质能资源的主要来源,是目前世界上仅次于煤炭、石油及天然气的第四大能源。与化石能源相比,生物质能具有清洁环保、可再生、分布分散、资源丰富等特点。因此,通过物理、化学等方法,对秸秆的纤维素、半纤维素等主要成分进行有目的的转化利用,是农作物秸秆污染防治与循环利用的重要课题。

(1)农作物秸秆直接燃烧技术

1)农作物秸秆燃烧供热技术

秸秆直接燃烧是秸秆能源利用的主要途径,也是传统的能量转换方式,成本低,容易推广。在我国,秸秆主要用于农户炊事用能,每年的直接燃烧量占农作物秸秆能源利用总量的99%以上。根据资料显示,秸秆热值约为15 000 kJ/kg,是标准煤的50%,不同秸秆的热值不同。

秸秆直接燃烧供热系统是以秸秆为燃料,以专用秸秆钢化炉为核心形成的供热系统。该系统由秸秆直燃热水锅炉、配套的秸秆收集与前处理系统和供热管路等组成,可以为乡镇机关、中小学校及相对集中的乡镇居民和经济比较发达的自然村提供热水及冬季采暖用能。但是这种方式具有效率低、污染环境等问题,因此推广起来较为困难。

2)农作物秸秆直接燃烧发电技术

农作物秸秆直接燃烧发电是指把秸秆原料送入锅炉中直接燃烧产出高压水蒸气,通过汽轮机的涡轮膨胀做功,驱动发电机发电。目前来讲,这种秸秆直接燃烧发电技术主要有水冷式振动炉排燃烧发电技术和流化床燃烧发电技术两类。

农作物秸秆直接燃烧发电需要注意的问题是解决床料烧结、受热面高温碱腐蚀及积灰问题。这是因为,秸秆燃烧后灰量很少,难以形成床料,所以多以河沙为床料,但是实际上会发生燃结现象,温度是主要因素。出现这种现象是因为生物质灰中富含碱金属(Na、K)氧化物和盐类。这些元素的化合物与沙子中的SO_2发生化学反应,生成低熔点的共晶体,熔化的晶体沿沙的缝隙流动,使沙粒结块,破坏流化。

（2）秸秆气化技术

1）秸秆气化集中供气技术

这项技术是我国十分重视的农村能源建设新技术，从1996年开始在全国发展起来。它是以农村丰富的秸秆为原料，经过热解和还原反应后生成可燃性气体，再通过管网送到农户家中，供炊事、采暖使用，改善了农民原有的以薪柴为主的能源消费结构。

由于生物质燃气在常温下不能液化，需要通过输气管网送至用户，因此农作物秸秆气化集中供热系统要以自然村为单位设置气化站（气柜设在气化站内），敷设管网，通过管网输送、分配生物质燃气到用户的家中，规模大小可以是数十户到几百户不等。集中供热系统包括原料处理机（粉碎机等）、上料装置、气化炉、净化装置、风机、储气柜、安全装置、管网和用户燃气系统等设备。因为燃气特性不同，生物质燃气的燃烧需要专用灶具。

现在我国农村规划多为整齐划一，因此有利于该技术的推广实施。因为项目规模小，省去了主干管网的敷设，输送管网距离短，降低了输送成本和对管材的要求，使项目投资成本大大降低。秸秆气化集中供气技术改善了农民的生活，提高了其生活舒适度，节省了用于炊事的劳动量和时间，而且使农作物秸秆得到了资源化利用，节省了木材的消耗，保护了环境。但是，农作物秸秆汽化后产生的粗燃气含有焦油等有害杂质，如处理不当，会造成周围空气、水和土壤的污染。

2）农作物秸秆气化发电技术

农作物秸秆气化发电是指生物质原料在缺氧状态下发生热化学反应转化为气体燃料（CO、H_2、CH_4），然后将转化后的可燃气体由风机抽出，经过冷却除尘、去焦油和杂质后，供给内燃机或者小型燃气轮机带动发电机发电。目前，秸秆气化发电主要应用于较小规模的发电项目。

（3）农作物秸秆发酵制沼技术

农作物秸秆发酵制沼主要是以秸秆为发酵原料，在隔绝空气并维持一定温度、湿度、酸碱度等条件下，经过沼气细菌的发酵作用生产沼气。沼气是一种可再生的清洁能源，以甲烷为主，其含量一般为55%~70%，沼气热值为20~25 MJ/m^3，燃烧热效率高，1座用户沼气池每年大约可代替0.8 t标准煤，节省农户50%以上的生活能源。

使用农作物秸秆、畜禽粪便等农业废弃物发酵制沼，不仅可以减少农户煤的使用，节约经济成本，而且使用沼渣、沼液种植蔬菜和其他农作物，能够提高农作物产量，改善农作物品质，降低生产成本，肥田效果佳，可以实现农作物秸秆的综合利用，体现了循环经济的效益。农作物秸秆发酵制沼技术从规模上来讲，可以是单独农户沼气利用，也可以敷设管道，集中供气。

农作物秸秆发酵制沼技术有两种方式。第一种是直接使用沼气池制取。该方法主要是把农作物秸秆及杂草、灌木枝条等农林废弃物混合牲畜的粪尿后，直接放入沼气池，在隔绝空气的条件下，调节合适的温度、湿度，经过微生物的发酵作用生产沼气。农作物秸秆制沼需要预处理，一般采取的是物理方法，即将秸秆粉碎或者是铡碎，然后在厌氧微生物的发酵作用下，生产沼气。为了保证沼气的正常生产，需要严格厌氧，沼气池必须密闭排空空气。由于秸秆不容易被微生物或者酶直接利用，因此需要在发酵时添加富含氮素的原料，来减少发酵启动时间，提高沼气产量。这就需要碳氮比正常，一般控制畜禽粪便与农作物秸秆的比例为2∶1。控制好温度，保证在25~40 ℃这个温度范围内，温度越高发酵越好，产生的沼气也越多。pH值为6.5~8，可以通过碳酸氢铵及石灰水等调节pH值。

此外，还有一种方式是对粉碎的农作物秸秆进行预处理，并按照产量需要量放入猪圈舍内，由猪来进行堆沤，干湿发酵，竹笼导气，还田循环。

（4）农作物秸秆固体生物质燃料技术

农作物秸秆固体生物质燃料技术是将农作物秸秆等农业废弃物粉碎，然后在一定压力和温度作用下，通过固体成型设备将农作物秸秆等压缩成型，主要有颗粒状、棒状和块状3种。该技术可以将秸秆中生物质的纤维素、半纤维素和木质素等在200~300 ℃温度下软化，用压缩成型机械将经过干燥和粉碎的松散生物质肥料在超高压的条件下，靠机械和生物质废料之间及生物质肥料相互之间摩擦产生的热量或外部的加热，使纤维素、木质素软化，经挤压成型后得到具有一定形状和规格的新型燃料。目前，这种技术能够提高生物质单位容积的质量和热值，其燃烧效率超过80%，产生的SO_2、氨氮化合物和灰尘少，减少了空气污染和CO_2的排放，方便运输和储存，可以实现商品化，具有较好的社会效益和环境效益。

从广义上讲，生物质燃料技术工艺可以划分为湿压成型技术、加热压缩成型技术和炭化成型技术3种形式。

1）湿压成型技术

这种技术常用含水量较高的原料，将原料水浸数日后将水挤走，或对原料喷水，加黏结剂搅拌混合均匀，利用简单的杠杆和木模等将腐化后的农业废弃物中的水分挤出，压缩成成型燃料。这样成型的燃料密度较低，湿压成型设备较为简单，容易操作，但是设备零部件磨损率高，烘干费用也高，产品的燃烧性差。

2）加热压缩成型技术

加热压缩成型的流程包括原料粉碎、干燥混合、挤压成型、冷却包装等环节。生产工艺流程不同，采用的压缩成型设备的工作原理也不同，主要采用螺旋挤压型成型设备、活塞冲压型成型设备和压模辗压式颗粒成型设备。

螺旋挤压型成型设备采用连续挤压的方式，生产的成型燃料通常为空心燃料棒。

活塞冲压型成型设备不用加热，采用冲压的方式把原料压紧成型，成型密度较大，对物料含水量要求较宽，但是生产率较低，产品质量不稳定，生产的成型燃料为实心燃料棒或燃料块。

压模辐压式颗粒成型设备多用于生产颗粒状的成型燃料,不需要外部加热,靠物料挤压成型时产生的摩擦热即可使物料软化、黏合。这种方式对原料的含水率要求为10%~40%。

3)炭化成型技术

炭化成型技术主要有两种形式:一是先炭化后成型,将生物质原料炭化成粉粒状木炭,然后再添加一定的黏结剂,用压缩成型机将其挤压成一定规格和形状的成品木炭;二是先成型后炭化,用压缩成型机将松散细碎的原料压成具有一定密度和形状的燃料棒,然后用炭化炉将燃料棒炭化成机制木炭。

(5)农作物秸秆液化技术

农作物秸秆液化技术是通过物理、化学或者生物方法,使秸秆的木质素、纤维素等转化为醇类、可燃性油或其他化工原料,目前有3种形式,分别为直接液化、高温高压液化和微波液化。

① 直接液化。这是指在中低温、高压并有催化剂参与的情况下,将生物质转化为液态的热化学反应过程,有 CO 等还原性气体参与,可以分为反应产物保留植物纤维原料的大分子结构和破坏原料的大分子结构两类。农作物秸秆生产乙醇需要提前进行预处理。

② 高温高压液化。这是在高压下发生热化学反应的过程,温度为 300~500 ℃,通过催化剂催化。此技术耗能较大,对设备耐压要求较高,主要应用于农作物秸秆制作柴油。

③ 微波液化。这是利用微波辐射使小分子极性物质产生物理效应,从而加速反应、改变反应机理或启动新的反应通道的技术。

4. 秸秆工业化技术

农作物秸秆的工业化广泛应用于建材工业、轻工业和纺织工业。目前,我国农作物秸秆主要用于造纸,占农作物秸秆总量的 2.3%;此外,还用于建材生产等。经过技术加工,秸秆还可以生产糠醛、酒和木糖醇等。

(1)农作物秸秆造纸技术

秸秆是中国造纸工业的重要原料,在 20 世纪 90 年代,秸秆造纸的应用量达到了 2000 万 t 以上。造纸技术在我国有 2000 多年的历史,是我国"四大发明"之一。目前,中国基本使用木材纸浆造纸,根据用途需要有机械制浆和化学制浆两种制浆方式。由于木材资源紧缺,近年来世界上积极研发非木材植物纤维的制浆造纸术。我国江南地区多数采用农作物秸秆中的纤维素部分作为原料制作纸浆。但是,在生产过程当中,秸秆制浆造纸会产生"黑液",这些生产废水未经处理排放到水循环中,不但会造成资源的巨大浪费,同时也会因"黑液"当中含有难以降解的物质而污染水源。农村的小造纸厂因资金紧缺,技术设备落后,无法建设并使用污水处理设备,因此国家禁止小型造纸厂采用秸秆制浆。以农作物秸秆为原料制浆造纸与废液处理技术是农作物秸秆循环利用技术需攻克的难题。

1）废液减量处理制浆造纸技术

这项技术的工艺过程为：提取农作物秸秆纤维，将纤维洗净、剪碎，进料、添加8%~15%的NaOH，缓压揉搓，制作纸浆。其原理是，将农作物秸秆纤维原料处理添加量增加30%，减少8%~12%的碱用量，利用缓压揉搓机械使纤维摩擦产生热量，在高温条件下制作纸浆。

此项技术可以使黑液废水量减少，而且在缓压揉搓过程中的升温也可以使纤维质分散，免除蒸解分散过程，同样减少废液的排放量。

2）生物酶处理制浆造纸技术

此技术主要是利用微生物和酶制剂处理农作物秸秆，然后再用简单的物理或者化学方法制浆造纸，其特点是成本低、环境污染风险低。

3）高速旋转纤维制浆造纸技术

这种造纸技术是将农作物秸秆纤维原料剪碎，把水和纤维混合后加入机器当中，通过机械的高速旋转使纤维分散，然后搓揉分散的纤维质，洗净制浆。

（2）农作物秸秆制造建筑、包装材料技术

农作物秸秆作为建筑材料自古就有。古代人们利用茅草作为瓦片的替代品，铺在屋顶上。在农村，人们将铡碎的农作物秸秆与泥土混合建造土坯房。农作物秸秆的纤维质既可以充当房屋的覆盖材料，也可以成为起加固作用的建筑材料。

1）农作物秸秆建材生产技术

农作物秸秆在建材领域应用广泛，可以做复合板材、纤维板等板材制品，还可以做石膏基、水泥基等。目前，我国将秸秆应用于节能、高强、利废、施工效率高、保护环境好等的新型建筑材料中。现在农作物秸秆制作的玻璃纤维增强复合材料、石膏板等新型建材已经成为主导的建筑材料。

当前，利用农作物秸秆生产建材，主要有模压秸秆墙体材料、挤压秸秆墙体材料、秸秆轻质保温内衬材料和定向结构板组合墙体材料等，主要工艺流程可以总结为集成工艺和碎料板工艺两种。

集成工艺主要是利用秸秆制造人造纤维板，这种工艺适用于小麦、水稻秸秆，加工过程中无须黏结剂，厚度为20~80 mm，可用于墙体材料。

碎料板工艺主要是利用秸秆制造人造纤维板，有秸秆硬质板材、秸秆轻质材料和秸秆复合材料3种。这种工艺主要应用于建筑材料、包装材料及家具、室内装修等。

2）可降解包装材料生产技术

塑料包装材料是环境污染源之一，每年消耗量很高。现在我国以可持续发展理念大力发展"绿色包装"。以农作物秸秆粉碎物和黏合剂作为原料，它们经过混合、交联、发泡等工艺加工后，可以做成具有减震缓冲作用的包装材料，具有可降解性，能够减少环境污染。例如，使用稻草为主要原料制成的新型无污染的植物纤维发泡包装，不仅可以在短时间内降解，在其腐烂降解后还可作为饲料原料，实现循环利用；

用农作物秸秆和玉米淀粉为原料,可以生产出可降解的聚乳酸包装材料,可用于包装材料、纤维和非织造物等领域,其强度、耐药性和缓冲性等都能与聚苯乙烯相媲美。

3）一次性可降解餐具生产技术

我国每年对一次性餐具的需求量极大,而利用农作物秸秆生产的一次性餐具具有物美价廉、性能好、无污染、可降解等优势,拥有广阔的市场,具有良好的经济效益、社会效益和环境效益。

一次性秸秆餐具主要是利用废弃农作物秸秆的天然植物纤维,将其粉碎成物料后,添加符合食用卫生标准的安全无毒的成型剂,经过加工制成可以完全降解的绿色环保餐具。此产品不仅无毒无害、防水耐高温、强度高、不变形,而且可以自行分解,之后循环利用成饲料或者肥料。

（3）秸秆生产工业原料技术

1）电子工业用高纯四氯化硅的生产技术

水稻、小麦、玉米等的秸秆富含硅元素,现已有采用水稻秸秆生产四氯化硅的生产技术,以满足电子工业对硅元素的需要。在400~1100 ℃的条件下,通过燃烧或炭化处理秸秆,所得的碳与含氯碳化合物（或盐酸）和含碳化合物混合物反应,在沸点56.8 ℃的条件下精馏得到超高纯的$SiCl_4$。

2）木质素黏合剂生产技术

以农作物秸秆为主要原料,可提取其木质素成分和甲醛发生接链反应制取黏合剂。此产品优势为黏合强度大、耐水、成本低、可生物降解等。

3）农作物秸秆陶瓷釉生产技术

为了防止陶瓷上釉烧制过程中出现釉质流动现象,通过添加秸秆灰、硅石和陶土等,尤其是水稻秸秆草木灰,富含硅元素,添加到釉料中可以得到优质的白色不透明的釉彩。

5. 农作物秸秆的其他利用技术

（1）菌类基料制作技术

农作物秸秆具有丰富的碳、氮、矿物质及激素等营养物质,十分适合食用菌的生长需要。农作物秸秆资源丰富、产量大、成本低,因此农作物秸秆适合食用菌基料的制作生产。

目前,我国利用农作物秸秆栽培食用菌技术十分成熟,而且投资少、见效快、技术含量低、不受客观因素限制,因此容易推广。现在我国已经可以利用农作物秸秆基料栽培20多种食用菌,如香菇、金针菇等一般产品及猴头菇、灵芝等名贵菌类。此外,秸秆栽培食用菌后的菇渣,由于菌体的生物降解作用,氮、磷等元素的含量也明显增加,因此还可以作为优质的肥料进行农业循环生产。

农作物秸秆用于食用菌基料制作时,应该严格按照《食品安全法》的要求进行卫生环境的消毒清理,保证良好的生产环境。此外,要选择良好的农作物秸秆原料,如选择新鲜、无霉变的秸秆,以保证食用菌的正常生长。还需要根据不同菌类认真选择配方,以保证食用菌的产量和质量。

(2)农作物秸秆工艺编织技术

利用秸秆手工编织制作工艺品,具有审美和实用价值,可以用于家具、室内装饰等领域。

以农作物秸秆,如麦秸、玉米皮编织的提篮、壁挂、蒲团、帘子等,具有回归自然的绿色风潮,深受城市消费者的喜爱,满足人们回归自然的理想。尤其是一些具有地方特色和少数民族特色的农作物秸秆工艺编织品,具有较高的文化价值,同时具有较高的经济价值。

第三节 畜禽养殖业废弃物循环利用

一、畜禽粪便的循环综合利用技术

畜禽粪便中含有丰富的有机质及氮、磷、钾等营养成分,是宝贵的资源,如果对其妥善处理,进行合理的资源化循环利用,可以对生态环境保护起到重要作用,同时也能为农业生产的经济效益提高带来巨大作用。目前,对畜禽粪便的资源化综合利用主要是从肥料化、能源化和饲料化3个方面进行。

1. 畜禽粪便的肥料化利用

畜禽粪便含有丰富的有机物和大量的氮、磷、钾等元素,用于农作物耕作中,可以改善土壤结构,提高土壤肥力,提高农作物的产值、改善其品质,是历来我国农民长期使用的"农家肥"。

(1)堆肥技术

我国自古以来就将畜禽粪便采用堆肥的方式进行处理,制成农家肥或者直接用于农作物种植。堆肥是在人工控制水分、碳氮比和通风的条件下,通过微生物作用,对固体粪便中的有机物进行降解,使之矿质化、腐殖化和无害化的过程。堆肥是处理各种畜禽粪便很有效的方法之一。在堆肥的过程中,高温可以使粪便中的营养成分释放出来,生成有利于提高土壤肥力的重要活性物质——腐殖质,它可以起到调节、改良土壤的作用;同时,高温还可以杀灭粪便中的各种病原微生物和杂草种子,使粪便达到无害化。由于粪便堆肥可以减少最后产物的臭气,且干燥、易于包装存储,因此被广泛使用。但是堆肥也存在诸多问题,如因为堆肥过程中的微生物活动程度直接决定堆肥的周期和产品质量,因此需要严格控制水分、碳氧比、温度、pH值等控制参数,故对人工操作要求较高。而且堆肥需要的场地大,时间较长(4~6个月),过程中有氨的释放,不能完全控制臭气挥发。

目前，堆肥技术主要有好氧堆肥技术和厌氧堆肥技术两种。

1）好氧堆肥技术

好氧堆肥是在有氧条件下，用好氧菌对废弃物进行吸收、氧化、分解。微生物通过自身的生命活动，把一部分吸收的有机物氧化成简单的无机物，同时释放出可供微生物生长、活动所需的能量；而另一部分有机物则被合成新的细胞质，使微生物不断生长繁殖，产生出更多生物体。好氧堆肥的温度一般为 50~65 ℃，最高可达到 80~90 ℃，因此也称高温堆肥。

好氧堆肥过程需要以下 3 个阶段。

A. 中温阶段（30~40 ℃）。这个阶段是堆肥过程的初期，也是产热阶段，嗜温性微生物较为活跃，并利用堆肥中可溶性有机物进行旺盛的生命活动，这个过程需要 1~3 天时间。

B. 高温阶段（45~65 ℃）。这个阶段，堆肥中残留的和新形成的可溶性有机物继续被氧化分解，复杂的有机物也开始强烈分解，这个过程需要 3~8 天。

C. 降温阶段。这个阶段剩余的较难分解的有机物进一步分解，腐殖质不断增多，进入稳定阶段，当进入到腐熟阶段，需氧量减少，含水率降低，堆肥孔隙度增大，氧扩散能力增强，需要自然通风，这个过程需要 20~30 天。堆肥发酵后可得到无臭、无虫（卵）及病原菌的优质有机肥料，具有运行费用低等特点。

堆肥过程的影响因素有：供氧量要适当，实际所需空气量应为理论空气量的 2~10 倍；物料含水量在 50%~60% 为宜，55% 最理想，此时微生物分解速度最快；碳氮比要适当。水的作用有二：一是溶解有机物，参与微生物的新陈代谢；二是调节堆肥温度，温度过高时水分蒸发可以带走一部分热量。

2）厌氧堆肥技术

厌氧堆肥以粪便原料中的原糖和氨基酸为养料，利用厌氧或者兼性微生物生长繁殖的特性进行乳酸发酵、乙醇发酵或沼气发酵。含水量高于 80% 的粪料主要以沼气发酵为主，含水量低于 80% 的粪料多为乳酸发酵。目前，也有用 EM 菌群对鸡粪进行发酵堆肥处理的。

厌氧发酵技术不需翻堆，不需要通气，节省能耗，费用也比较低，操作方便。同时，经过厌氧处理，可除去大量可溶性有机物，杀死大量的传染性病菌，较为安全。

（2）生物有机肥技术

生物有机肥是将有益微生物与有机肥结合形成的一种新型、高效的微生物有机肥料。生物有机肥的原料主要有鸡鸭等禽粪、猪牛羊等畜粪、其他动物粪便、秸秆、农产品加工废弃物等。其生产工艺一般包括原料前处理、接种微生物、发酵、干燥、粉碎、筛分、包装、计量等，配料方法因原料来源、发酵方法、微生物种类及设备不同有所差异。

生物有机肥富含有益微生物菌群,营养功能强,适应性好,同时富含各种养分,且体积小便于施用,适合规模化生产。将畜禽粪便生产成有机肥可以克服粪便含水量高以及运输、储存和使用不便的缺点,相比其他方法,安全无害。又因生物有机肥能够提高肥料利用率,改善土壤肥力,增加农作物产量,提高其品质,非常适合无公害农产品生产的需求,具有良好的经济效益。畜禽粪便生产有机肥的市场前景光明,经济效益好,同时也具有很好的社会、生态价值。

(3) 生物转化利用技术

生物转化利用技术是利用蚯蚓消化处理畜禽粪便,一般是处理含水量在85%以上、有机质含量高的粪便,如猪粪、牛粪、羊粪及其他禽类粪便。相关研究表明:在土壤中施用蚯蚓堆肥与不施用情况对照比较,土壤中速效氮、磷、钾分别增加15.68、10.71、24.30 mg/kg。蚯蚓堆腐处理的猪粪,有机氮更多地转化为无机氮,减少了氮的挥发;接种蚯蚓处理的未腐熟牛粪比不接种蚯蚓的未腐熟牛粪或自然堆制的腐熟牛粪显著增加了矿质氮和速效磷的含量,提高了碱性磷酸酶的活性,降低了微生物中碳、氮的含量和脲酶的活性。由此可见,畜禽粪便可以经过蚯蚓、蝇蛆处理后再施用,这样能提高粪便的肥效,改良土壤结构,增加土壤透水性,防止土壤表面板结,提高土壤的保肥性。这种从整个生态系统考虑,使畜禽粪便资源化、无害化增值利用的生物方法,不仅可以解决污染问题,还能提高养殖业的经济效益。

除了上述畜禽粪便处理技术外,还能采用快速烘干法、膨化法、微波法等技术生产高效优质的肥料。

2. 畜禽粪便的能源化

主要用直接燃烧、沼气技术和发电等方式实现。

(1) 直接燃烧

直接燃烧方式主要是在草原地区使用,牧民们收集晾干的牛、马等动物的粪便,作为燃料直接燃烧,用来烧饭取暖。吉林省利用畜禽粪便与秸秆、煤灰等物质加工制成牛粪煤作为能源。这是畜禽粪便能源化的最简单的方法,但这种方式容易产生大量浓烟,产生空气污染,又存在卫生问题,随着人民生活水平的提高,这种燃料逐渐被其他燃料取代。

(2) 沼气技术

目前,我国规模化畜禽养殖场逐年增加,且大多数养殖场都采用水冲式清粪方式,造成粪便含水量高。对于这种畜禽粪便,目前较多采用沼气技术进行处理。

沼气法主要采用厌氧发酵技术将粪便、垃圾、杂草与污水等按照一定比例,在一定的温度、水分、酸碱度等条件下,经过沼气细菌厌氧发酵生产出以甲烷为主的可燃气体。这种方法可以将畜禽粪便中的微生物病原体杀死,减少生物污泥量,实现无害化生产,达到净化环境的目的;还可达到资源的多级利用,即"三沼"产品的综合利用,沼气可直接为农户提供能源、气肥等;沼液可直接用来肥田、养鱼等;沼渣可制作高效优质的有机肥等。"三沼"完全可以当成一种农业生产资料，作为肥料、饲料、饵料，用于农作物浸种，防治病虫害，提高农作物、果品的产量与质量，储存保鲜农产品。更主要的是，通过"沼气"这一环节，把种和养联系起来，形成一个物质多层次高效利用的生态农业良性循环系统。此外，作为新兴能源，沼气的用途广泛，除了用作生活燃料使用外，还可以用于生产能源使用。

我国目前利用沼气技术处理畜禽养殖废弃物，主要有农用户沼气生态工程模式和规模化沼气工程模式两种。

1）农用户沼气生态工程模式

农用户沼气生态工程模式在全国各地农村都有成功的经验，如以沼气为纽带的适应北方冬天寒冷的特定环境的沼气池、猪圈、厕所、太阳能温室"四位一体"的生态模式、"猪—沼—作物"能源生态工程、"器—气—池"生态家园工程等。

农村地区建立以沼气池为主的能源生态工程模式，可以使畜禽养殖废弃物得到良好处理，不仅能增加农民收入，还可极大地改善农民的生产生活条件，有利于农村生态环境的改善，对促进农村经济的可持续发展具有深刻意义。

2）规模化沼气工程模式

沼气工程模式主要有小型沼气工程模式和大中型沼气工程模式，是根据养殖方式、养殖场规模来确定、优化不同区域和不同养殖方式的畜禽粪便处理方式。此方式能使废弃物合理利用，变废为宝。

虽然目前我国利用沼气技术处理畜禽养殖废弃物技术已经成熟，但是在实际操作上，因为建造沼气池及其配套设备的投资巨大，沼气池的运行又受温度、季节的影响较大，小型沼气工程比较成功，而大中型沼气工程运行状况并不成功，没有发挥其最佳的经济效益和环境效益。

（3）发电

畜禽粪便以无污染方式焚烧，然后发电利用，焚烧过程中产生的灰还可以作为优质肥料。英国Fibrowatt公司用鸡粪作燃料发电。我国福建圣农集团将谷壳与鸡粪的混合物进行燃烧发电，年消耗鸡粪和谷壳混合物约25万t，相当于节省煤约8.8万t，既创造了经济效益，减少了环境污染，又节约了煤炭、天然气等不可再生资源。

（4）畜禽粪便热化学转化

畜禽粪便作为一种生物质能源属于可再生能源，可再生能源的开发与利用日益受到国际社会的重视。我国是能源消费大国，常规能源储备相对不足，因此，多元化的能源配置是解决我国能源问题的必由之路。可再生能源在我国蕴藏量丰富，开发利用新能源对我国的能源战略安全和环境、经济的可持续发展意义重大。热化学转化法是当前开发生物质能源的主要技术，也是各国研究的重点。其基本原理是将生物质原料加热，使其在高温下裂解（热解），热解后的气体与供入的气化介质（空气、氧气、水蒸气等）发生氧化反应并燃烧，最终生成含有一定量固体可燃物（如木炭）与液化油、生物油或生物质燃气（CO、H_2、CH_4）等的混合气体。

3. 畜禽粪便的饲料化利用

畜禽粪便不仅是优质的有机肥料，而且也是畜禽本身较好的饲料资源。畜禽粪便中的粗蛋白含量几乎比畜禽采食饲料中的粗蛋白含量高50%；畜禽粪便中含17种氨基酸，其占比达到8%~10%。此外，粪便还含有粗脂肪、粗纤维、磷、钙、镁、钠、铁、铜、猛、锌等多种营养物质。其中，鸡粪的营养成分最为丰富，粗蛋白含量占鸡粪干物质的25%，相当于豆饼的57%~66%，而且氨基酸的种类齐全，并含有丰富的矿物质和微量元素，因此，鸡粪可以成为优质高效的饲料资源。在美国，农场主用混入鸡粪和垫草的饲料直接饲喂奶牛，其结果与饲喂豆饼效果相同。

畜禽粪便中存在重金属元素、病原体、寄生虫等有害物质，因此需要经过适当处理，杀死病原菌，提高蛋白质的消化率和代谢，改善适口性，才可作为饲料用。国外畜禽粪便饲料已经商品化多年，我国畜禽粪便饲料化研究工作也已开展多年，现在已经有部分地区实现了商品化。目前，畜禽粪便饲料化的方法有以下几种。

（1）直接用作饲料

这种方法是最简单的，仅需要用化学药剂对粪便进行杀菌处理后即直接用于动物饲料。该方法的原料主要是鸡粪，原料来源广泛。因为鸡的肠道较短，对饲料的消化吸收能力差，饲料中约有70%的营养成分未被消化吸收就被排出体外，故鸡粪营养物质丰富，可以用作猪、牛的饲料。但是鸡粪中含有非常复杂的成分，包括寄生虫、尿素、病原体等，因此需要用化学药剂提前进行处理，防止畜禽间的交叉感染及传染病传播。处理烘干鸡粪的步骤简单，操作方便，处理成本低，而且经烘干的鸡粪所含营养成分丰富，完全可以替代部分精、粗饲料和钙、磷等添加剂，可较大程度地降低饲料成本，提高经济效益，促进养殖业的发展。

（2）青贮法

畜禽粪便中的碳水化合物含量低，为了调整饲料和粪的比例，掌握好水分含量，不能单独青贮，需要与禾本青饲料一起青贮，来防止粪便中的粗蛋白损失过多。这种饲料具有酸香味，适口性高，而且青贮法可以杀死粪便中的微生物、病原体等，提高了饲料的安全性。

（3）干燥法

干燥法是畜禽粪便循环利用最常用的方法。这种方法可以使畜禽粪便干燥脱水，能够除臭和彻杀虫卵，可以达到卫生防疫和生产商品肥料的要求。该法主要利用热效应和喷放机械。干燥法主要用来处理鸡粪，优势是对粪便的处理效率高，设备简单，投资小，便于推广。但是鸡粪在夏天保鲜困难且具有臭味，因此，需要在加工时添加乳酸菌等除臭效果好的添加剂进行臭气处理，因此使成本增加。目前采用的干燥技术主要有日光自然干燥、高温快速干燥、微波烘干、烘干膨化处理等。

1）日光自然干燥

这种方法主要是在自然条件下或在大棚内，对粪便粉碎、过筛、除杂后将其放置在干燥的地方，利用阳光照晒进行干燥处理，经干燥后的粪便可作为饲料或者肥料。这种方法投资小、成本低，易于操作，但是处理规模小、时间长、占地大、受天气因素影响大，且处理过程中氨气易大量挥发，臭气较大，不但影响肥效，还会对环境造成威胁，不适于集约化的畜禽养殖场采用。

2）高温快速干燥法

这种方法利用机械对粪尿进行固液分离、烘干，通过高温、高压、热化、灭菌、除臭等处理过程生产有机肥料，主要是对鸡粪进行处理，同时也是我国处理畜禽粪便较为广泛使用的方法之一。干燥机主要是回转式滚筒烘干机，鲜鸡粪含水量为70%~75%，经过高速烘干，可达到干燥、除臭、灭菌、耐储存的效果。高温快速干燥法的优点是不受天气影响、能大批量生产、干燥快速等，适合大型畜禽养殖场使用。但是其具有一次性投入大、能耗较大、在烘干过程中产生大量的臭气、耗水量大等缺点。

3）微波烘干

微波烘干是利用微波产生高温，迅速使湿畜禽粪含水量降到13%以下的处理方法，在干燥过程中可以达到消毒、杀灭细菌、消除臭味的效果，但是这种方法的养分损失较大，成本较高。

4）烘干膨化处理

烘干膨化处理是利用热效应和喷放机械效应两方面的作用，使畜禽粪便膨化、疏松，既除臭又能彻底杀菌、灭虫卵，达到卫生防疫和商品肥料化、饲料化的要求。该方法的缺点是一次性投资较大，烘干膨化时耗能较多，特别是夏季保持鸡粪新鲜较困难，大批量处理时仍有臭气产生，且成本较高等，从而导致该项技术的应用受到限制。据报道，一个饲养10万只蛋鸡的农户购置一台日处理10 t的鸡粪膨化烘干机，7~8个月可以收回成本，以后每年可以获纯利50万~80万元。

（4）分解法

分解法是利用蚯蚓、苍蝇等低等动物来分解粪便，达到提供运动蛋白质和处理畜禽粪便的目的。蚯蚓和蝇蛆是非常好的动物性蛋白质饲料。蚯蚓的蛋白质含量为10%~14%，可以作为水产养殖的活饵料，也可以作为猪牛羊的饲料，同时蛆粪可作为肥料。这种方法比较经济，生态效益显著，但是操作技术难度较大，同时对温度的要求苛刻，难以全年生产，推广不易。

二、养殖场废水处理和综合利用技术

畜禽养殖场的废水主要包括畜禽的粪尿和冲洗水，固形物含量高，含有大量的氮、磷等有机物、悬浮物和微生物致病菌，有机质浓度高，易于生化处理。但是各地因饲养方式、管理水平、畜舍结构、清粪方式等的不同，畜禽养殖场的污水排放量差异很大。其中大规模养猪场废水处理难度很大，原因为：①由于大多数养猪场都是采用漏缝板式的栏舍和水冲式清粪，排水量大；②冲洗栏舍的时间相对集中，冲击负荷很大；③粪便和污水量大且集中，而农业生产是季节性的，周围农田无法全部消纳；④废水固液混杂，有机质浓度较高，而且黏稠度很高。据相关数据表明：年出栏1万头育肥猪的猪场，每天产生的污水量为73 t，粪尿量约为1.05 t。

畜禽养殖场废水的主要处理方式有以下几种。

1. 生态还田

将畜禽养殖废水作为肥料直接还田用于农业种植，这是传统的处理方法，方式简单，经济有效，广被使用。该法不仅能使畜禽废水不排往外环境，达到污染物的零排放，还能将废水中有用的营养成分循环利用于土壤—植物生态系统，减少土壤化肥的施用量，实现养殖废水的循环利用。但是这种方式也具有一定的风险，如存在传播微生物病原体的危险，当施用方式不当或者施用量过多时，可能会导致土壤污染等。

2. 物理处理技术

（1）固液分离技术

固液分离是畜禽养殖废水的预处理步骤，主要是对水清粪工艺清理出的畜禽养殖废弃物通过沉降、过滤、压缩、离心等方法进行固液分离。先是通过沉降将废弃物中的固体废弃物和废水分离，然后对其过滤进一步处理废水中的固体废弃物，降低后续处理负荷和成本。

一般利用滤网等固液分离设施可去除其中40%~65%的固体悬浮物，降低25%~35%的生化需氧量（BOD）。

目前，常用的分离设备有转动筛、斜板筛、带式滤机和挤出式分离机等。规模化畜禽养殖场固液分离主要采用机械式分离，通过筛分和挤压方式实现。固液分离技术的成本和运行费用低、工艺和设备结构简单、维修方便。但是分离后的固体和液体均需进一步处理，才能满足相关要求。

（2）介质吸附法

这种方法主要是采用吸附容量较大的吸附介质材料对畜禽养殖废水中的氮、磷等进行吸附预处理，根据废水中的污染物种类不同选择不同的吸附剂，可以达到处理某种污染物的目的。王雅萍等将凹凸棒石黏土应用于畜禽养殖废水处理中，氨、氮去除率达到75.1%。杭小帅等（2012）研究了3种红色黏土对畜禽养殖废水中磷的吸附去除性能，3种红色土对含磷量为35 mg/L的养殖废水中磷的去除率均达到90%，对含磷量50 mg/L养殖废水中磷的去除率均达到85%，均显著优于活性炭。于鹄鹏等（2009）按凹凸棒土∶稻壳为9∶1的比例制成新型凹土吸附剂用于吸附处理某养殖场废水中的NH_4^+-N，NH_4^+-N的最高去除率可达87%。高萌等研究了改性壳聚糖对畜禽

废水中 Cu^{2+}、Zn^{2+} 的捕集，结果表明，改性壳聚糖对复杂体系下的实际废水有很好的处理效果，出水中的 Cu^{2+}、Zn^{2+} 的残余浓度能达到国家排放标准。

（3）化学氧化法

化学氧化是利用氧化势能较高的氧化剂产生强氧化性的自由基，将水中有机物、无机物等氧化分解。这是一种新兴的水处理技术，有研究显示，采用 Fenton 氧化法处理 COD 浓度高达 5000～5700 mg/L 的畜禽养殖废水，当 Fe^{2+} 投加浓度为 4700 mg/L、H_2O_2 投加浓度为废水初始 COD 浓度的 1.05 倍时，反应 30 min 后，COD 的去除率可达 80% 以上，甚至可达到 95%。采用 Fenton 氧化法处理经过 UASB（升流式厌氧污泥床）消化的畜禽废水，其对厌氧出水中 COD 和色度的去除率分别达到 95% 和 96%。

同时，利用点氧化和电还原，能够有效去除有机物和重金属。欧阳超等（2010）采用电化学氧化法处理养猪废水，在反应 3 h 后，养猪废水中的 NH_4^+-N 去除率可达 98.22%，但 COD 的去除率仅为 14.04%。此外，电化学方法还能同步去除养殖废水中的抗生素、激素和重金属等污染物。李文君等（2011）采用 UV/H_2O_2 联合氧化法处理含抗生素（磺胺甲恶唑、磺胺二甲氧嘧啶、磺胺二甲嘧啶等）的畜禽养殖废水，在紫外波长 254 nm、抗生素浓度 2.0 mg/L、H_2O_2 投加量 7.0 mmol/L、pH 值 5.0 条件下，反应 1 h 后，废水中 5 种抗生素去除率均可达 95% 以上。

（4）生物处理技术

生物处理技术主要包括自然处理技术、厌氧生物处理技术、好氧生物处理技术和厌氧—好氧组合处理技术。

1）自然处理技术

自然处理技术是畜禽养殖废水传统的处理方法，主要利用天然水体、土壤和生物的物理、化学与生物的综合作用来净化污水。其净化机理主要包括过滤、截留、沉淀、物理和化学吸附、化学分解、生物氧化及生物吸收等。其原理涉及生态系统中物种共生、物质循环再生原理、结构与功能协调原则及分层多级截留、储藏、利用和转化营养物质机制等。这种方法投资省、工艺简单、动力消耗少，但净化功能受自然条件制约。

自然处理的主要模式有氧化塘处理法、土壤处理法、人工湿地处理法等。

氧化塘又称生物稳定塘，是一种利用天然或人工整修的池塘进行污水生物处理的构筑物。其对污水的净化过程和天然水体的自净过程相似，污水在塘内停留时间长，有机污染物通过水中微生物的代谢活动而被降解，溶解氧则由藻类通过光合作用和塘面的复氧作用提供，亦可通过人工曝气法提供。氧化塘主要用来降低水体的有机污染物，提高溶解氧的含量，并适当去除水中的氮和磷，减轻水体富营养化的程度。

土壤处理法不同于季节性的污水灌溉，是常年性的污水处理方法。将污水施于土地上，利用土壤、微生物、植物组成的生态系统对废水中的污染物进行一系列物理、化学和生物净化过程，使废水的水质得到净化，并通过系统的营养物质和水分的循环利用使绿色植物生长繁殖，从而实现废水的循环、无害化和稳定化。

人工湿地处理法可通过沉淀、吸附、阻隔、微生物同化分解、硝化、反硝化及植物吸收等途径去除废水中的悬浮物、有机物、氮、磷和重金属等。近年来，人工湿地的研究越来越受到重视，叶勇等利用红树植物木榄和秋茄处理畜禽废水中的氮、磷，结果表明，两种植物对氮、磷的去除效果较好。廖新梯、骆世明分别以香根草和风车草为植被，建立人工湿地，随季节不同，湿地对污染物的去除率不同，COD_{Cr} 去除率可达 90% 以上，BOD_5 去除率可达 80% 以上。

由于自然处理法投资少，运行费用低，在有足够土地可利用的条件下，是一种较为经济的处理方法。畜禽废水的自然生态处理技术适合我国国情，特别适用于小型畜禽养殖场的废水处理，具有广阔的应用前景。

2）厌氧生物处理技术

畜禽养殖污水厌氧生物处理是利用厌氧微生物在无氧条件下的降解作用，使污水中有机物质达到净化的处理方法。在无氧的条件下，污水中的厌氧细菌把碳水化合物、蛋白质、脂肪等有机物分解生成有机酸，然后在甲烷菌的作用下，进一步发酵形成甲烷、二氧化碳和氢等，从而使污水得到净化。

厌氧生物处理系统主要由厌氧反应器、沼气收集系统、净化系统、储存系统、使用系统及配套管线、沼液和沼渣收集、处理系统组成。厌氧反应器类型的选择和设计可根据畜禽养殖污染物的种类和工艺路线确定。畜禽养殖废水厌氧生物处理的 BOD 负荷较高，一般为 3.5 kg/($m^3 \cdot d$)，去除率可达 90% 以上。厌氧生物处理通过厌氧发酵产生沼气，在降低污水中 COD、BOD 含量的同时，实现循环利用。该技术投资少、耗能少，不需要专门进行管理，运行费用低，因此在畜禽养殖场废水处理中得到了广泛的应用，尤其是在处理高浓度有机废水处理领域备受青睐。但是由于这种技术厌氧出水很难达到排放标准，必须与其他技术联合使用。

利用厌氧生物处理技术进行畜禽养殖场沼气工程的建设模式主要是"生态模式"。这种模式可以使畜禽粪便污水全部进入处理系统，进料 TS 浓度可达到 10% 以上，可以根据具体情况采用全混合厌氧池等厌氧消化器，产生的沼液和沼渣都可以进行综合利用，当作有机肥料用于种植业。产生的沼气可用于农户的生活燃料或者用于发电。该法是以沼气为纽带的良性生态系统，值得推广。但是这种"生态模式"工程建设要求畜禽养殖场周围有足够的农田来消纳大量的沼渣、沼液。

3）好氧生物处理技术

好氧生物处理技术的基本原理是利用微生物在好氧条件下分解有机物，同时合成自身细胞。好氧微生物以污水中的有机污染物为底物进行好氧代谢，经过一系列的生化反应，逐级释放能量，最终以低能位的无机物稳定下来，达到无害化的要求。目前，畜禽养殖废水处理常用的好氧生物处理方法主要有活性污泥法、生物滤池、生物转盘、SBR 和 A/O 等一系列方法。

采用好氧技术对畜禽废水进行生物处理,研究较多的是水解与SBR结合的工艺。SBR（Sequencing BatchReactor）工艺,即序批式活性污泥法,该法是基于传统的FillDraw系统改进并发展起来的一种间歇式活性污泥工艺,它把污水处理构筑物从空间系列转化为时间系列,在同一构筑物内进行进水、反应、沉淀、排水、静置等周期循环。SBR与水解方式结合处理畜禽废水时,水解过程对COD_{Cr}有较高的去除率,SBR对总磷去除率为74.1%,高浓度氨氮去除率达97%以上。此外,其他好氧处理技术也逐渐应用于畜禽废水处理中,如间歇式排水延时曝气法（IDEA）、循环式活性污泥系统（CASS）、间歇式循环延时曝气活性污泥法（ICEAS）。

通过好氧生物处理,可以去除废水中的大部分氮、磷、有机物等,一般用于废水厌氧消化处理的后续步骤。

4）厌氧—好氧组合处理技术

由于畜禽养殖废水性质复杂、成分多变、有机负荷及氮磷含量均较高,因此,采用单一的处理工艺在经济成本和处理效果上往往不够理想,所以一般会采用组合方法对其进行系统处理,以弥补单一方法的不足,其中常用的是厌氧—好氧组合处理技术。无论是厌氧技术还是好氧技术,单独处理的时候,均无法实现畜禽养殖废水的达标外排。结合它们各自的优势,厌氧—好氧联合处理的时候,既克服了好氧处理能耗大和占地面积大的不足,又克服了厌氧处理达不到排放要求的缺陷,具有投资少、运行费用低、净化效果好、能源环境综合效益高等优点,特别适合规模化畜禽养殖场污水的处理。因此,大多数经济发达、集约化规模的畜禽养殖场均采用厌氧（缺氧）—好氧组合处理工艺。

例如,杭州西子养殖场采用了厌氧—好氧组合处理工艺,养殖场废水经处理后,水中COD_{Cr}含量约为400 mg/L,BOD_5含量为140 mg/L,基本达到废水排放标准。李金秀等采用ASBR-SBR组合反应器系统,ASBR作为预处理器（厌氧）主要用于去除有机物,SBR（好氧）用于生物脱氮处理。

膜生物反应器是由膜分离技术与生物反应器相结合的新型生物化学反应系统,它用膜取代了传统的二沉池,具有出水稳定、活性污泥浓度高、抗冲击负荷能力强、剩余污泥少、装置结构紧凑、占地少等优点。

利用组合处理方式的代表是"环保模式",这种模式要求较高,养殖场必须实行严格的清洁生产,干湿分离,冲洗污水和尿进入系统。污水进行严格预处理,强化固液分离、沉淀,控制SS浓度。厌氧消化器采用上流式厌氧污泥床反应器（UASB）等,厌氧出水COD控制在1000 mg/L。好氧处理可采用SBR等,处理过程中产生的污泥可以制作有机肥或者用作菌种出售。后处理则可采用氧化塘、人工湿地等自然处理方法。这样投资少,运行管理费用低,耗能少,污泥量少,对周围环境影响较小；但是对土地占地要求较大,对气温要求较高。

畜禽养殖废水是比较难处理的有机废水，主要是因为其排量大、温度较低、废水中固液混杂、有机物含量较高、固形物体积较小，很难进行分离，而且冲洗时间相对集中，使得处理过程无法连续进行。由于废水中的 COD 和 BOD 等指标严重超标，悬浮物较大，氮磷含量丰富，氨氮含量高且不易去除，单纯采用物理、化学或者生物处理方法都很难达到排放要求。因此，一般养殖场的废水处理都需要使用多种处理方法相结合的工艺。根据畜禽废水的特点和利用途径，可采用不同的处理技术。

第四章　生态循环农业资源利用模式

第一节　猪—沼—粮循环模式

一、系统生态学原理及模式内涵

猪—沼—粮循环模式是指按照生态学和生态经济学原理，应用系统工程方法，因地制宜地规划、设计、组织、调整和管理畜禽生产，以保持和改善生态环境质量，维持生态平衡，保持畜禽养殖业协调、可持续发展的生产形式。在合理安排粮食生产的情况下，种草养畜，以畜禽的粪便养地，种养结合，实现我国养殖业可持续发展。

二、模式特点与关键技术

土壤结构破坏、地力下降与水资源、肥源、能源的短缺和失调密切相关，成为"高产、高效、优质"农业发展的制约因素。猪—沼—粮生态循环模式将传统的粮食生产一元结构向粮食作物和畜禽养殖二元结构发展，实施"种、养、沼、肥"四环产业并举，其关键是变单环技术为组合链条技术，集大成于一体，提高农业生产的环境质量，合理调整生产过程中的相互关系。猪—沼—粮循环模式使一个生产过程中的排泄物（废弃物）转变为另一个生产过程中的输入物（原料资源），从而实现农业生产的无废弃物过程（零排放目标），即废弃物资源化过程。粮食秸秆还田及粗制品作为饲料进入规模化养殖场，养殖场畜禽粪便进入发酵池产生沼气用于能量循环，解决燃料问题；沼液还田减少肥料用量，沼渣用于生产专用有机肥，进入农业生产循环系统，充分利用循环系统内各个环节的资源，减少农药化肥用量、生产出绿色优质稻米。建立以规模集约化养殖场为单元的生态农业产业体系，是以粮食作物生产为基础、养殖业为龙头、沼气能源开发为纽带、有机肥料生产为驱动，形成饲料、肥料、能源及生态环境的良性循环。

1. 猪—沼—粮模式中的养殖关键技术

① 养殖及生物环境建设。畜禽养殖过程中利用先进的养殖技术和生物环境建设，达到禽畜生产的优质、无污染，通过禽畜舍干清粪技术和疫病控制技术，使畜禽生长环境优良，从而达到无病或少生病。

② 固液分离技术。对于用水冲洗的规模化畜禽养殖场，其粪尿采用水冲洗方法排放，既污染环境又浪费水资源，也不利于养分资源利用。采用固液分离设备，首先进行固液分离，固体部分进行高温堆肥，液体部分进行沼气发酵。同时，为减少用水量，尽可能采用干清粪技术。

③ 污水处理与综合利用技术。采用先进的固液分离技术，液体部分利用污水处理技术，如氧化塘、沼气发酵及其他好氧和厌氧处理技术，在非种植季节进行处理，达到排放标准后排放；在作物生长季节可以充分利用污水中水肥资源进行农田灌溉。

④ 畜牧业粪便无害化高温堆肥技术。采用先进的固液分离技术，固体部分利用高温堆肥技术和设备，生产优质有机肥和商品化有机无机复混肥。

2. 猪—沼—粮模式中的沼液高效利用关键技术

① 沼液浸种杀菌技术。针对种植领域不同的作物品种，采用不同的技术措施，主要从浸种时间、沼液稀释倍数、沼液用量等方面进行浸种消毒杀菌。

② 稻田沼液施用技术。根据水稻生长需肥、需水规律，分时间、分批次控制沼液浓度进行沼液灌溉。

三、技术操作规范

规模化养殖场建成，沼气池完成后，稻麦轮作条件下沼液利用主要有以下几个步骤。

1. 管道架设措施

由于沼液浓度较高，需稀释施用，根据稻田现有水渠情况布设沼液管道，安装控制阀，配合灌溉用水进行施用。

2. 沼液水稻浸种措施

沼液浸种杀菌使用上年或当年生产的新鲜种子，浸种前应对种子进行筛选，清除杂物、秕粒并翻晒，翻晒时间不得低于 24 小时。选用正常发酵产气 2 个月以上的沼液，将种子装在能滤水的袋子里，并将袋子悬挂在沼气池出料间的上清液中，要求沼液温度 10 ℃以上，pH 为 7.2~7.6。

常规稻品种采用一次性浸种。在沼液中浸种时间：早稻 48 小时，中稻 36 小时，晚稻 36 小时，粳、糯稻可延长 6 小时，然后清水洗净，破胸催芽。

抗逆性较差的常规稻品种应将沼液用清水稀释一倍后进行浸种，浸种时间为 36~48 小时，然后清水洗净，破胸催芽。

杂交稻品种采用间歇法沼液浸种，三浸三晾，清水洗净，破胸催芽。浸种时间：杂交早稻为 42 小时，每次浸 14 小时，晾 6 小时；杂交中稻为 36 小时，每次浸 12 小时，晾 6 小时；杂交晚稻为 24 小时，每次浸 8 小时，晾 6 小时。

3. 稻田沼液肥料配施灌溉管理

水稻适时移栽、合理密植。一般于 5 月中旬播种，6 月上旬移栽，秧龄控制在 20 天左右，叶龄 3~4 叶。采用机插秧方式，株行距控制在 12 cm×30 cm，每 667 m^2 移栽约 1.85 万穴，基本苗达 7 万~8 万株。

全程施用沼液替代化肥，配合灌溉用水稀释后进行沼液施用，整个水稻生长季沼液施用量在 21~27 t（折合纯氮 16~20 kg/667 m^2）。不同生育期施用量比例为：基肥：苗肥：秆肥：穗肥＝（3∶3∶2∶2）~（4∶2∶3∶1）较为适宜。

沼液替代 25% 化肥施用，配合灌溉用水稀释后进行沼液施用，整个生长季沼液用量在 15~20 t（折合纯氮 12~15 kg/666.7 m^2，不同生育期施用量比例为：基肥：苗肥：秆肥：穗肥＝4∶2∶3∶1 较为适宜。

4. 稻田病虫害防治管理

① 稻瘟病（穗瘟）。在 5% 破苞及齐穗期，每公顷用 75% 三环唑 900 g，兑水 900 kg 均匀喷雾，预防穗瘟。

② 稻曲病。圆苞期及齐穗期，每公顷用 40% 菌核净 1500 g 或 20% 井冈霉素 1500 g，兑水 900 kg，均匀喷雾防治稻曲病，共 2 次。

③ 白叶枯病。在白叶枯病常发地区，孕穗期用叶枯宁 100 g，兑水 60 kg 均匀喷雾防治；水淹没的田块及暴风雨袭击过的田块，在稻叶水干及天晴后及时用叶枯宁 100 g，兑水 60 kg 均匀喷雾，预防白叶枯病。

④ 稻飞虱。田间百丛虫口密度达 800 头的防治指标，可选用以下药剂进行防治：75% 艾美乐，8000 倍液进行喷雾防治；25% 稻虱净，800~1000 倍液进行喷雾防治；15% 金好年（吡虫啉＋丁硫克百威）30~40 mL，兑水 60 kg 均匀喷雾防治；50%~70% 吡虫啉 3~5 g，兑水 60 kg 均匀喷雾防治。

5. 麦田沼液施用管理

在浸种前，选择晴天将麦种晒 2~3 天，提高种子的吸水性能。选用发酵时间长、腐熟较好并正常使用的沼气池发酵液浸种。在播种前一天进行小麦浸种，浸泡时间根据水温而定，一般 17~20 ℃ 浸 6~8 小时。麦种浸 6~8 小时后取出种子袋，用清水洗净，并沥干袋里的水分，然后把种子摊平，待种子表面水分晾干后即可播种。

沼液作为基肥在犁地前均匀洒施田面，然后用犁翻入底层，拔节期用尿素追肥。小麦 11 月上旬播种，播种粒 15 kg/667 m^2，麦田沼液施用量一般为 4 t 左右，配施 10 kg 尿素，主要采用机械沟施或者喷洒的方式进行。

6. 小麦病虫害防治管理

① 沼液防治小麦蚜虫。用沼液喷施小麦可防治小麦蚜虫侵害，方法如下：用沼液 14 kg，洗衣粉溶液 0.5 kg（溶液按洗衣粉和清水 0.1∶1 的比例配制），配制成沼液复方治虫剂，用喷雾器喷施。每公顷麦田一次喷施 450 kg，第二天再喷 1 次。喷施时间最好在晴天的上午进行。

② 沼液防治小麦赤霉病。每公顷麦田喷施 750 kg 沼液，原液效果最好，盛花期喷 1 次，隔 3~5 天再喷 1 次。

第二节　草—兔—粪—稻循环模式

一、系统生态学原理及模式内涵

草—兔—粪—稻循环模式是一个循环的农业生态系统，牧草、兔、兔粪、水稻作为系统的生态因子，组成一个食物链和循环链。按照"养殖集中化、粪便资源化、污染减量化、治理生态化"思路，通过对牧草品种、兔种、水稻品种的选择，从茬口布局、水稻牧草栽培技术措施、兔粪废弃物综合利用等关键技术的搭配与实施，实现整个循环圈内的能量循环和物质流动，建立起一种动态的、平衡的、能够充分实现高效、优质、高产的生态循环模式，最终生产出优质的农牧产品（水稻、兔等）。

二、模式特点

该模式最大的特点就是充分实现了草—兔—粪—稻的物质循环，没有任何的流出。养殖兔子作为循环模式的核心，通过对粪尿进行干湿分离，兔尿进入发酵处理池和氧化塘，处理达标后作为液体肥直接还田；兔粪配合稻草直接堆肥养殖蚯蚓，生产蚯蚓粪及有机肥，还田肥田；水稻和牧草进行轮作，牧草养殖兔子，形成一个小的生态循环链。

三、操作技术规范

1. 兔子养殖关键技术

① 品种选择。选择适宜的兔优良品种，如新西兰大耳兔等。肉兔活动量小、采食均匀、易管理、生长快。采用笼养，兔笼一般用铁丝制成平列式或立柱式，并设有料槽、水槽和草架，笼底板采用竹条或木条等，宽度要求 2.0~2.5 cm，条间距 1 cm。在兔笼规格上，种兔笼比商品兔笼大些，一般公母兔和后备种兔每只所需面积为 0.25~0.40 m²，育肥兔为 0.12~0.15 m²，要求舍内布局整齐紧凑，又符合卫生防疫规定。

② 投喂方法。在投喂青绿饲料时，适当补饲精饲料，可使肉兔生长快且不易生病，饲料利用率高，缩短饲养周期，提高出栏率，增加经济效益。连续投喂精饲料，并供给充足的清水，饲养 70~80 天，肉兔体重便可达 2.3~2.7 kg。

③ 定期驱虫和投药预防。肝片吸虫、蛔虫、线虫、球虫、肠炎等常发疾病，都会影响肉兔生长，在肉兔不同的生长阶段，定期用药驱虫和预防，可以保证肉兔的健康生长。

④ 定期做好免疫注射、消毒卫生工作。兔瘟、巴氏杆菌病等流行病对肉兔生产影响很大，在断奶、4月龄等不同生长期注意做好兔瘟、巴氏杆菌病等流行病的疫苗免疫注射，同时经常洗刷兔舍，保持清洁卫生。每隔15天或30天对兔场进行定期消毒，建立严格的消毒制度，一旦发生疫情，须及时采取隔离与封锁措施。定期做好免疫注射和消毒卫生工作，对减少疾病发生、稳定养殖生产非常关键。

2. 兔粪尿无害化处理技术

对兔粪尿采取固液干湿分离技术，兔尿直接由管道进入污水处理池进行发酵处理，达标后混合雨水直接排入农田作为液体肥料；兔粪运至塑料大棚，混合水稻秸秆进行堆置，养殖蚯蚓，生产有机肥。

3. 水稻绿色生产管理技术

① 品种选择。选择农艺性状好、抗病抗虫性强、生育期适中的优质粳稻品种。

② 施足基肥。结合土地耕翻，根据不同水稻品种一生对养分的需求，每667 m^2 施兔粪蚓粪有机肥1000 kg左右，再辅以水稻BB肥15~25 kg作基肥，同时追施碳酸氢铵15~20 kg作分蘖肥。

③ 适时移栽、合理密植。根据蛙苗产出期，确定水稻播种期。一般于5月中旬播种，6月上旬移栽，秧龄控制在20天左右，叶龄3~4叶。采用机插秧方式，株行距控制在12 cm×30 cm，每667 m^2 栽约1.85万穴，基本苗达7万~8万株。

④ 合理灌溉。水稻移栽至分蘖期，浅水勤灌，水深2~3 cm，达到穗数时适度搁田，孕穗期保持浅水层，抽穗扬花灌浆期干湿交替。

4. 牧草绿色生产管理技术

牧草品种选择适宜饲喂兔的牧草：黑麦草、紫花苜蓿等。

（1）黑麦草

① 播种期选择：黑麦草喜温暖湿润的气候，种子发芽适宜温度13 ℃以上，幼苗在10 ℃以上就能较好地生长。因此，黑麦草的播种期较长，既可秋播，又能春播。秋播一般在9月中下旬至11月上旬均可。随着播期推迟，由于播后气温下降、出苗迟，分蘖发生迟而少，鲜草收割次数减少，产量降低。

② 播种量选择：在一定面积范围内，播种量少，个体发育较好，但密度过小，就会影响单位面积内鲜草总产量，特别是前期的鲜草产量；相反，播种量过大，鲜草产量未必高，且个体生长发育也受到影响。因此，只有合理密植，才能充分发挥黑麦草的个体群体生产潜力，才能提高单位面积产量，一般每公顷播种量15~22.5 kg最适宜。

③ 播种方法：黑麦草种子细小，要求浅播，稻茬田土壤含水量高、土质黏重，秋季播种时往往连续阴雨，或者因秋收季节劳力紧张推迟播种而影响出苗。为了使黑麦

草出苗快而整齐，有条件的地方，用钙镁磷肥 150 kg/hm^2、细土 300 kg/hm^2 与种子一起拌匀后播种。

④ 黑麦草的收割：黑麦草再生能力强、可以反复收割，因此，当黑麦草作为饲料时，就应该适时收割。黑麦草收割次数的多少主要受播种期、生育期间气温、施肥水平影响。秋播的黑麦草生长良好，可以多次收割。另外，施肥水平高，黑麦草生长快，可以提前收割，同时增加收割次数；相反，肥力差，黑麦草生长慢，不能在短时间内达到一定的生物量，也就无法收割利用。适时收割，当黑麦草长到 25 cm 以上时就收割，若植株太矮，鲜草产量不高，收割作业也困难。每次收割时留茬高度约 5 cm 左右，以利于黑麦草残茬的再生。

⑤ 黑麦草的留种：成熟后的黑麦草种子落粒性较强，因此，当黑麦草穗子由绿转黄，中上部的小穗发黄，而小穗下面的颖还是黄绿色时就应及时收割。为了防止收割时落粒，最好在早晨有露水或阴雨时收割，要做到轻割、轻放、即时摊晒、脱粒、晒干扬净。若农活紧张、劳力不足，或天气不佳，不能及时脱粒、晒干时，应将黑麦草挂放在干燥通风的地方，以防霉烂，保证种子质量。

（2）紫花苜蓿

① 播种前准备：播种前晒种 2~3 天，可以打破休眠，提高发芽率和幼苗整齐度。从未种过紫花苜蓿的田地应用根瘤菌剂拌种，可提高紫花苜蓿的成苗率、幼苗结瘤率、增加产草量，同时还可以增加土壤肥力，促进后茬作物生长。常用拌种方法有老土拌种、根瘤菌剂拌种等。

② 播种量及播种期：如采用机械条播，每公顷用种量 15~22.5 kg，撒播时每公顷用种量 30 kg 左右。根据需要，紫花苜蓿可以分春播、夏播、秋播。春播：在春季土地解冻后，与春播作物同时播种，春播紫花苜蓿当年发育好、产量高，种子田宜春播。夏播：如果春季干旱，土壤墒情差时，可在夏季雨后抢墒播种。秋播：秋播不能迟于 8 月中旬，否则会降低幼苗越冬率。

③ 播种方法：有条播、撒播，可根据具体情况选用。种子田要单播，采用穴播或条播的方法播种，一般行距 50 cm，穴距 50~70 cm；产草田可条播或撒播，条播一般行距 30 cm 左右。撒播时要先浅耕后撒种，再耙糖，最后镇压保墒。播种后要根据土壤墒情及时镇压，确保种子与土壤充分接触，有利于种子吸水发芽。镇压还具有提墒的作用，有利于种子所处的浅表土壤保持湿润状态。

④ 合理施肥：播种前结合耕翻，每公顷施农家肥 7.5 t 以上、优质过磷酸钙 750 kg、硫酸钾 150 kg、尿素 225 kg 做底肥。每次收割后，每公顷施用过磷酸钙 300 kg 或磷二铵 75 kg。追肥一般在春季返青后、分枝期、现蕾期结合灌水进行。第一年在苗期根瘤菌未形成前结合浇水，每公顷追施尿素 75~150 kg。

⑤ 病虫防治：夏季是紫花苜蓿病虫害的高发期，病害主要有锈病、霜霉病，可用 25% 粉锈宁可湿性粉剂 1000~1500 倍液喷雾防治；虫害主要有蚜虫、黏虫、潜叶蝇、

甜菜夜蛾、蓟马、盲椿象等，可针对性选用4.5%氯氰菊酯乳油、2.5%溴氰菊酯乳油、10%吡虫啉乳油等杀虫剂喷雾防治。化学防治时严禁使用剧毒、高残留农药。依据收割时间，确定合理的安全间隔期，防止环境污染和植株内有害残留物超标而引起牲畜中毒。

⑥ 适时收获：紫花苜蓿前两茬的产量约占全年产量的70%，且品质优良，商品性好。产草田宜在始花期至盛花期及时收割，收割时留茬4～5 cm，以保证充足的营养积累。收割后尽量减少在地里的晾晒时间。打捆后选择通风避雨处自然风干。

第三节　草—羊—粪—粮循环模式

一、系统生态学原理及模式内涵

草—羊—粪—粮循环模式是一个循环的农业生态系统，牧草、羊、羊粪尿、水稻作为系统的生态因子，组成一个食物链和循环链。按照"养殖标准化、粪便资源化、污染减量化、治理生态化"思路，通过种养结合、对牧草品种、羊种、水稻品种的选择，从茬口布局、水稻和牧草栽培生态种植、羊粪尿废弃物综合利用等关键技术的搭配与实施，实现整个循环圈内的能量循环和物质流动，建立一种动态平衡的，能够充分实现高效、优质、高产的生态循环农业模式，生产出优质的产品（水稻、湖羊等），达到生态、优质、安全、高效，具有良好的生态效益和社会效益。

二、模式特点

该模式最大的特点是通过草—羊—粪—粮的物质循环链，形成一种环境友好型的农业生态模式。该模式以标准化生态养殖湖羊作为循环模式的核心，采用先进的高床漏缝板和粪尿收集等生产工艺，通过对粪尿进行干湿分离，羊尿由污水管道统一收集流入污水收集池，用泵抽入三级处理池进行厌氧发酵无害化处理，然后泵入污水储存池暂存，根据作物需要直接还田。在黑麦草（秋季播种夏季收割）收割翻耕后，在播种水稻前，将处理好的污水作液肥用粪泵抽入地下管道与水混合后灌入农田作基肥，在水稻分蘖阶段再进行1～2次淌灌作追肥。

羊粪配合稻草经高温发酵腐熟后，制成优质有机肥料还田种植牧草或水稻，为种植提供养分，牧草又可养殖湖羊。通过牧草或者秸秆养羊—羊粪尿还田—牧草和水稻种植，产出优质湖羊和稻米，形成具有较高经济效益和良好社会效益的生态循环链。

三、操作技术规范

1. 湖羊养殖关键技术

在选择适宜的湖羊优良品种基础上要做好以下工作。

① 建舍。羊舍应选择在地势高、向阳背风的地方,通风保温兼顾。羊舍采用高床漏缝板和粪尿收集等生产工艺,进行分栏养殖,圈舍中要有足够的槽位和活动空间,每只羊应有 1.5～2.5 m² 的活动场地,栏间留 2～3 m 饲养区。舍内布局整齐紧凑,符合卫生防疫规定。

② 分舍。舍饲养羊应按照工厂化生产模式,把不同年龄、不同品种、不同体况的羊分舍饲养,设立专门的产房和羔羊舍、肉羊舍、母羊舍、公羊舍、病羊隔离舍等,并配以相应的饲养管理措施。

③ 配种。选择理想的杂交组合和适宜的配种时期进行配种,发情期受配率达到 90% 以上,总受胎率达到 95% 以上,母羊初配体重应达到成年体重的七成以上为宜。妊娠母羊应做好保胎工作,并使胎儿发育良好。不得饲喂发霉、变质、冰冻或其他异常饲料,不得空腹饮水和饮冰渣水。日常管理中,不得有惊吓、驱赶等剧烈动作,特别是羊在出入圈门或补饲时,要防止相互挤压,避免流产。妊娠后期的母羊要给予补饲,不宜进行防疫注射。

④ 接羔。根据当地气候,确定接羔是否在暖棚舍内进行,用肥皂水或 2%～3% 的来苏水清洗母羊乳房和后躯,接羔后及时断脐带,称初生重、喂初乳,观察排出胎粪。

⑤ 饲喂。采用粗料为主,精料为辅。规模饲养最好是种植紫花苜蓿、黑麦草进行"种草养羊",这样一年四季均可吃上鲜青草。青草秋季收割后还可以晒制成青干草或制成青贮饲料喂羊。配制精饲料时,除要有一定量的玉米外,还要按比例搭配豆粕、麸皮、鱼粉、骨粉等蛋白质饲料。羔羊出生后 30 分钟内需吃上初乳;5～7 日龄开始训练采食精料和干草;1 月龄内,每日补饲精料 0.05～0.1 kg、干草 0.1 kg;1～2 月龄,每日补饲精料 0.15～0.2 kg、干草 0.3～0.5 kg、青贮饲料 0.2 kg;3 月龄,每日补饲精料 0.2～0.25 kg、干草 0.5～0.8 kg、青贮饲料 0.2～0.3 kg。一般应在 1～2 月龄断奶。育成羊,采用粗、精结合,选择市售精饲料配置粗饲料。肉羊饲料的日喂量应根据羊的不同生长阶段进行调整,一般为每天 2.3～2.7 kg。要做到饲喂量的分配调整以饲槽内基本不剩料为标准,做到饲喂定时、定量,并有专人负责。混合精料组成为玉米 55%～75%、麦麸 5%～15%、豆粕(饼)10%～25%、石粉 0.5%～1.5%、食盐 1.0%～1.5%、预混料 0.5%～1%,含硒微量元素和维生素 A;粗饲料组成为青草、干草或者青贮玉米。育肥期第一个月,混合精料占 60%,粗饲料占 40%,以后精粗比为 1∶1。6 月龄体重达到品种标准要求,一般公羊应达到 45～50 kg,母羊 35～40 kg。

⑥ 防病。群养湖羊易发病，主要有羔羊痢疾、链球菌病、传染性脓疱、线虫、羊虱等。要贯彻"防重于治"的方针，应做好重点疾病的预防接种和综合防治工作。每年3月、6月、9月、12月各进行1次全群驱虫。体外寄生虫可使用阿维菌素注射液（0.2 mg/kg体重，皮下注射），体内寄生虫可使用丙硫苯咪唑（口服剂量为15～20 mg/kg体重）；每年春、秋各注射四联苗（快疫、猝疽、羔羊痢疾、肠毒血症）1次，肌肉或皮下注射5 mL，对肠毒血症免疫期6个月，其他病为1年。

⑦ 综合管理。羔羊生后3天内，打耳号或耳标；出生后10天内，在第三、第四尾椎处采取结扎法进行断尾。非种用公羔，生后1～2周采取结扎或手术法进行去势，利于提高肉的品质（减少腥膻气），并使之性情温顺，便于管理，快速育肥。此外，种公羊、母羊还要定期浴蹄和修蹄，春季剪毛后药浴1次，对寄生虫病发病较重地区则可适当增加次数。要定期（1周左右）轮换选用不同类型的消毒剂，如20%石灰乳、2%～5%火碱、10%的漂白粉溶液、3%福尔马林、10%百毒杀等对羊舍、运动场、饲槽、饮水器皿、饲养工具及圈舍等进行消毒，尽量做到羊栏净、羊体净、食槽净、用具净。病死羊的尸体要深埋或焚烧，严防传染病的流行。每年春、秋两季应进行1次大型的消毒，场门、场区入口处消毒池的药液要经常更换，保持有效浓度，并谢绝无关人员入场。

2. 羊粪尿无害化处理技术

湖羊养殖采用高床漏缝板技术实现粪尿干湿分离，干粪收集运至公司有机肥厂混合水稻秸秆进行好氧发酵堆肥，生产优质有机肥料。羊尿（尿泡液）和清洗污水漏入污水收集池，用泵抽入三级处理池进行厌氧发酵无害化处理，放入污水储存池暂存，根据作物需要直接还田。在黑麦草（秋季播种，夏季收割）收割或翻耕后、播种水稻前，将无害化处理好的尿液泵入农田作基肥；在水稻分蘖阶段，视实际情况再进行1～2次淌灌追肥。

3. 水稻绿色生产管理技术

① 品种选择。选用农艺性状好、抗病抗虫性强、生育期适中的优质粳稻品种。

② 施足基肥。结合土地耕翻，根据不同水稻品种一生对养分的需求，每667 m^2 施羊粪有机肥1000～1500 kg，辅以水稻BB肥（氨：磷：钾=24：8：10）20～25 kg作基肥。分蘖期追施尿素10～15 kg或者碳酸氢铵15～20 kg，穗肥根据叶色情况追施尿素5 kg左右，如叶色过浓不宜施用穗肥，以免贪青晚熟或倒伏。追肥期间也可以适当减少化肥用量，采用经无害化处理的粪尿液进行淌灌，每公顷粪尿液用量120～150 t。

③ 适时移栽、合理密植。根据茬口安排，确定水稻播种期。一般5月中旬播种，6月上旬移栽，秧龄控制在20天左右，叶龄3～4叶，采用机插秧方式，株行距控制在12 cm×30 cm，每667 m^2 约栽1.85万穴，基本苗达7万～8万株。

④ 合理灌溉。水稻移栽至分蘖期，浅水勤灌，水深2～3 cm，达到穗数时适度搁田，孕穗期保持浅水层，抽穗扬花灌浆期干湿交替。

4. 牧草绿色生产管理技术

牧草品种选择适宜湖羊饲喂的黑麦草。

① 播种期选择。黑麦草喜温暖湿润的气候，种子发芽适宜温度为 13 ℃以上，幼苗在 10 ℃以上就能较好地生长。因此，黑麦草的播种期较长，既可秋播，又能春播。秋播在 9 月中下旬至 11 月上旬均可。随着播期推迟，由于播后气温下降，出苗迟，分蘖发生迟而少，鲜草收割次数减少，产量降低。

② 播种量选择。每公顷适宜播种量 15～22.5 kg，合理密植，保证个体发育良好，充分发挥黑麦草的个体和群体生产潜力，提高单位面积产量。

③ 播种方法。黑麦草种子细小，要求浅播。稻茬田土壤含水量高、土质黏性重，秋季播种时往往连续阴雨，而且秋收季节劳力紧张容易影响播种质量。为了使黑麦草出苗快而整齐，有条件的地方可用钙镁磷肥 150 kg/hm²、细土 300 kg/hm² 与种子一起拌匀后播种。

④ 黑麦草的收割。黑麦草再生能力强，可以反复收割，因此，当黑麦草作为饲料时，应该适时收割。麦草收割次数的多少，主要受播种期、生育期间气温、施肥水平等影响。秋播的黑麦草生长良好，可以多次收割。适时收割，也就是当黑麦草长到 25 cm 以上时就收割；若植株太矮，不仅鲜草产量不高，而且收割作业也困难。每次收割时留茬高度约 5 cm、以利于黑麦草的再生。

⑤ 黑麦草的施肥。黑麦草一般不需单独使用肥料，可结合湖羊养殖经无害化处理的粪尿液，在每次鲜草收割后、用泵将粪尿液加压后淌灌或者喷灌，每公顷用量 150～225 t。

⑥ 黑麦草的留种。成熟后的黑麦草种子落粒性较强，因此，当黑麦草穗子由绿转黄，中上部的小穗发黄，而小穗下面的颖还是黄绿色时，就应及时收割。为了防止收割时落粒，最好在早晨有露水或阴雨天收割。要做到轻割、轻放，随时摊晒、脱粒、晒干扬净。若农活紧张、劳力不足，或天气不佳，不能及时脱粒、晒干时，应将黑麦草挂放在干燥通风的地方，以防霉烂，保证种子质量。

第四节 稻—菜—蚓循环模式

一、系统生态学原理及模式内涵

稻—菜—蚓循环模式即将稻麦轮作改为稻菜的水旱轮作，通过水稻和耐寒蔬菜完成一个水旱轮作的生产周期，秸秆还田的同时施用一定量的有机肥，并添加蚯蚓这一地球上最有价值的土壤动物，使得水稻秸秆、蔬菜茎叶、有机肥和土壤腐殖质综合起来成为蚯蚓的饵料，利用蚯蚓生命活动的取食，调节土壤微生物等性能，此举不仅可以利用蚯蚓对水稻秸秆和蔬菜残叶的吞食加速了秸秆还田处理速度，还可以通过蚯蚓活动改变土壤物理性质，调节土壤微生物活性，加快秸秆腐解，实现秸秆的原位还田快速处理。此模式可改变稻麦茬口转换期短而造成的秸秆无法充分腐解归田，从而影响稻麦种植过程的诸多问题，尤其是水稻种植前泡田对小麦秸秆的影响，既实现稻—菜—蚓3收，又可原位处理秸秆和残叶，减少化肥用量，提高单位土地产出效益，达到增收、节支和绿色的可持续发展目的。

二、模式特点

1. 更加丰富的生产模式，增加经济效益

结合当地实际，建立"冬闲利用、春荒填补、冬春兼顾、综合发展、常年栽培"的稻—菜—蚓循环模式，增加了单位土地产出，提高了农民收入。

2. 改善农田环境，增强示范带动作用

秸秆等农业废弃物得到循环再利用，提高了资源利用率，改善了生态环境，培育了一批规模主体，发挥了积极的示范带动作用，社会生态效益显著。

三、技术操作规范

1. 水稻技术操作规范

（1）品种选择

选用高产、优质、矮秆、抗逆性强的水稻品种或组合，如"花优14""寒优香晴"等。

（2）种子处理

以日平均气温稳定超过15 ℃的初日为始播期指标，每年3月20日左右播种，播种前对种子进行精选、晾晒，然后间歇浸种催芽24~48小时。催芽分为3个阶段：第一阶段高温破胸，32~35 ℃，保持10小时；第二阶段适温催根长芽，24~30 ℃，保持24小时；第三阶段摊晾催芽。

（3）秧田准备

秧田应选择土质疏松、地势平坦、地下水位低、排水方便、水源清洁的地块，并施腐熟人粪尿15 t/hm²，复合肥300 kg/hm²作基肥。采取湿法育秧，畦宽100~130 cm，沟宽30~40 cm，沟深约15 cm，要求平、直、细。

（4）秧田肥水管理

一叶包心期，施尿素30 kg/hm²；二叶包心期，施尿素37.5 kg/hm²；移栽前4~5天，施尿素6075 kg/hm²作送嫁肥。出叶前畦面不上水，以湿润为主；出叶后浅水勤泄，以水调温、调气、调肥，促分蘖。如有抛秧，抛秧前2~3天要断水。

（5）插秧

前茬作物采收后，机耕深翻，灌足水，使蔬菜叶、茎等充分淹没腐化，插秧前1~2天再耕翻耙平耙匀。秧龄20~25天插秧，要求浅插（2~3 cm）。稳（不浮秧）、匀（苗数均匀）、直（不斜插）。栽插密度为"寒优香晴"30万穴/hm² 左右，"花优14"约22.5万穴/hm²。

（6）田间管理

① 追肥：以"稳头、顾中、保尾"为原则，菜地少施氮肥，多施钾肥。插秧后6~10天，施用水稻专用肥225 kg/hm² 及氯化钾180~225 kg/hm²；抽穗期可结合喷药施0.2%磷酸二氢钾。

② 大田水管：分蘖期应灌浅水，排灌结合，适时烤田，并分次轻烤。当田间总苗数达到预期穗数的80%时开始烤田。在孕穗期，则采用浅水勤灌、干湿交替的方法，以后水不见前水为宜。抽穗至齐穗期保持浅水，后期保持湿润状态，收割前7天断水，以防止脱水过早，影响灌浆结实。

③ 病虫害防治：以防治为主，采取水稻群体优化调控技术，增强水稻植株自身抗性。当季主要病虫害有水稻纹枯病、水稻二化螟、三化螟和稻纵卷叶螟等，水稻纹枯病可用30%苯醚甲环唑、丙环唑300 mL/hm² 兑水600~750 kg/hm² 喷雾，或用6%井冈·A蛇床子素可湿性粉剂800 g/hm² 防治；水稻二化螟、三化螟和稻丛卷叶螟可选用10%阿维·氟酰胺（稻腾）悬浮剂、5%氟虫腈（锐劲特）悬浮剂450~600 mL/hm²，5%氟虫腈（锐劲特）悬浮剂300 mL/hm²+99绿颖乳油500 mL/hm²，40%氯虫·噻虫嗪120~150 g/hm² 喷雾防治。

④ 收获。水稻籽粒成熟度达95%~100%时便可收割。

2. 花椰菜技术操作规范

（1）品种选择

选用抗病、优质、丰产、耐寒的长胜90天等花椰菜品种。

（2）育苗

① 苗地整理：选土壤肥沃、排水方便、富含有机质，与生产田隔离的沙质土作育苗田，撒施优质商品有机肥3 t/hm²、硫酸钾复合肥（氮：磷：钾＝15：15：15）150 kg/hm²、钙镁磷肥375 kg/hm²，然后深翻20~25 cm，播前3天左右按畦面宽100 cm、畦高30 cm、畦沟宽30 cm的规格起畦及平整畦面。

② 播种：每隔3~4 cm压一条深1 cm的条沟，然后沿沟均匀播种，粒距约3 cm。播完后覆盖厚1~1.5 cm的细砂（土），并淋透水，再用2层遮阳网覆盖畦面，以后每天早晚直接在遮阳网上各淋1次水。

③ 苗期管理：待70%左右的种子出苗时，及时去掉覆盖物，齐苗后控制水分，防立枯病和徒长。当苗长至2~3片真叶时，若出现苗弱、苗小、叶淡黄色，可追施尿素37.5 kg/hm²。苗期主要防止菜青虫、小菜蛾、跳甲等危害，可用10%除尽3000倍液，

或15%安打4000倍液喷雾防治。4~5片真叶时移苗定植。

（3）定植

水稻收割后，及时深翻，并结合翻地施足基肥。一般撒施商品有机肥4.5 t/hm²、钙镁磷肥300 kg/hm²、硫酸钾复合肥300 kg/hm²和硼砂15 kg/hm²，条施在畦中央，并按畦带沟120~130 cm作畦。低的种植基地宜做成高畦，以防涝降渍。大田整理好后，选择苗龄25天左右的无病虫壮苗，在午后阳光温和时，采用双行定植，定植密度3.00万~3.75万株/hm²，株距为40~50 cm。

（4）大田管理

① 灌溉与排水：定植后每天傍晚浇水1次，保持土壤湿润。缓苗后，可以结合淋水进行追肥。开花需要充足的水分，但是在中耕及施肥时，土壤应稍干燥，以便中耕除草及肥料的吸收。如遇降雨过多时，应该及时清沟排水。

② 追肥：花椰菜生长初期和抽蔓期需要较多的氮素和适量的磷、钾肥。生长盛期，除施足氮肥外，还必须配合增施磷、钾肥及补充硼、镁、锰、钼等微量元素。第一次追肥在定植后7~10天进行，浇施尿素75 kg/hm²；栽后30天中耕除草，结合培土追施硫酸钾复合肥150 kg/hm²；第三次追肥在栽后60天，约开始现蕾前施尿素75 kg/hm²、硫酸钾复合肥150 kg/hm²；后期配合喷药，根外喷施0.2%硼砂1~2次。

③ 返叶：花椰菜开花期如不遮叶，则花色容易发黄，商品性比较差。因此，在花球直径7~8 cm时，折断老叶把花球全部遮住。

④ 病虫害防治：主要病害有霜霉病、菌核病、黑腐病、软腐病等。霜霉病发病初期用64%杀毒矾，或58%雷多米尔，或72%克露500~600倍液喷雾；菌核病在发病初及多雨天用50%扑海因1000倍液，或50%速克灵1500倍喷雾；黑腐病、软腐病在移栽成活后用80%必备500倍液，或77%可杀得500倍液，或47%加瑞农800倍液喷雾。主要虫害有蚜虫、白粉虱、小菜蛾、斜纹夜蛾等，可用10%毗虫啉2500倍液，或3%农不老5000倍液防治；选用10%除尽3000倍液，或15%安打4000倍液防治小菜蛾和斜纹夜蛾。

⑤ 收获：采收太早花球未充分长大，产量低；太迟则花球松散，花蕾粒粗松，影响其品质和价值，故应在花球发育适量大小、各小花蕾尚未松开之前采收。用刀在花球基部砍下，去掉多余的外叶、套上泡沫网袋，并做好标签后放进塑料筐，及时送进附近工厂进行预冷，并做好记录，以便后期加工或销售。

3. 蚯蚓技术操作规范

（1）品种选择

赤子爱胜蚓选育的太平2号、北星2号、北京条纹蚓等。

（2）养殖条件及准备

饲料准备。搞好"三性"饲料的组合搭配。"三性"饲料，即动物性饲料：猪、牛粪等；植物性饲料：各种秸秆、树叶、杂草等；水果性饲料：西瓜皮、烂水果、橘子等。第一、第三类饲料大体占 70%。

室外养殖法。择向阳潮湿、能灌排水的地方，箱宽、箱间沟宽、深，箱面平整、稍压实。先在中央填上 1 m 厚的发酵饵料，再放上含有幼蚓的饵料，使总厚度达 23 cm 深，最后用麦秸或草帘覆盖。气温达到 15 ℃以上开始养殖，气温降至 10 ℃时应转入室内保种。

温湿度条件。①最适温度 10~30 ℃。冬季稍加遮盖即可，不让蚯蚓冬眠，但不可暴晒及雨淋。②蚯蚓对湿度要求不高，相对湿度为 60%~70%，一般发酵新鲜牛粪直接投入即可，如果堆放太久偏干，可稍喷些水，水分掌握在用手握料，手指间见水珠但不滴下为宜。酸碱度 pH 值 6~8。

（3）管理工作

蚯蚓食性广，几乎所有的植物残体、腐殖质、腐烂动物和生活垃圾都是蚯蚓的食物。蚯蚓的饵料尽可能多样化，避免单一。饵料须经处理，先去杂质，然后将秸秆、杂草、甘蔗渣等切碎，加上猪牛鸡粪，堆成 1 m 高的圆锥形，用草帘或塑料薄膜覆盖发酵。假如原料过干，应喷水浇湿；也不要太湿，用手紧握时，能挤出少量水滴为宜。经过 5 天发酵，料温一般升到 70 ℃左右，倒翻 1 次，有利于发酵均匀。经过几天温度降下来，再喷水浇湿，再翻几次。经过 3 周左右饵料发酵完毕。理想的发酵饵料应是黑褐色，无恶臭，松散，不粘手。

一般每 20 天观察一次，并加料一次，方法同上。每 40 天可成倍扩大饲养面积。一般每 40 天为一周期，一年可养 9 批。

第五节　菜—鱼共作模式

一、系统生态学原理及模式内涵

菜—鱼共作模式是把水产养殖与旱耕蔬菜栽培这两种原本完全不同的农耕技术，通过巧妙的生态设计，达到科学的协同共生，从而实现养鱼不换水而无水质忧患，种菜不施肥而正常成长的生态共生效应。菜—鱼共生系统是一种涉及鱼类和植物的营养生理、环境、理化、机电等学科的农业技术。在传统的水产养殖中，随着鱼的排泄物积累，水体的氨氮增加，毒性逐步增大。而在菜—鱼共生系统中，水产养殖的水被输送到水培系统，由硝化菌将水中的氨氮亚硝化成亚硝酸盐，然后继续被硝化菌硝化成硝酸盐，硝酸盐则可以直接被植物作为营养吸收利用。菜—鱼共生让动物、植物、微生物三者之间达到一种和谐的生态平衡关系，从而形成小区域良性循环，最大限度地提高了水产品和蔬菜产量，还能把水质污染程度降至最低限度，是可持续循环型零排放的低碳生产模式，更是有效解决农业生态危机的最有效方法。共生系统是高密度水产养殖技术与蔬菜无土栽培有机结合的产物。

二、模式特点

1. 病害防治，菜—鱼模式少见鱼病发生

除设施本身的隔离条件有利于病害的防治外，植物根系会分泌一种排异性的微毒素，对动物某些病菌有抑制作用，可减少农药的使用。

2. 菜鱼产量及品质有所提升

蔬菜吸收了水中的氮、磷等养分，净化了水质，利于鱼的生长，促进产量提升；同时鱼的粪便作为蔬菜的优质肥料，减少了肥料的使用，提升了蔬菜品质。

3. 自证清白

菜—鱼共生脱离了土壤栽培模式，避免土壤的重金属污染，因此，菜—鱼共生系统中蔬菜和水产品的重金属残留都远低于传统土壤栽培。且因为菜—鱼共生系统中鱼的存在，故不能使用任何农药，否则会造成鱼和有益微生物种群的死亡和系统的崩溃。如果菜—鱼共生农场生产的蔬菜带着根配送的话，消费者很容易识别蔬菜的来源，避免消费者产生这个菜是不是来自批发市场的疑虑。

三、技术操作规范

1. 浮架制作工艺

① PVC 管浮架制作方法。用 PVC 管（直径 110 mm）按照 1 m×2 m 或者 1 m×4 m 规格，四端用弯头连接并密封，上下两层用上疏下密两种聚乙烯网片拴牢制成浮床，主要作用是使蔬菜漂浮水面生长，隔断食草性鱼类吃菜和控制茎叶生长方向。下层网片密度规格适中，既可固定菜根、防止鱼吃菜，又有利于水体交换。养殖户可根据自己池塘的实际条件，按照移动、清理、制作、收割方便的原则，选择合适规格的浮架。

② 竹子浮架制作方法。选用直径在 6 cm 以上的竹子，首尾相连，按照竹子的长短通过竹篾固定做成四边形、三角形（防止变形），上下层用竹篾编制成大小不同规格网格，既便于栽种和固定菜苗，又防止鱼吃菜苗。具体形状可根据池塘条件、材料大小、操作方便灵活而定。

2. 栽培蔬菜种类选择

水上栽种的作物并不单单指吃的蔬菜，还应包括观赏用的花卉、食用的饲草等。栽培蔬菜种类应选择根系发达、耐水、净水能力强的蔬菜、瓜果、花卉植株，利用其根系发达与庞大的吸收表面积，进行水质的净化处理。一般选择的品种有空心菜、水芹菜、水白菜、青菜、生菜、菱角、丝瓜、南瓜、西瓜等。经实验，栽种空心菜、水芹菜、菱角、丝瓜等效果较好，因为其生长旺盛，产量高，净水效果佳。

3. 种植面积的选择

种植和养殖的面积与池塘面积的比例，关系到生物间的生态平衡和物质能量的循环利用。比如，多少鱼排出的粪便能为多少菜提供养分，多少菜能够对水质净化产生最佳的生态效果，这是菜—鱼共生技术的基础。1.5～2 m 水深的精养池塘鱼单产 1200 kg/hm^2 以上，养殖周期 3 年以上，种植空心菜等蔬菜面积占池塘的 5% 左右，可取得较好的结果。蔬菜种植的比例应根据池塘水质的肥瘦、水体的大小、养殖鱼类的多少合理确定，一般养殖密度比较大、水体比较小、养殖周期长的池塘，种植比例可适当提高，但应控制在 10% 以内。

4. 蔬菜栽培技术

主要采用移植的方式栽种，如 PVC 标准浮架可采用直接栽培法、营养杯栽培法和泥团栽培法。直接栽培法指直接将植物茎秆按 20～30 cm 株距插入下层较密网目（空心菜在 5 月后气温基本稳定、水温在 15 ℃ 以上时，水芹菜在 10 月后，可在岸上定植好后将浮架放入池中），此后注意浮苗的补栽，此法操作快捷、简单，成活率高。营养杯栽培法主要采用花草培育杯，在杯内置入营养液或泥土（塘泥），按 20～30 cm 株距（瓜类株距 1 m 以上）放入浮架。此法成活率较高，但相对繁琐，成本也较高，适宜于瓜类定植，瓜类藤蔓附着在池塘周围或者用绳网固定在池塘上面。泥团栽培法主要是指将植物茎秆直接插入做好的小泥团（塘泥即可），按 20～30 cm 株距放入浮架。此法成活率相对较高，操作也简单方便，成本低。

各养殖单位可根据自己的劳动力数量灵活选择合适的种植方法，各种植方法成活率依次为：营养杯栽培法＞泥团栽培法＞直接栽培法；劳动力花费量依次为：营养杯栽培法＞泥团栽培法＞直接栽培法。

5. 病虫草害防治策略

养鱼菜田的病虫草害防治策略同养鳝菜田。菜田养鱼后，鱼能吃掉一些菜田害虫，但仍然需要结合化学防治。农药对鱼类的毒性可分为高、中、低 3 级。呋喃丹、敌杀死、速灭杀丁、三唑磷、灭扫利、DDT、六六六、鱼藤精、五氯酚钠、敌稗、去草胺等属高毒农药；敌敌畏、敌百虫、1605、久效磷、叶蝉散、稻瘟净、异稻瘟净、稻瘟灵、除草醚、杀草丹、甲草胺、扑草净等为中毒农药；杀虫双、甲胺磷、多菌灵、井冈霉素、扑虱灵、草甘膦等为低毒农药。

第五章 现代生态循环农业发展模式与技术

现代生态循环农业在实践过程中创造出了许多不同类型的发展模式与配套技术，涌现出了一大批好的典型与案例，发挥了积极的示范带动作用。例如，农业种养空间立体结构生态系统模式与技术、农业资源节约与物质循环利用系统模式与技术、庭院现代生态循环农业模式与技术、绿色休闲农业与农业科技（生态）园区模式与技术等。总结推广这些不同类型的发展模式与技术，让广大农民生产者在实施现代生态循环农业的过程中得以借鉴、应用和推广，对促进现代生态循环农业的发展有着十分重要的意义与作用。同时，这些模式与技术还需要在实践中不断创新、不断完善，开发出更多的新型发展模式与技术，引领与推动现代生态循环农业的可持续发展。

第一节 种养空间立体结构模式与技术

一、农田立体种植模式与技术

农田立体种植模式是在继承和发扬中国优良传统种植经验的基础上，广泛利用现有的自然资源、生产条件和现代农业科学技术，将不同作物及品种按照农田生态规律科学地进行间作、套种、混种、复种等，使之有效地提高单位土地、单位时间和空间内光、热、水、肥、气等自然资源的利用率，切实做到地尽其力、物尽其用，建立良好的农业生态经济复合体，从而获得较好的经济效益、社会效益和生态效益。

1. 粮菜（粮）高效立体种植模式与技术

这类模式大多是在水肥条件较好的精耕细作地区和城镇郊区开发推广。主要以小麦套种玉米、小麦套种谷子、小麦套种甘薯、小麦套种水稻等一年两熟的粮食作物为主体作物，在保证粮食作物持续稳定增产的前提下，参与经济价值较高的经济作物如瓜类、蔬菜作物，提高单位面积的经济效益。这类模式在耕作制度上多是从小麦播种开始，将播种带的宽窄、畦面的大小、留套种行的大小及参与后茬作物的时间一次性确定下来，一年之前即将两茬主体作物间、套、混、复种的瓜类、蔬菜、豆类等经济效益较高的作物列入种植模式计划，一般结合小麦秋种，间作越冬性蔬菜，如菠菜、蒿菜、大蒜、洋葱、大葱等。早春多在麦田内套种小白菜、小油菜、小红萝卜、早熟甘蓝、地芸豆等速生蔬菜。麦收之前，除了套种（或移栽）第二季主体作物玉米、谷子、甘薯、大豆等粮食作物外，还可在麦田套种（或移栽）西瓜、生姜、马铃薯等高效益作物。麦收后夏播作物间作大豆、绿豆、秋菜等。例如，小麦套种玉米间作大豆、小麦套种玉米间作夏谷、小麦间作甘蓝套种玉米间作大豆、小麦间作越冬菜套种玉米间作蘑菇、小麦间作越冬菜（指菠菜、蔓菜、大蒜、洋葱下同）套种玉米套种大白菜、小麦间作越冬菜套种玉米套种黄瓜、小麦间作越冬菜套种玉米套种番茄、小麦间作越冬菜套种生姜、小麦间作大蒜（或洋葱）套种玉米间作秋菜（萝卜、芥菜、大白菜等）、小麦套种玉米间作菜葫芦、小麦套种马铃薯套种玉米间作大白菜等。此类模式能在保证粮食不减产或少减产的情况下，多收获1~2季蔬菜（或其他作物），每亩可增加经济效益500元以上，实现高产高效。

这类模式主要技术包括：小麦、玉米和萝卜、芥菜、大白菜、菠菜、蔓菜、大蒜、洋葱等蔬菜种植与田间管理技术，农作物之间的空间合理配置与套种技术，施肥技术，行株距比例设计等。

2. 粮棉菜立体种植模式与技术

这类模式主要是在春棉一熟制地区，通过改革耕作制度，利用其冬闲季节增种一季小麦或间套越冬菜，或者在小麦套种棉花一年二熟制的基础上，利用其棉花套种行套种越冬菜或瓜类等，正进行开发推广。在保证棉花增产或小麦、棉花双增产的前提下，通过间套瓜类、蔬菜等，增加经济收益。在耕作制度上，多在拔棉花柴以后，整地、施肥，确定好种植规格，播种晚茬小麦，留出棉花的套种行，在套种行上结合种麦间作洋葱、大蒜、菠菜等越冬菜，或第二年早春间作小白菜、小油菜、小红萝卜等速生菜，收菜后于4月中下旬套种棉花。例如，小麦间作大蒜套种棉花，小麦间作洋葱套种棉花，小麦间作越冬菜（或秋芥菜）套种棉花间作萝卜，小麦间作越冬菜套种西瓜间作棉花，小麦间作越冬菜套种棉花间作绿豆，小麦间作芥菜套种棉花间作辣椒等。

这类模式主要技术包括：小麦、棉花和蔬菜、瓜类种植与田间管理技术，农作物之间的空间合理配置与套种技术，施肥技术，行株距比例设计等。

3. 粮油菜立体种植模式与技术

这类模式多是在水肥条件较好的花生产区开发推广。在小麦套种花生一年二熟制的基础上，利用其套种花生的套作行间套种瓜类或蔬菜，一般每亩增加经济效益300元以上。在耕作制度上，主要是从秋种开始，扶大垄，在垄沟内播种小麦，在垄上间作菠菜、蔓菜等越冬菜，第二年春季收菜后，套种春花生或半夏花生，并间作西瓜、辣椒等秋菜。例如，小麦间作越冬菜套种花生套种西瓜、小麦间作大蒜苗套种西瓜套种花生间作秋菜、小麦套种西瓜套种花生间作玉米、小麦间作越冬菜（或油菜）套种花生间作芝麻。

这类模式主要技术包括：小麦、花生、油菜、芝麻、蔬菜、瓜类种植与田间管理技术，施肥技术，农作物之间的空间合理配置与套种技术，行株距比例设计等。

4. 水稻高效立体种（养）植（殖）模式与技术

这类模式以水稻为主体作物，冬春季种植各类经济作物或在水稻田中养殖各种高档水产品，既可保持水稻较高的产量，又可比稻麦两熟制大幅度提高经济效益。一般选用的模式有"洋葱（荷兰豆、大蒜、马铃薯）—单季稻""简易大棚栽培甜瓜（草莓、黄瓜、辣椒、茄子）—单季稻""冬春蔬菜—青糯玉米—后季稻""冬春蔬菜—西瓜、甜瓜（四季豆）—后季稻""冬春蔬菜—单季稻+螃蟹（青虾、鲫鱼）"等。

这类模式的主要技术包括：水稻、马铃薯和荷兰豆、大蒜、黄瓜、辣椒等蔬菜种植与田间管理技术，螃蟹、青虾、鲫鱼等养殖管理技术。

5. 棉菜高效立体种植模式与技术

该类模式以棉花为主体作物，冬春秋季间套复种各类经济作物，既保持棉花的高产，又可大幅提高棉田经济效益。一般选用的模式有"马铃薯（榨菜）套种棉花""春毛豆（蔬菜）套种棉花套种荷兰豆（萝卜、大蒜）""棉花套种大蒜（西葫芦、西瓜）"等。

这类模式主要技术包括：棉花和蔬菜、马铃薯等种植与田间管理技术，施肥技术，农作物之间的空间合理配置与套种技术，行株距比例设计等。

6. 农业综合开发生态模式与技术

该模式是集种植、养殖、加工、经营与生态保护为一体的综合发展模式，主要技术是立体种植技术、养殖技术、加工技术、沼气技术、资源综合利用技术等。

以上立体种植模式是在总结传统的间、套、复种增产经验的基础上，经过大量的试验、示范和生产实践，研究总结出的一门新兴的、综合性的应用农业科学技术。农田立体种植模式是增加农业收入，农民科学致富的重要途径，符合中国人多地少、精耕细作、集约经营等国情特点，具有强大的生命力和广阔的发展前景。广大农民可以通过立体种植，高效栽培，利用当地的自然资源、生产条件，因地制宜地改革和运用各种各样的立体种植模式，充分发挥自己的智慧和才能，不断地挖掘土地的增产潜力，加快现代生态循环农业发展。

二、水域立体养殖模式与技术

池塘立体养殖模式主要有莲藕塘立体养殖、鱼鸭立体养殖、青虾河蟹混合立体养殖等。

1. 池塘莲藕—鱼立体养殖模式与技术

池塘莲藕立体养殖是选择水源充足，水体无污染，进排水比较方便，面积为10~20亩，池深1~1.5 m，池底淤泥厚20~40 cm的池塘，塘基高70 cm以上。池塘水深一般保持在15~50 cm。春季在池塘中栽植莲藕，在莲藕池塘中养殖黄鳝、泥鳅、鲢鱼等，形成藕鱼共生共养。高温季节要注意及时换水或者是加注新水，冬季要做好防冻工作。一般大面积养殖进入9月即可采用鳝笼等工具捕获鳝鱼、泥鳅，也可在越冬前用双手逐块翻泥捕获鳝鱼、泥鳅，还可留待春节前后出售。每亩莲藕塘可产鳝鱼、泥鳅 200 kg 左右，产藕 2000~3000 kg。

这种模式的主要技术有黄鳝、泥鳅、鲤鱼等养殖技术，莲藕的种植与管理技术，池塘管理技术等。

2. 池塘鱼—鸭立体养殖模式与技术

池塘鱼、鸭立体养殖就是在池塘中养鱼和养鸭组合的配套高效养殖模式。鱼、鸭同塘是一种特有的共生现象，塘中的水面可以供鸭子在水中活动，为鱼增氧，水底的浮游生物供鸭子食用，鸭粪供鱼食用，是一种共生共养立体养殖技术。不仅节约了养鱼的饵料投入，提高了养鱼产量和收益，而且水面为养鸭提供了活动场所，鸭在水中洗浴，可以促进鸭的新陈代谢，提高体质和健康水平，鸭蛋品质较高，还解决了养鸭的排泄物污染的问题，有利于改善养鸭环境。实施鸭、鱼配套养殖，二者相得益彰，互为促进。

这种模式的主要技术有鱼、鸭的立体养殖技术，池塘的管理技术等。

3. 池塘青虾、河蟹混合立体养殖与技术

池塘青虾、河蟹混合立体养殖，一般要求池塘面积在 5~10 亩，水深 1.5 m，水源充足，水质优良，无污染源，塘口进排水方便，进排水口要独立，成对角线排列，要设有滤水装置，冬春季应清塘，晾晒鱼塘消毒处理。根据天气情况及时种植水草，以苦草为主，适当移植黑藻、水花生、水葫芦等，使水面覆盖率达 60%~70%。幼蟹、仔虾一般是正月前后放养，幼蟹规格不小于 5 g，放养密度每亩 400~500 只，仔虾规格每千克 400~800 尾，放养密度每亩 7~8 kg，苗种下池时应做消毒处理。一般使用质量安全渔用饲料，根据河蟹养殖密度规格的大小、饵料成本高低及河蟹市场等因素及时捕蟹上市，年底一次性捕完并做好清塘工作，为明年的养殖做好准备。

这种模式的主要技术有水草、黑藻、水花生、水葫芦等的种植技术，青虾、河蟹混合立体养殖技术，池塘管理技术等。

4. "水面种菜，水中养鱼"的立体循环养殖模式与技术

"水里养鱼，水面种菜"立体循环养殖模式，是在鱼池上搭起竹筏，选择空心菜等水生叶菜，用浮床作为载体栽种。在养殖塘内种出的空心菜不需施肥也不需打农药，水里养殖的鱼虾粪便是空心菜最好的肥料。浮床上种出来的空心菜不仅口感、鲜嫩度等比旱地上种植的要好，而且产量也比旱地上种植的高出好几倍。最重要的是，水面上种空心菜可以改善养殖塘内的水质环境，既节省养殖成本，又增加了效益，不仅可以收获有利于人体健康的绿色、安全水产品，而且还可以获得鲜嫩可口的蔬菜，可谓是一举两得的立体种养技术。

这种模式的主要技术有水生叶菜的种植技术、鱼的养殖技术、池塘的管理技术等。

5. 淡水珍珠与本地草鸭立体循环养殖模式与技术

按水域为单元，在一个湖或水库边建猪舍养猪，利用斜坡地建鸭舍栏网养鸭，

水中立体养鱼，水底育蚌，水面吊蚌育珍珠，珍珠加工成项链、水解珍珠粉、珍珠护肤霜，贝壳加工成工艺品和矿物饲料，形成循环综合利用立体养殖加工模式。养殖项目按比例配套：每亩水面养育肥猪 5 头、母猪 0.3 头，0.5 m^2 幼蚌池，0.67 hm^2 鱼池，20 羽鸭。猪粪、鸭粪入池养鱼、育珠，每亩水面可生产珍珠 1kg、鲜鱼 50kg。可形成在人工控制和调节下，以食物链为纽带的水陆物质流和能量循环利用的水面立体生产体系，这一综合立体养殖模式大大提高了经济效益。

这种模式的主要技术有猪、鸭、鱼、蚌、珍珠的养殖技术，珍珠的加工技术，水域的科学管理等。

6. 鸭—鱼—鳖立体养殖模式与技术

鸭—鱼—鳖立体养殖模式是按水域为单元，渔业与畜牧业相结合的养殖方式。鸭在水面养，把鸭子放养在水中，水里的青草、昆虫就成了鸭的天然饲料，每只鸭的饲养成本比饲养喂养少 4~5 元，由于放养，鸭的肉质鲜美，销售价格比一般养的鸭要高。同时，每只鸭又是一位"施肥员"，用鸭粪当鱼的饵料，鱼产量可提高 20% 左右。鳖还可以把水里多余的、质量不好的小鱼小虾吃掉，提高了鱼的质量。鱼鳖养殖要采取多品种混养，做到不同食性、不同层次的鱼类混养，提高水体利用率和饵料的合理利用。

这种模式的主要技术有鸭、鱼、鳖养殖技术，水域的科学管理等。

7. 海水池塘立体养殖模式与技术

海水池塘立体养殖模式，就是水中养虾、水底养蟹、底泥养贝的立体养殖方式，具体讲，就是在一个海水池塘水体中，上层水养虾、池塘底部养梭子蟹、池塘的底泥里养殖杂色蛤或者蛏子。这种养殖模式不仅有利于生态互补，而且可综合利用水域，提高经济效益。

这种模式的主要技术有虾、蟹、贝立体养殖技术与疾病防治技术，池塘科学管理等。

8. 淡水水域多品种综合立体养殖模式与技术

该模式是在水库、池塘养殖鲤鱼、草鱼、鲫鱼等多品种鱼类。不同的鱼种可根据其生理特性在不同的水层繁养，可充分利用水体的空间、时间和饵料资源进行综合主体养殖，实现高产、高效、生态环保的目的，是淡水养殖的理想模式。

这种模式的主要技术是不同鱼种的养殖技术、水域的科学管理技术等。

三、林果地立体模式与技术

1. 林果—粮经立体生态模式与技术

这一模式国际上统称农林复合系统，林果—粮经立体生态模式是利用林果和农作物之间在时空上利用资源的差异和互补关系，在林果株行距中间开阔地带种植粮食、蔬菜、药材乃至瓜类经济作物，形成不同类型的农林复合种植模式，也是立体种植的主要生产形式,一般能获得比单一种植更高的综合效益。

这一模式的主要技术有立体种植、间作技术等,配套技术包括合理密植栽培技术、节水技术、平衡施肥技术、病虫害综合防治技术。

2."林果—畜禽"复合生态模式与技术

这类模式的基本结构是"林果业+畜禽业"。在林地或果园内放养各种经济动物,以野生取食为主,辅以必要的人工饲养,较集约化养殖更为优质、安全的多种畜禽产品,接近有机食品。一般在丘陵山区,在山坡地发展林果业或林草业,在林地中或果园里建立禽舍,将畜禽粪便直接投放林地或果园,从而形成了"林(果)、草、禽养殖单元"相互联系的立体生态农业系统。牧草种在林间,利用林间空地生长,一方面可以保持水土,另一方面还可以减少杂树、杂草的生长。为避免牧草种植过程中受野草侵害,选择在林木落叶季节进行秋播牧草。畜禽在林间轮放或圈养,用刈割牧草来饲喂,禽畜在林间放养过程中,采食牧草的同时也将粪便施到了林间,这样有利于牧草与林间苗木的生长。主要有"林—鱼—鸭""胶林养牛(鸡)""山林养鸡""果园养鸡(兔)"等模式。这类模式,畜禽为树木提供有机肥、啄食害虫,树木又为畜禽创造适宜的生长环境,起到了种植、养殖协同发展的良好效果。

这类模式的主要技术包括林果种植技术、动物养殖技术和种养搭配比例等,配套技术包括饲料配方技术、疫病防治技术、草生栽培技术和地力培肥技术等。

3."农业林网"生态模式与技术

这是一种农林复合系统模式。主要是平原区为确保种植业的稳定生产,减少农业气象灾害,改善农田生态环境条件,通过标准化统一规划设计,利用路、渠、沟、河进行网格化农田林网建设及部分林带或片林建设,一般以速生杨树为主,辅以柳树、银杏等树种,并通过间伐保证合理密度和林木覆盖率,这样便逐步形成了与农田生态系统相配套的林网体系,如黄淮海地区的农田林网。

这种模式的主要技术包括树木栽培技术、网格布设技术,配套技术包括病虫害防治技术、间伐技术等。

第二节　农业资源节约和物质循环利用系统模式与技术

一、农业资源节约型模式与技术

中国农业属于资源约束型农业,虽然农业自然资源总量大,但人均拥有量小;山地丘陵分布广,土地难利用比重大,耕地面积不足;水资源短缺,干旱、半干旱区域广;水土流失严重、面源污染不断加剧,农业生态环境日趋恶化;农村人口素质不高,成为农业发展的严重制约因素。这一现实既凸显了建设节约型农业的紧迫性,又成为研究如何建设节约型农业的出发点和依据。为此,大力发展资源节约型农业意义深远。

农业资源节约型模式是以提高资源利用效率为核心,以节地、节水、节能、节肥、节农药、节种子和农业资源综合循环利用为重点的农业生产方式。节地,要建立和推广立体种植模式,充分利用地域空间发展节地农业。节水,要推广普及管道输水、膜下滴灌、水肥一体化等高效节水灌溉技术,建设旱作农业示范基地,加大旱作节水农业技术推广力度。节能,要淘汰老旧农业机械,推广使用节能型农业机械,推进抽水泵站节能改造,推广普及节能型太阳能蔬菜大棚。节肥,要大力推广测土配方施肥技术,科学使用化肥,鼓励农民增施有机肥,减少不合理化肥施用量。节农药,要淘汰落后施药机械,推广使用高效、低毒、低残留农药,开展有机农产品基地建设。节种子,要大力推广优良品种,实施定量精播。由于各地农业自然条件的不同,具体可根据各地农业资源状况、自然经济技术条件、区域特点和发展优势,发展各具特色的农业资源节约型模式,如农业的"节地—节时—节水"模式、畜牧业的"节粮—食草型"模式、渔业的"节饵—多层型"模式、林业的"速生—木本粮油—立体型"模式。

这一模式的根本在于依靠科技、突出节约、重在效益。主要技术是节地、节水、节能、节农药、节种子的相关技术。

二、种养加销模式与技术

"种—养—加—销"一体化的生态农业循环模式,是在农田种植农作物,农作物为发展养殖业提供饲料;农作物秸秆、养殖业粪污作沼气,经无害化处理后返回到农田;利用沼气作能源,利用种养业提供的产品为原料,发展加工业,生产净菜、豆腐、特色肉食品等产品,副料豆腐渣等作饲料,又返回养殖业;经过深加工的食品提供给配送中心,进入销售渠道;其收入返回到种养业,为扩大再生产提供资金支持。这种模式实现了农业生态系统内的资源循环利用,提高了资源转化率,节约了资源、保护了生态环境,构建了高产、优质、高效、生态、安全良性循环的农业生产体系。

主要技术是种植、养殖、加工技术,物质循环利用技术,营销管理技术,"种—养—加—销"一体化的生态农业循环模式的设计技术。

三、农业产业化现代生态循环模式与技术

该模式按照自然生态系统中物质循环共生的原理设计农业产业化生产体系,以绿色低碳循环(生态)农业"龙头"企业为依托,立足于当地农业资源的综合开发来确定农业产业化的主导产业和主导产品,把农副产品生产、加工、销售有机结合起来,推进绿色低碳循环(生态)农业向商品化、专业化、现代化转变。在农业产业化的过

程中尽量延长生物链和加工链，形成"自然资源—产品—资源再生利用"的物质循环过程，使投入的自然资源量最少，废物的排放量最小，对生态环境的危害程度最低，实现经济效益最大。

主要技术是农业产业链设计技术，农产品加工技术、资源循环利用技术、清洁生产技术等。

第三节　农家庭院现代生态循环农业模式与技术

农家庭院是指农村中农家的院落。庭院绿色低碳循环生态农业模式是农民利用自家的庭院区域从事集约化生产的一种经营形式，主要是发挥庭院资源优势，依据生态学、循环经济学的原理，运用系统工程方法，建立种植、养殖、沼气、综合利用的生态循环农业模式。有的种养并举，综合利用；有的利用有限空间发展立体种养业。由于庭院面积的大小、地域环境及资源条件的不同，庭院现代生态循环农业主要有以下模式。

一、"厕所、猪栏、沼气池"三位一体生态循环农业模式与技术

该模式形成以沼气促养殖，以养殖促种植，以种植促养殖的良性循环，实现庭院经济效益、生态效益和社会效益的统一。

该模式以沼气池为纽带，充分利用生物链和生物的共生、互生作用，通过能量多层转化和物质循环利用，将种植业、养殖业、能源及生态环境保护有机结合起来，形成系统。人、畜粪尿及其他有机物直接进入沼气池，沼渣、沼液为种植业提供肥料，沼气为农户提供能源，种植业为养殖业提供饲料，形成生物链条，实现系统整体协调、物质循环再生的生态循环农业模式。

在设计与技术方面，选择猪栏、厕所等排污较为便利的适当位置，按沼气建池技术标准，建上流式浮罩沼气池。用暗管与沼气池进料口相连接，将人、畜排泄物及其他污染有机物排入沼气池发酵，使"厕所、猪栏、沼气池"形成三位一体。沼气池出料口设有一个出料池，便于种植业随时提取沼渣、沼液。产出的沼气入室做饭、照明。

建设规模。三位一体户用沼气池大致需占地 30～42 m^2，一栏一厕需约 30 m^2，两栏一厕需 42 m^2。其原则是，结构布局合理，与主体构筑物（如住房）协调大方，建筑地平标高应低于主房 10 cm 以上，通行作业方便，各部尺寸协调。

沼气池建筑按照"三位一体"布置。尽量在土质坚实、地下水位低和地势较高处建池。注意避风向阳，综合考虑其他配套工程和"三位一体"的方便，以及运输车辆的畅通。根据家庭需气量和养猪多少确定适宜的体积。以 14 m^3 为例，先准备水泥 1.7 t、

河砂（粗）4 t、砾石2 t、红砖800块、直径6 mm的钢筋15 kg、石灰30 kg、直径300 mm、长1.5 m混凝土管或优质胶管1根作进料管，直径110 mm、长3 m下水道优质胶管1根作出料管。池身为圆柱形，底部为锅底形，高为2.05 m，土方严格按标准开挖，直径误差不超过5 cm。

厕所建筑。总高度应达2.3~2.4 m，内部空间净高度应达2.2 m以上，以便兼作浴室；内部面积2~3 m^2为宜；厕所门高度应达1.85 m，宽0.8 m；厕所应设通气窗；厕所地平可比猪栏地平高10 cm以上；厕所内外应粉饰平整光滑，线条横平竖直。

猪栏（舍）建筑。猪舍高度与跨度应与厕所一致；地平应比猪栏高出5~8 cm，以利干燥；猪舍内外墙应粉刷平整光滑，线条横平竖直；猪栏栏圈高度为80 cm左右，面积视养猪规模及场地面积而定。

厕所与猪舍平顶可盖预制板，也可支模板现浇。若为现浇，应适当加铁丝、钢筋、竹条等作筋，平顶上下面均作粉饰，光滑平整，并作防渗处理。

在开挖沼气池坑时，凡遇地下水位高和流沙土质时，要按一定坡度开挖，挖出的土要堆放在距池坑较远的地方。池坑周围不能堆放重物，以免发生塌方。在雨季施工建池，应在坑周围挖好排水沟，避免雨水冲垮坑壁；支模现浇时，模架要牢固平稳，安全可靠，装拆方便。拆模不能太早，一定要等水泥强度达70%以上时方能拆模。拆模时要小心，池盖上禁止站人。初次投料时，接种物的量最好是总发酵料液的10%~30%。一般来说，接种物的投量越大，沼气发酵启动越快。在沼气池使用管理过程中，应避免混入重金属盐类、农药、有毒植物等。

该模式的主要技术是猪饲养技术、菜果种植技术、沼气技术、模式的系统设计等。

二、"大棚、猪（鸡）舍、沼气池"种养生态循环农业模式与技术

这种模式是合理高效利用有限的农家庭院，在农家庭院内建一个塑料大棚，大棚内一端种植蔬菜，另一端建猪舍或在棚内搭建鸡舍。用菜叶配合饲料养鸡；将鸡粪和落地鸡料再加入部分精料和青菜、牧草喂猪；猪粪投入地下沼气池，再加入部分杂草及秸秆，产生沼气，利用沼气点灯照明，作燃料煮饭取暖；还可利用沼气渣养蚯蚓，蚯蚓可用来喂鸡，充当鸡的蛋白质饲料；剩下废渣作蔬菜的肥料。这种模式利用蔬菜及动物之间营养需求的互补关系，形成一条良性循环的生物食物链，保持生态平衡，生产出安全的绿色蔬菜和肉蛋。

在设计与技术方面，庭院建设主要分为地下、地面和空间3个部分，即空间的塑料大棚、地面的猪舍及地面下的沼气池。把发展养猪、沼气和建塑料蔬菜大棚有机地结合在一起，形成相互之间良性循环。塑料大棚不仅能提高蔬菜生长环境的温度，解决了冬季蔬菜不能生长的问题，还能提高猪舍及沼气池的湿度。猪的粪尿进入沼气池产生沼气，沼气可用于照明和做饭，沼气渣又可作为优质肥料，还可提高地表温度，

有利于猪床保暖，猪呼出的二氧化碳能满足大棚内蔬菜的需要，而蔬菜产生的氧气又可供猪利用，这样可保持猪舍内空气的新鲜，在冬季蔬菜大棚内饲养的猪生长快、疾病少。

该模式的主要技术是猪鸡饲养技术、蔬菜种植技术、沼气技术、模式的系统设计等。

三、"畜—沼—鱼、鸭、果、菜"绿色生态循环农业模式与技术

这种模式是在有水源条件的地区，农户修筑猪舍养猪，在猪舍旁建沼气池，利用猪屎猪尿生产沼气，供农户作燃气，用来做饭和照明。同时，开挖鱼塘，在水下养鱼，水上养鸭。用沼液、沼渣、鸭粪作鱼饲料，或用沼液、沼渣种果种菜。这样既解决了猪粪的污染环境问题，又可以生产沼气，作为群众生活所需的能源，还可以利用沼渣、沼液为发展无公害种植、养殖业提供绿色肥料、饲料，形成物质循环利用的绿色生态农业发展模式。

该模式的主要技术是猪鸡鸭鱼饲养技术、蔬菜果树种植技术、沼气技术、模式的系统设计等。

四、农家庭院立体种植养殖模式与技术

在农家庭院，充分利用土地和光能资源，种植苹果、石榴、杏、枣、木瓜、柿、梨、葡萄、猕猴桃等。在果树行间、棚下养殖蚯蚓或筑塘栽藕养鱼，修圈舍养畜（猪）禽（鸡）等。

这种模式的主要技术有果树栽植、栽后管理、整形修剪技术，病虫害防治技术，畜禽养殖技术，畜禽疫病防治技术等。

第四节 绿色休闲农业和农业科技（生态）产业园区模式及技术

绿色休闲农业和农业科技（生态）园区模式是以农业生产的系列开发与农业休闲观光有机结合为特色，来提升农业自身的附加值与旅游价值的新型现代生态循环农业。它是利用农业生产活动（过程）、田园与园林景观、农业科技园区、农业生态园区、现代农业园区、设施农业、农耕文化、农家生活等资源与环境，为游客提供休闲、观光、度假、娱乐、健身、教育等多项旅游活动的一种集生态功能、生产功能、生活功能、科普教育功能和旅游娱乐功能于一体的多功能农业。这种农业生产和旅游产业的结合，实现了互利双赢，休闲观光农业带动了旅游产业发展，旅游产业又促进了休闲观光农业的发展。同时，休闲观光农业促进了农业生产方式、经营方式及人们

消费方式的创新，是现代农业未来发展的一种新思路、新模式，市场潜力大，发展前景广阔。

绿色休闲观光农业技术主要有各种园区的景观与功能的规划设计、休闲观光农业项目（旅游产品）设计开发、植物（蔬菜、花卉、果树、观赏树等）的栽培（立体、无土等）技术、农业高新技术示范与展示、农耕文化传承、旅游商品设计开发等。

由于中国自然资源、人文资源、农业资源和经济状况等条件差异，休闲观光农业发展类型与模式呈现多样性。从休闲内容、观光对象和活动项目综合考虑，可分为以下几种类型。

一、绿色休闲观光农业模式与技术

绿色休闲观光农业是利用农业生产活动（过程）、田园景观、园林景观、农业设施等资源与环境，为游客提供休闲、观光、体验、采摘、品尝和购置瓜果、花卉等活动，从而提升农业价值和效益。

1. 田园农业观光模式与技术

田园农业观光模式以大田农业生产为本，通过设施农业、立体种植、生态养殖、种养加销一体化建设，与休闲观光功能有机结合，引进适合休闲观光农业发展的品质优良特种蔬菜品种、水果、花卉和其他观赏植物，以先进的农业种植模式和栽培技术，提高农产品科技含量、产量、质量和销售量，提高农业效益。综上所述，开发欣赏田园风光、观看农业生产活动、品尝和购置绿色食品、学习农业技术知识等旅游活动，让游客在休闲观光中了解农业、体验农业，回归自然生态并感受快乐。

主要技术是设施农业、立体种植、生态养殖等相关技术及旅游管理技术。

2. 园林观光模式与技术

园林观光模式以果园和林园生产为重点，建造果园和林园自然生态观光景观，使自然景观、人文景观与农业园林景观和谐统一，生态环境良性循环。依托自然优美的园林风光、舒适怡人的清新空气、良好生态环境的绿色空间，让游客回归自然，尽享生态自然之美。以绿色、生态、自然的果园和林园为载体，让游客观光赏景；开发采摘、观景、赏花、踏青、购置果品等旅游活动，让游客观看绿色景观，体验美好园林自然风光。

主要技术是果园、林园栽培与管理技术，以及果园、林园规划与设计技术。

3. 花卉观光模式与技术

花卉观光模式是以各种花卉种植、栽培、观赏、销售为一体的花园为载体，开发看花、赏花、采花（购花）等旅游活动，让游客观看花的海洋，体验五光十色、艳丽多姿的花卉风景。

主要技术是花卉种植与管理技术、花园的规划与设计技术。

4. 都市型现代农业休闲观光模式与技术

都市型现代农业休闲观光模式是依托现代农业设施和现代农业栽培工程技术，通过对农作物设施、园艺作物的创意型栽培，实现设施栽培的生产性和观赏性，例如，利用墙体、圆柱、多层栽培床等装置和工程技术，采用叶菜无土栽培技术，形成栽培平面，充分利用蔬菜的种类和颜色，形成观赏点；设施蔬菜立体栽培利用营养液槽等装置和工程技术，采用果菜无土栽培技术，利用果菜（辣椒、茄子和番茄）无限生长的生物学特性，运用设施环境可控的优势，充分延长辣椒、茄子和番茄等的生长期，形成高大的树形体态，形成观赏点；植物工厂的设施栽培集现代农业栽培工程技术、智能控制技术、传感技术和环境调控技术于一体，实现了育苗、蔬菜栽培、蔬菜产品销售，形成观赏点。

都市型现代农业休闲观光模式将作物栽培技术与植物的观赏性、景观艺术性有机结合，演绎出新奇瑰丽、异彩纷呈的景观，使都市型设施园艺与观光休闲农业中的元素相得益彰，提高和丰富了都市型设施园艺观光休闲农业的内涵和外延，从而使城市居民和游客在都市型现代农业园区，通过观赏、采摘、休闲散步，满足休闲观光的精神需求。这不仅为城市居民和游客提供了良好的休闲观光农业景观，而且为生产经营者提供了丰厚的农产品销售收入和旅游收入，综合效益可观。近年来，在北京、上海、天津、广州、深圳等一些大城市的郊区，都市型现代农业已成为休闲观光的亮点和城市新的经济增长点。

主要技术是植物工厂的设计技术、现代农业栽培工程技术、智能控制技术、传感技术和环境调控技术，以及都市型现代农业园区的规划与设计技术。

二、生活型休闲农业模式与技术

生活型休闲观光农业模式是人们通过参与农业生产活动、体验农事生活，通过"吃农家饭、住农家屋、做农家活、看农家景"等活动，让游客尽享农业、农村生态自然之美，农家风情之乐，绿色有机农产品之味，从而增加农业效益。

1. 农事生活体验模式与技术

该模式为城市居民和学生提供一种全新的旅游环境和休闲方式，通过参加农业生产活动，与农民同吃、同住、同劳动，让游客接触实际的农业生产、农耕文化和特殊的乡土气息，提高城市居民的休闲生活品位，改变学生认知社会和学习农业科技的方式，有利于社会的和谐与协调发展。

主要技术是农田生产管理技术，游客吃、住、行管理。

2. 农家乐休闲模式与技术

农家乐模式是农民利用自家庭院、自己生产的农产品及周围的田园风光、自然景观，为游客提供舒适、卫生、安全的农家居住环境和可口的农家特色饭菜，让游客体验农家生活，以及参与观赏、娱乐、休闲、购物等旅游活动。"吃农家饭、住农家屋、做农家活、看农家景"，让游客尽享农村生态自然之美，农家风情之乐。

主要技术是农家乐旅游品种设计，吃、住、行的科学管理。

三、贸易型休闲农业模式与技术

农业贸易模式是利用各类大中型农副产品集散市场、商务会展中心及农产品加工园等，把休闲观光内容与农业经贸活动有机结合起来，为游客提供休闲观光、优质农副产品采购等服务活动的休闲观光农业模式，如各种农贸会、农产品交易会等。

主要技术是商务会展设计，农产品加工园、农副产品集散市场管理等。

四、农业科技园区休闲模式与技术

1. 农业科技园区观光模式与技术

该模式是以农业科技示范园区为载体，以开展农业高科技生产、农业立体种植与无土栽培、生态农业、科普教育等和农业休闲观光为一体，向游人展示现代高新农业科技和新奇特植物、农作物、农产品和农业高科技观光的创意，以及现代农业与休闲观光农业结合的魅力。让游人观看农业科技园区的各种蔬菜、花卉种植、栽培的展示，设施农业和生态农业的景观，了解农业和农业科技知识。

主要技术是农业高科技生产、农业（花卉）立体种植、无土栽培、生态农业技术，园区的规划与设计技术，园区的管理技术。

2. 农业生态园休闲观光模式与技术

农业生态园休闲观光模式是围绕生态农业生产，利用田园景观、自然生态及环境资源，开发具有区域特色的农副产品及旅游产品，以供游客进行观光、游览、品尝、购物、参与农作、休闲、度假等多项活动，是一个集生态农业与科技示范、休闲观光农业、绿色有机农业生产、科普教育为一体，实现生态效益、经济效益和社会效益统一的多功能的新型产业园区。

主要技术是生态农业生产与管理技术，生态园区的规划与设计技术，园区的管理技术。

五、农耕文化休闲观光模式与技术

该模式是利用农耕技艺、农耕用具、农耕节气、农耕展示、农产品加工活动等，开展农业文化休闲观光活动。就是把农村特有的生活文化、产业文化及许多民俗文化等，通过休闲观光农业的发展得以继承，并创造出具有特殊风格的农村文化。中国传统农耕文明文化源远流长，博大精深，与休闲娱乐思想文化息息相关，通过二者融合赋予了新的含义。同时，中国的农村风土人情、民俗文化资源丰富。利用居住民俗、服饰民俗、饮食民俗、礼仪民俗、节令民俗、游艺民俗等，开展民俗文化休闲观光活动；利用民俗歌舞、民间技艺、民间戏剧、民间表演等，开展乡土文化休闲观光活动；利用民族风俗、民族习惯、民族村落、民族歌舞、民族节日、民族宗教等开展民族文化休闲观光活动。还可以利用农村二月二闹春社、清明前后到田野乡村去踏青、九月九重阳节登高赏菊花等，以及新农村建设等富有时代特色的新内容融入农耕文化休闲观光活动，原汁原味的农耕文化发展前景广阔。

主要技术是农耕文化、民俗文化资源的挖掘、收集、整理，以及休闲观光的产品开发、设计、管理。

六、教育农园模式与技术

教育农园是农业生产与科普教育相结合的农业经营形式，是农业园区的一种特殊类型，是新型的素质教育和科普教育的基地。教育农园利用农业园中栽植的作物、饲养的动物及配备的设施，如特色植物、热带植物、农耕设施栽培、传统农具展示等，进行农业科技示范、生态农业示范，传授农业知识，集科技、科普、教育、休闲、观光为一体，富有浓郁乡土文化特色，是开展农业科普教育的好形式。教育农园具有以下功能：

① 科普教育功能。利用园内优美而独特的植物、动物、自然景观组成科普教育基地和综合实践教育基地等，成为学生春游、秋游、夏令营、军训、写生等课外教育的好场所。

② 生产与生活功能。园内种植的各类树木、花卉、瓜果、蔬菜，饲养的家禽、家畜等可以为游客提供新鲜、卫生、安全的食品。加上无土栽培、立体栽培、嫁接、杂交等的应用为游客提供更多新、优、特的食品和花卉，如彩色甜椒、香蕉、西葫芦、魔豆、各种花卉等。

③ 观光休闲娱乐功能。园内的树木园、花卉园、蔬菜园、瓜果园等绿色植物，园内的园林小品、建筑、水体等形成了奇异的田园景观，可吸引游客前来观光、品尝、体验、娱乐，举行各种社团活动，如单位团建、家庭亲友聚会、老年人休闲度假等。

目前，教育农园在世界各地已经有了很大发展。20世纪90年代以来，欧美很多国家已经发展了大量的教育农园。大多数是私人农场为了增加收入，整合传统的农场生产生活资源，为城市居民提供农业和农村的生活体验。也有政府机构、教会和其他社会团体从社会福利或生态教育等角度设立的教育农园，这种农园不以营利为目的，主要进行农业技术推广和激发人们对自然环境和生物的爱心。日本政府每年都投入大量的资金来组织和实施青少年教育活动，其创办的"学童农园"是对青少年进行道德教育、心理健康教育、环保教育的示范基地。

中国的教育农园（农业科普教育园区）是利用农业观光园、农业科技园区为游客提供了解农业历史、学习农业技术、增长农业知识的教育场所。例如，利用农业科研试验场，把农业生产、科技示范、科技教育融为一体，向农民和学生进行农业科技教育；利用农业园区的资源环境、现代农业设施、农业经营活动、农业生产过程、优质农产品等，开展农业观光、参与体验、DIY教育活动；利用当地农业种植、畜牧饲养、农耕文化、农业技术等，让学生参与休闲农业活动，接受农业技术知识的教育。

开发建设教育农园可以通过科普活动向人们提供认识农业、了解农业生产过程、体验农村生活、认识农村文化及生态环境的重要性与意义；可以通过农业科技示范、生态农业示范，让人们更加珍惜农业和农村的生态环境、自然资源，增加保护自然与环境的自觉性，特别是懂得粮食和农副产品的来之不易，从而更加爱惜粮食。

教育农园的主要技术是教育农园的规划、设计，科普教育活动形式、休闲观光新型产品的开发、设计及管理。

前面已经论述，发展绿色休闲观光农业不仅是调整农业产业结构、促进农业农民增收的重要手段，也是提升城乡居民生活质量的有效途径。中国休闲观光农业资源独特丰富、品位高雅、区域特色明显、民族风格多样，各地应根据本地资源特点，因地制宜，统筹规划，科学布局，不断创新，开发适合本地发展的休闲观光农业类型，推进本地区休闲观光农业的发展。

随着城乡居民休闲旅游消费需求的增长，对休闲活动内容质量要求也会提升。在发展休闲农业中，既要注重休闲农业产品供给的规模与数量，更要把开发差异化产品与不断提升产品质量放在重要位置。要充分发挥政府引导、企业参与、市场推动的机制，合理调整和把握旅游需求的发展方向，开发多种多样具有现代生态循环农业与高科技特色、浓郁乡村景观、淳朴风俗民情、浓厚乡村文化的休闲观光农业项目，把绿色休闲观光农业做大做强。

第六章 现代生态循环农业园区规划基本理论

第一节 现代生态循环农业园区概念、类型和意义

一、现代生态循环农业园区概念

循环农业生态园是在循环经济理论指导下，进行园内农业的布局和生产，以资源高效和循环利用为核心，以低消耗、低排放、高效率为特征，在一定区域生态系统中进行农业生产活动、文化活动、休闲娱乐活动等，且和环境之间通过能量流动、物质循环和信息传递的方式，使园区各种活动达到高度适应、协调和统一的状态，是将资源化、减量化和再利用技术按照优化的生产模式以产业链的模式结合在一起，是循环经济基本原理、技术、模式在园区空间的高度集成与展示，以此实现节能减排与增收的目的，促进现代农业园区的可持续发展。

通俗地讲，现代生态循环农业园区就是运用物质循环再生原理和物质多层次利用技术，实现较少废弃物的生产和提高资源利用率的农业生产方式，是转变农业增长方式、提高资源利用效率、改善农村生态环境的重要途径。现代生态循环农业园区以密集组装农业清洁生产和废弃物综合利用技术为主要特征，成为循环农业科技成果转化和现代循环农业生产的重要示范载体。循环农业作为一种环境友好型农作方式，具有广泛的社会效益、经济效益和生态效益。只有不断输入技术、信息、资金，使之成为充满活力的系统工程，才能更好地推进农村资源循环利用和现代农业可持续发展。

现代生态循环农业园区在生态上能够自我维持、低输入，经济上有着较强的生命力，是在环境、伦理和审美方面都可接受的新型农业。它在实践上力图克服"石油农业"所带来的危机，其目的是将农业建立在生态学基础上而不是建立在石油化学基础上。实际上，生态农业源于生态学思想对农业生产的指导，要求人们充分利用当地的自然资源，利用动物、植物、微生物之间的相互依存关系，并利用现代科学技术，实行无废物生产和无污染生产，提供尽可能多的清洁产品，满足人民生产、生活需要，推动乡镇规模经济的发展，同时创造一个优美的生态环境。

二、现代生态循环农业园区的主要类型

循环农业生态园是未来规模化、现代化和标准化农业发展的趋势。现代生态循环农业园区规划建设正是以对局部区域农业生态系统进行整合协调的一项技术密集型工作，使区域农业生产建设与发展充分发挥地方主导产业优势和产品特色；充分挖掘地方文化、绿化美化环境；充分发挥企业或个人资金优势，大力加强农村生产基础设施建设，建立农业科技成果转化常态化链接纽带；充分发挥科技人才、现代化技术装备的先进性作用，使现代生态循环农业园区对所在地区的农业产业经济发展具有示范、指导和推动意义。

根据我国农业发展实际，以及区域交通、经济、环境等的影响，我国现代生态循环农业园区发展类型一般有以下几种方式：

① 按主导产业分：种植业主导型、养殖业主导型、加工业主导型、旅游业主导型；
② 按主导功能分：农业科技园区、农业观光休闲园区、休闲农场；
③ 按地域类型分：都市郊区型、景区周边型、特色村寨型；
④ 按组织类型分：社区集体经营管理型、企业自主经营管理型、农户自主经营管理型。

不同类型的现代生态循环农业园区，皆是根据区域实际、场地特征等因地制宜的发展规划模式，各园区基于良好的环境、经济、技术及产业等优势，更加利于现代生态循环农业园区的建设与发展。

三、建设生态循环农业园的意义

1. 提高综合生产能力

生态循环农业园的建设着重调动和发挥农业生态系统整体功能，从规模化种植业、养殖业着手，按照整体考虑、全面协调、循环利用、持续发展的原则，进行全面的协调规划，最大限度地调整和优化农业产业结构，使涉及其中的农业形态与各产业综合协调发展，使各行业之间能相互支持、相互配合，从而达到提高综合生产能力的目的。

2. 发挥多样的产业优势

考虑到我国农业生产的实际情况，由于我国幅员辽阔，自然条件、资源基础、经济和社会发展水平存在较大的差异，生态循环农业充分吸收了传统农业的精华，结合现代科学技术，把多种生态模式、生态工程建设和实用的技术应用到农业生产中，通过不断地、深入地分析、对比，发掘不同地区的农业生产优势，改造落后的生产方式，进而整合不同的、多样的发展模式和产业，突出发展的多样性。

3. 提高产业竞争力

中国加入 WTO 后，对农业的机遇大于挑战，影响是巨大的，将使我国农业长期存在的一些痼疾，如基础设施落后、规模小、结构单一、效益低下、竞争乏力等跃然纸上，如何迎接 WTO 的挑战已成为我国农业面临的重大问题。要解决基础设施落后、规模小、结构单一、效益低下、竞争乏力等一系列问题，就应通过建立生态循环农业园的方式，提升我国的现代农业发展水平，优化农业结构，实行规模化、产业化经营，把一些效益较低的产业整合连接到一起，建立起一个循环发展的农业平台，实现产品生产与市场需求的高度契合，同时保护好农业环境，以期提高产业竞争力。

4. 促进效益持续发展

建设生态循环农业园，促进生态循环农业的科学发展，能够有效保护农业生态环境，防止农业生产中的化肥、农药等有害物质对土地、作物、牲畜、农产品的污染，不仅维护了生态平衡，而且提高了农产品的质量，保障其安全性，把环境保护、农业生产和经济发展紧密结合起来，在保证农产品的供应、满足人们消费需求的同时，使我国现代农业朝着更加稳定、和谐、持续的方向发展。

5. 起到示范带动作用

我国农业生产区，特别是西部山区，由于农户分散、交通不便、通信技术落后、信息来源渠道不畅，农户掌握不到更多的农业实用新技术，对资源的利用率也比较低，农民增收困难，从而导致农产品市场疲软，形成种植业、养殖业低水平重复的不良局面，因此迫切需要建立产业优势突出、技术新、品种新、管理新、思路新的集合园区。建设生态循环农业园，一是可以有效地引进农业新品种进行培育，促进农业高新技术的发展，大力推广先进适用技术，促进农业科学技术的进步；二是有利于科研单位、科技人员、产业实体和农民的有机结合，集聚各种力量，发挥综合优势；三是可以探索出一条适合于区域农村经济发展，便于操作，而且科学、高效、高质的无公害农业发展之路；四是可以对农民进行培训，提高农民整体农作素质。

第二节 现代生态循环农业园区的理论及功能产业定位

一、现代生态循环农业园区的理论

现代生态循环农业园区从本质上是一种带动地方经济发展的集生产、研究示范、休闲观光、科普教育于一体的新型农业组织形式。因此，现代生态循环农业园区所承载的责任是比较重大的，它不仅关乎政府、企业、农村集体和农户等的经济利益，更关乎地区农业发展的前景及农业产业结构优化升级等问题。现代生态循环农业园区的规划，总体方针是为人民服务、为生产服务。要从实际出发，遵循因地制宜、合理布局的原则，既要符合国家、地方农业发展水平，又要创造出具有地方特色和创新精神且具备引导性、示范性的现代生态循环农业园区系统。

规划理念是对现代生态循环农业园区整个规划的概括性、条款性的限定和指导，直接影响园区发展定位和战略目标的确定、重点项目的选择和园区未来的发展方向等，是规划过程中必须遵循的一些基本理念。规划理念是规划单位和规划编写人员在现代生态循环农业园区规划方案编写过程中所确立的主导思想，它赋予方案独特的文化内涵和特点。科学、因地制宜的规划理念在现代生态循环农业园区建设中起到至关重要的作用，它不仅是规划的精髓所在，而且能使规划方案更具地方独特性、专业性。当前，现代生态循环农业园区总体规划主要包括以下几个规划理念。

1. 生态理念

生态理念在现代生态循环农业园区的应用，是从规划的角度出发，依据城乡生态环境的特点及平衡开发与生态环境的交错关系，注意环境容量和土地开发利用的限度，强调环境的总体协调性、资源的综合利用性，最大限度地保护和改善生态环境，达到建筑与环境共生、建筑与环境互融、建筑与环境相映生辉，使现代生态循环农业园区呈现出生态、生产、生活共生互融的一派和谐景象，成为具有浓郁地方文化特色和生态景观纹理的城市后花园。

2. 节能理念

现代生态循环农业园区规划布局应当充分体现节能的设计理念，办公与生产区等生产性建筑要保证充分的日照间距，采用保温性能好的外墙材料等，既保证工作、生产、生活的舒适，又体现现代节能的特点。连栋温室、日光温室、拱棚等农业设施，按照当地太阳高度角以设计最经济的采光要求，科学设计栋与栋的间距。同时，尽可能使建筑群之间所形成的小区院落与周边的绿化、水体、阳光、自然通风条件等紧密结合，达到人与自然的和谐统一。另外，在生产中，要充分利用节能生产工艺，优化人工建筑环境，节约人工能源，避免能源浪费。

3. 节水理念

现代生态循环农业园区的规划布局将充分体现节水的理念，生产、生活以及加工区的污水，处理后达到排放标准可作农田灌溉用水使用。园区的低洼区域及温室区尽可能采用雨水集蓄利用工程设施，避免雨水的流失。在管理区、加工区设集蓄雨水沟，使雨水流到蓄水池后，备用于农田的灌溉。露地栽培尽可能采用滴灌，温室内的灌溉全部采用膜下滴灌技术。

4. 可持续发展理念

现代生态循环农业园区的规划应当同近期项目和近期发展相结合，应建立具有弹性的长远发展构架，描绘出未来现代生态循环农业园区的清晰轮廓。由于现代科技的迅速发展，高科技产业、生产工艺的改建与更新，以及生产规模扩建的节奏加快，要求基地的建设项目必须具有弹性，建筑设计要留有灵活变动空间。同时，在土地、水、配电等资源的综合利用上也应体现可持续发展的理念，应预留未来发展区域空间，以适应未来科技与社会经济发展的需求。

5. 人性化设计理念

当今时代，以人为本的理念越来越凸显，物质生产和社会经济的发展越来越体现人性化要求，工作、生活、生产环境的规划与设计需要具备良好的高效性和舒适性。因此，现代生态循环农业园区的规划应当处理好人与建筑、人与设施、人与环境之间的和谐统一。无论是示范区、生产区、科研区还是观光休闲区，都应该体现以人为本的人性化设计理念。

6. 个性化设计理念

现代生态循环农业园区规划在整个建筑风格上应有主基调，但体现在具体设计时，又要有各自的个性特征。园区的各功能区，在布局上既要符合总体格局的基本要求，同时又要体现其特点，在建筑形体和空间的组合、企业的标志等方面都要有鲜明的个性，以丰富整个项目区的视觉效果和文化环境。

7. 园林化设计理念

在满足生产农业高新技术示范、产业开发的前提下，园区的建设力求按照园林化的思路进行规划和布局，通过名特优瓜果蔬菜、花卉、苗木、牧草等展示项目的安排，设施工程、生态观光园、不同建筑风格的特色设计，以及道路、绿化、水面、桥涵、沟渠、绿色长廊等配套项目的精心设计与规划，逐步形成一个优质的现代生态循环农业园区和与周边气候、环境和人文景观相适应的农业观光休闲点。

8. 智能化设计理念

随着现代科技的发展，建筑与生产系统需要智能性、社会性、结构清晰性等特征，充分体现现代生态循环农业园区的智能建筑特点。在满足生产和生活功能的前提下，园区的规划设计应优先考虑各独立功能区和各单位建筑之间的管理与安全、消防、节能等方面自动化及电力、通信、网络服务等方面的功能设置，引进先进的设施设备，以体现出现代生态循环农业园区的智能化特色。

二、现代生态循环农业园区的功能产业定位及功能区分析

现代生态循环农业园区的功能定位必须基于当地产业发展实际进行确定，这样，在后期的发展过程中才会具备更好的发展优势和竞争优势。一般以当地的主导产业为基础性分析资料，产业定位要根据区域性主导产业的优势和自身的特点来进行，突出优势，挖掘潜力，形成特色。通过功能定位，解决园区产品结构模式、生产运作模式的低技术含量现状，运用区域经济学、产业经济学原理，以区域资源优势为基础，以发挥优势比较为原则，分层次对园区功能定位、分区域和分时序对产业发展定位进行研究。功能定位是园区规划的基础，是园区规划的起点和主体思路，围绕功能定位展

开园区的产业规划、产品规划、项目总平面规划等，并符合《农业标准化示范区建设》标准要求。

1. 技术创新与科技转化功能

现代生态循环农业园区与以往其他的生态循环农业园区不同，必须以科技为先导、项目为载体、企业为依托、人才为基础、效益为目标，以此形成良好的自循环生态可持续发展模式，是社会主义市场经济条件下促进农业科技转化为现实生产力的综合平台，可有效提升农产品的市场竞争能力，既保障国家粮食安全，又推动农业走出去和园区国际化发展，成为区域农业经济的增长极。现代生态循环农业园区具有生态效益好、科技含量高、科技成果转化率高、综合经济效益高等特点。园区围绕新的科技革命，以农业科研、教育和技术推广单位作为技术依托，以科技创新为核心，以体制机制创新为动力，以科技开发与示范、辐射、推广为主要内容，实现了现代科学技术和生态循环农业产业的高效结合，促进了产城融合、园镇融合、三产融合。

2. 企业化培育功能

现代生态循环农业园区是农业新技术集成创新的重要基地。一方面，园区可以吸引高校和科研院所来园区建立科研基地、研发中心，进行技术创新和试验，研发和培育科技含量高、市场需求高、附加值高的新品种，开发农产品保鲜、精深加工及相关配套技术；另一方面，科技创新成果转化是支撑现代农业发展的关键环节，因此，园区还需要扮演"孵化器"的角色，促进科研成果转化，不断孵化出现代农业高新技术企业。

3. 农业科技人才集聚功能

农业科技园区的建设可以吸引农业科技企业和科研机构集聚，促进信息交流和知识共享，加速产品和技术创新。另外，各种生产要素的集聚使得企业能够与合作伙伴进行资源互补和降低交易成本，逐渐形成产业网络，吸引外部大量的人才、技术和资金，进一步促进农业科技人员集聚。

4. 生产加工功能

现代生态循环农业园区本质上是一个经济实体，承担着应用现代农业科技成果的重任，对高新技术产品的生产加工是其基本功能。园区通过吸引相关的农业科技企业入驻，实现规模化生产和产业化经营。同时，园区内企业作为农产品生产加工的关键载体，可以将农业科技与生产、加工相结合，通过精、深加工，提高产品的附加值，增强市场竞争力。

5. 生态和旅游观光功能

着重特色发挥，充分展示现代农业科学技术与地域传统农业文化，通过旅游业带来巨大的人流、物流和信息流，促进现代生态循环农业园区的发展。园区在建设过程中，除保持农业的自然属性外，还可通过新型农业设施和高新技术的展示，配合园林化的整体设计和名特优瓜果蔬菜、花卉、特种珍禽、鱼的生产与示范，形成融科学性、艺术性、文化性为一体，天人合一的现代农业休闲观光景点，扩展园区的功能。

6. 辐射带动功能

新品种引进、技术示范及带动周边经济发展是农业科技园区的主要任务。通过这些技术成果的展示、示范和推广，有助于对传统农业的改造升级，推进农业现代化，促进农业增长方式由"资源依存型"向"科技依存型"转变。作为农业经济发展的"增长极"，现代生态循环农业园区能够带动周边地区的经济发展和科技创新能力的提升，而且有助于进一步提高当地农民养殖、种植的热情和创业意愿，促进农民增收。

7. 教育示范功能

近年来，失地农民成为国家改革开放和城市化进程中所产生的一个隐形弱势群体，开展失地农民职业技能培训，提升失地农民的综合素质和就业能力，是解决其再就业问题的有效途径之一。现代生态循环农业园区可以基于人才和科技资源的优势，为技术人员和当地农民开展教育培训。例如，通过技术培训，提高他们的知识技能；通过创业教育，提升他们的创业能力。园区建立完善的农业技术推广体系和健全的培训机制，能够有效提高培训人员的文化素质、技术水平及就业创业能力，使其适应农业发展和乡村振兴的需要。

第三节　现代生态循环农业园区规划原则和内容

现代生态循环农业园区规划与设计是一项系统工程，从学科的角度分析，综合了循环农业与生态工程、园林景观、农业经营管理等方面的基本原理和方法，属于多学科交叉领域。其中，循环农业与生态工程原理包括减量化、资源化、再利用、废弃物饲料化、废弃物材料化、废弃物能源化、废弃物肥料化等工程技术；现代化农业科学技术原理包括设施农业、现代农业装备、畜牧养殖、农业产业化与信息化等农业技术；园林景观学原理包括景观生态学、园林工程技术；农业经营管理科学包括组织管理、产品营销、物业与物流管理、农产品质量管理等。

规划设计的相关理论包括农业区位优化理论、资源合理匹配理论、复合生态农业理论、农业转型升级理论、农业工程系统理论、高新技术产业理论、农业环境保护理论、农业资源利用理论等。规范框架包括农业系统规范、规划及建筑体系规范及相关的政策、法律。

一、现代生态循环农业园区规划背景与指导思想

1. 规划背景

近年来，国家越来越重视并规范现代农业产业园的发展，2016年10月，国务院印发《全国农业现代化规划（2016—2020年）》。其中提出，创新第一、第二、第三产业融合机制，以产业为依托，发展农业产业化，建设一批农村一二三产业融合先导区和农业产业化示范基地，推动农民合作社、家庭农场与龙头企业、配套服务组织集群集聚；以产城融合为依托，引导二三产业向县域重点乡镇及产业园区集中，推动农村产业发展与新型城镇化相结合。2016年12月，中央农村工作会议提出"现代农业产业园是优化农业产业结构、促进三产深度融合的重要载体"。2017年中央"一号文件"首次提出，以规模化种养基地为基础，依托农业产业化龙头企业带动，聚集现代生产要素，建设"生产＋加工＋科技"的现代农业产业园，推动我国现代农业产业园快速发展。2017年3月，农业部联合财政部发布了《关于开展国家现代农业产业园创建工作的通知》，要求通过创建国家现代农业产业园，突出现代农业产业园产业融合、农户带动、技术集成、就业增收等功能，使其成为引领农业供给侧结构性改革的新平台，为培育农业农村经济发展新动能探索新经验，为促进农民增收开辟新途径，为加快推进农业现代化提供新载体。2017年4—5月，农业部和财政部先后召开"全国集中部署现代农业产业园创建工作会议"，举办"国家现代农业产业园创建方案编制完善培训班"，联合报送"关于拟创建国家现代农业产业园遴选工作情况的报告"，推动了国家现代农业产业园的创建。2017年，财政部、农业部公示了第一批11个国家现代农业产业园名单。2018年，农业农村部和财政部共同下发了《关于开展2018年国家现代农业产业园创建工作的通知》，规范了创建国家现代农业产业园的条件及任务。

2. 指导思想

规划按照高质量发展要求，围绕实施乡村振兴战略，以推进农业供给侧结构性改革为主线，立足优势特色产业，聚力建设规模化种养基地为依托、产业化龙头企业带动、现代生产要素聚集、"生产＋加工＋科技"的现代农业产业集群，促进一二三产业融合发展，创新农民增收利益联结机制，培育农业农村经济发展新动能，打造高起点、高标准的现代农业建设样板区和乡村产业兴旺引领区，示范带动省（自治区、直辖市）、市、县，形成梯次推进的现代农业产业园建设体系，为农业农村现代化建设和乡村振兴提供有力支撑。

二、现代生态循环农业园区规划原则

规划原则以前瞻性原则、示范性原则、市场性原则、特色性原则、因地制宜原则、可操作原则、可持续发展原则为核心，科学合理指导园区规划。

1. 政府引导，市场主导原则

现代生态循环农业园区发展的显著特点是与市场发展紧密结合，规划要在政府调控的同时，以市场为导向，把握最新的市场需求，调整农业产业结构，结合市场细分和市场需求的发展趋势，科学合理地规划。因此，现代生态循环农业园区规划要在政府引导下，准确及时把握市场情况，分析现代农业产业化发展趋势，研究农业生产条件及生产潜力，农业产业的市场容量及前景预测，确定园区的目标市场，充分发挥市场主体在产业发展、投资建设、产品营销等方面的主导作用。

2. 因地制宜原则

我国幅员辽阔，东西南北跨度较大，地势、土壤、水、植被条件，以及交通条件、气候条件、能源供应条件有很大的不同，国家现代生态循环农业园区因选址的区位条件和地理环境现状存在差异，导致其资源条件、气候条件和经济发展水平也千差万别，因而具有不同的规划建设方式。因此，在规划设计方案的制订和特色产业项目的选择上，应与地区特征相结合，因地制宜地考虑当地资源条件和生态要素构成，以及产业优势和经济基础、生产条件，科学地选择适宜的主导产业和农业生产项目，围绕当地主导产业，发展名优产品，保证粮食等大宗农产品产量不断增长，多样性地开发利用农业资源，以期更好地综合利用区域资源，降低建设成本，实现园区效益最大化。

3. 多目标协同，可持续发展原则

应当坚持经济效益与社会效益、生态效益的统一，走可持续发展道路。注重园区内部产业结构和景观生态要素的协同共进，构建绿色、低碳、循环发展长效机制。在规划设计的过程中，既要满足园区产业基本运行的功能要求，又要在最大程度上提升景观美感；既要设定短期目标，又要兼顾长远利益。力求在土地开发利用和环境容量允许的最大限度范围内，保护和改善生态环境，促进园区的可持续发展。在科技水平高、经济发达的地区建立高起点、高标准的科技和资金密集结合为主的现代生态循环农业园区。对自然资源丰富、经济发展水平相对滞后的地区，应首先建立科技先导、劳动密集、资源实力型现代生态循环农业园区。随着经济的发展，逐步改善生产手段和条件，使现代生态循环农业园区的建设呈现多层次、多类型、多形式的格局。从产业层次建立多种链条，有利于充分利用资源，使园区内各产业之间促进、互补、共生。

4. 以人为本原则

园区的规划设计引入"人本性"原则。现代生态循环农业园区建设的目标是要解决好"三农"问题，增加农民收入，提高农民生活水平。以人为本的考虑是有层次的，其中心不能是片面地考虑个体，而是综合地考虑不同群体，同地域、文化、社会、效益等结合起来，体现以人为本的规划理念。为此，现代生态循环农业园区的规划不仅要符合生产需要，还要满足消费者和旅游者的物质文化需求。规划的过程中必须从宏观到微观、从整体到局部进行全面的人性化的设计，注重政府、企业、科研、服务等机构围绕产业开发和建设的协调工作和利益共享。

5. 前瞻性和可操作性原则

现代生态循环农业园区规划项目的选择，必须结合园区发展目标和消费群体的范围，考虑市场供求和未来市场饱和度情况，项目设置在技术和产品生产上要有前瞻性。在生产方式和技术上，应遵循先进实用和适度超前的原则，避免无法将规划措施落实到具体建设步骤上。现代生态循环农业园区规划必须要切实可行，要符合当地农业生产条件、经济社会条件和国家的政策，避免不能落实到实际工作的抽象化、形象化的规划。可操作性和前瞻性是相对的，需要在动态发展中不断进行适时调整，以确保相对合理状态。总的来说，规划内容一方面要满足相当一段时间的领先要求；另一方面要参照现有的规划设计规范要求，为当地长远发展考虑，规划要凸显开放性、科学性和生态性。

三、现代生态循环农业园区的规划内容

现代生态循环农业园区规划设计的第一步，应对待建区的实地条件进行基础性资料分析，包括相关资料的搜集分析和实地调研。

1. 相关资料的搜集分析

搜集规划园区的地域、基地、政策条件，园区所在区域所做的该区域农业宏观布局规划的成果，相似型现代生态循环农业园区的研究成果。地域研究是对规划园区所在地域的宏观区位条件进行研究，包括地理环境、自然条件（包括环境污染严重程度）、气候特征、资源分布、社会条件、社会经济发展现状、农业发展现状（生产力水平、技术水平、主要产业等）、农业产业（主导产业、特色产业、潜在产业）结构发展方向及市场需求（包括地区、全国乃至相关的国际市场）等。基地研究则是针对规划园区的地理位置、交通区位条件、用地规定、基础设施建设及种植类型、企业建设现状等进行研究，如果有上位规划，还应对其进行解析。政策研究则是对国家和地方政府出台的有关统筹城乡发展、农业产业发展、现代生态循环农业园区规划建设方针政策和规范要求等进行研究，确保规划设计符合国家规定。

通过基础资料的分析，能够确定现代生态循环农业园区所在地的农业资源情况及农业生产的整体水平，为现代生态循环农业园区的定位提供依据。对园区发展方向有整体把握，有助于确定园区的指导思想和战略目标，主要包括园区的性质与规模、园区的主要功能与发展方向、园区的发展阶段及每阶段的发展目标。

调研当地的农业现状条件、农业整体水平、经济社会发展水平、农业特色产业发展现状等；调研规划园区的农业现状条件，包括地理区位、自然条件、环境污染严重程度等。

2. 指导思想和战略目标确定

现代生态循环农业园区规划的指导思想是实现园区发展所遵循的根本依据和行动指南，也是确立发展定位、建设目标、项目方案的理论基础。规划的指导思想和目标应当立足园区的实际情况，大到国情省情，小到村情户情，因地制宜，根据具体环境条件，选择规划建设内容，指导产业园区发展优势农业，实现战略目标。内容主要有：确定实现园区目标的可能途径，找出提高农产品竞争力的核心因素，制定园区的发展战略。

3. 功能分区与布局

国家现代生态循环农业园区的基本功能包括农业生产、农产品加工与销售、科技研发、示范辐射、观光游憩、配套服务、生态保护等，将这些基本功能贯穿于园区的规划中，在园区具备营利性基本功能的基础上，兼具科技创新、机制创新、示范辐射、服务农民、带动产业发展的社会公益性功能，以及显著生态保护效应。功能布局是国家现代生态循环农业园区规划的重要内容。影响园区功能布局的因素很多，但从产业为园区规划核心的角度考虑，主要影响因素包括土地利用现状及效益水平、产业关联程度、功能相似性、总体规划要求及园区的定性定位等方面。根据产业布局情况，确定功能分区及经济和景观轴线，划定核心区、示范区和辐射区的范围。功能分区应依据园区规划的指导思想、发展目标、规划原则进行规划，坚持突出重点、体现特色的原则，打造功能齐全、空间联系紧密、合理高效的功能分区布局形式。园区功能分区与产业布局结合，产业布局与项目规模相适宜，既要考虑全局统筹，又要兼顾局部协调，还要充分确保各区域内所引进的项目规模、技术选择等方案的先进性、科学性和可操作性。农业产业在园区中具有重要的基础性地位，产业设置与布局需要相对具体和详尽的项目建设方案。要根据园区的目标定位和区内资源性质，确立若干个功能区或产业带；确立核心区、示范区及辐射区的范围；要在围绕种植、养殖和农产品加工等第一、第二产业的同时，结合实际，充分发挥旅游观光、科普教育等第三产业在园区规划中的作用。

核心区集中了园区行政管理、科技开发、产业机构，是园区生产与建设的核心和重点，是农业高新技术的科研和产业开发基地。核心区有完整的边界和明确的范围。核心区是园区的主体，园区的管理机构、大部分经济实体组织和生物技术组培中心、新品种引种区、设施园艺区、科研开发中心、培训展示中心和信息中心都集聚于此。核心区的主要功能包括：引进和应用现代农业高新技术和科技成果，进行农业科技成果的试验和示范；培育优势产业的龙头企业，形成有市场竞争力的主导产业及企业或企业集团，实施高科技农业和设施农业项目的开发，运用高新技术改造传统农业，实行农业产业化经营，提高园区农产品的市场竞争力；对整个园区进行管理、指挥和协调，为农业产业开发和农业生产提供技术服务、生产资料运销服务、农产品运销加工服务；引进国内外农业科技专家，对园区的农业科技产业开发进行攻关和咨询；农业科技培训，向该区域的农户及农业企业传播和推广先进农业技术，提高农业劳动者的素质。

示范区是现代生态循环农业园区的农产品生产基地和农业科技成果的试验基地，还是核心区的农业产业化带动基地、现代生态循环农业园区的主要示范平台、核心区的直接作用对象。示范区具备完整的边界和明确的范围，一般示范区紧靠核心区，面积是核心区面积的3~5倍。示范区通过吸收核心区传播的新技术和新品种，在核心区的技术、品种、人才、资金、培训等要素的带动下，通过"公司+农户"的运作形式，进行农产品标准化生产和示范，探索和开拓现代农业产业化经营模式，并通过核心区的技术支撑，在示范区重点展示生物技术、节约高效栽培技术、现代设施农业技术、节水灌溉技术、集约化种养技术、农畜产品精加工技术、信息管理技术、农产品加工技术和农业产业化成果；展示人工与自然相结合的环境改良成果，展示现代化乡村改造与建设成果，从而拉动和带动周边地区农业经济的发展。

辐射区与核心区距离比较远，它并不在现代生态循环农业园区的界域内，没有完整的边界和明确的范围，只能划出大致的影响层面或作用半径。辐射区是现代生态循环农业园区核心区主导产业涉及和影响到的周边的农业生产与农村经济区域，或是地理环境、资源特点、生产与经济特征相近的同类型的农业区域。辐射区一般紧靠示范区，面积为示范区的10~15倍。按照"发展极"理论，现代生态循环农业园区的先进技术、新品种和新产品通过示范区种植企业的示范作用，对周边区域的农户产生扩散效应和牵引效应。

以国家现代生态循环农业园区作为示范基地，将聚集到园区的先进生产力要素进行有机结合，并通过现代农业科技企业的集约化生产过程，转变为具有竞争力的农产品，通过市场销售获取利润。

4. 产业规划

产业规划是国家现代生态循环农业园区规划的核心内容和设计重点，通过对区域适宜的优势农业产业进行选择与分析，确定园区的产业发展内容，是实现园区产业化发展，获得经济、社会、生态协同的重要举措。产业决定着园区的战略发展方向，其功能定位要立足于当地社会、经济的实际条件，因地制宜，选择恰当的建设内容和技术路线。国家现代生态循环农业园区在产业规划前，要分析产前技术支撑能力和农资服务业务水平，并充分挖掘潜在产业的发展潜力，在明确产前相关信息后规划产业布局，以一产为基础、二产为支柱、三产为纽带，实现农业多产业的融合发展。

第一产业即农业生产，是园区存在与发展的基础，占地面积最大。空间布局应以片式或环状围绕园区的核心区辐射展开，规划布局于农业生产基础实力较好、农村人口分布合理、农业生产特色鲜明的农业用地上。第二产业是以农产品加工、储存、销售为一体的产业链，其空间布局应以土地集约、产业集群为原则，应集中成块规划在建设用地或远期建设的备用地上。第三产业则是以农业高新科技研发、科教宣传、培训推广、游览观光、休闲娱乐为主要功能的产业形态，空间规划布局形式多样化。第三产业可与一产、二产的空间规划形式相结合，开展产业园区农业生产体验、休闲观光采摘等休闲娱乐活动；可与农业文化、民俗文化相结合，规划建设农业特色文化主题园；亦可将农业高新科技进行展示，独立建设科技研发中心、技术推广中心、农业科技观光温室、生态温室休闲项目等科普游乐相结合的项目。

农村一二三产业融合可分为5种类型：一是产业整合型，如种植与养殖相结合；二是产业链延伸型，即以农业为中心向前后链条延伸；三是产业交叉型，如农业与旅游业、文化融合；四是技术渗透型，如信息技术的快速推广应用，使得网络营销、在线租赁托管等成为可能；五是综合型，即综合运用现代工程技术、生物技术、信息技术等技术成果，最大限度地摆脱自然条件对农业生产经营活动的束缚，在相对可控的环境条件下，实现农业的周年性、全天候、反季节的企业化生产。

现代生态循环农业园区的项目规划主要指产业规划及与产业规划内容密切相关的其他项目设计。产业规划需要明确如何通过选择适宜的优势农业产业，来实现园区产业化发展，实现经济、社会、生态三效和谐，主要包括支柱产业选择、产业规模、产业链发展思路、产业组织等内容，涉及种植业、养殖业、加工业、销售业、研发、物流及观光旅游业等领域。

① 种植业。主要涉及蔬菜、果树、茶叶、大田作物、观赏植物种植及种子种苗生产等。

② 养殖业。根据区域的资源优势、地域优势及产业优势，发展以肉牛、奶牛、羊、猪、禽等常规畜禽养殖；因地制宜规划水域滩涂养殖，根据区域水生生物资源、生态环境状况，推进环境友好型水产养殖；根据园区规划的内容及地域特色，进行珍稀动物的养殖。养殖产品优势区主要有：东北、中部、西南的生猪主产区，中原、东北、西北、西南的肉牛主产区，中原、中东部、西北、西南的肉羊主产区，东北、内蒙古、华北、西北、南方和大城市郊区的奶业主产区，华北、长江中下游、华南、西南、东北等肉禽优势产区，华东、华北、华中、华南、西南禽蛋主产区。在沿海地区，积极保护滩涂生态环境，规划发展生态养殖、深水抗风浪网箱养殖和工厂化循环水养殖，开展海洋牧场建设。在沿海和长江中下游地区建设优质淡水产品加工产业带。

③ 农产品。加工园科技创新平台凭借龙头企业，带动围绕区域的主导产业和特色农业，着力建设环境友好、资源集约、市场导向的农产品加工区，有效地融合农业一二产业。主要项目有农产品初加工、粮油精深加工、果蔬茶加工、畜禽加工、水产品加工、其他项目的加工、农产品副产物综合利用。

在国家现代生态循环农业园区的加工区，发展相互配套、功能互补、联系紧密的与农业生产相关的加工企业，依据园区发展配套及规划愿景需求，选择发展工业饲料、畜禽粪污资源化利用、有机肥生产、农业机械生产、温室制造及温室配套设备生产加工。

④ 农业休闲观光、餐饮等产业。以园区农业元素及当地文化为背景，以生态景观为出发点，根据场地现状因素，总体采用组团式布局，运用动静分区的原则，利用水系或园路进行功能区域划分，兼顾生态保护、休闲观光、体验参与和教育示范等多种功能，满足人们亲近大自然、学习农业知识、感受民俗风情、体验农耕乐趣和享受农家生活的需求。规划时，应注意动态游览与静态观赏相结合，有效地保护和开发资源，保护农业环境，尽量把握生产和旅游的平衡点。除保持农业的自然属性外，还可利用新型农业设施和高新技术展示的现代气息，加上园林化的整体设计和常年进行的名特优瓜果、蔬菜、花卉及特种珍禽、观赏动物养殖的生产与示范，形成融科学性、艺术性、文化性为一体的、天人合一的现代化农业休闲观光景点。

5. 项目规划

现代生态循环农业园区项目规划主要是针对园区产业的配套、服务项目的设计，主要包括建设目的、项目规模、关键技术、工艺流程、保障措施、风险评估、效益分析等内容，通常依据园区功能设置项目类别，如设施蔬菜栽培项目、粗粮精细加工项目、基础设施建设项目等。园区项目规划工作在开展过程中还必须与土地利用指标、人口发展指标及生态循环规划相协调。

6. 园区运行模式规划

确定优势产业在园区的根本地位只是规划的一方面，产业的正常运行和发展，还离不开一整套的运行机制和管理模式。在对园区的资源、投资、技术、管理制度及产业特点等分析后，提出适合园区发展与建设的运行模式，是园区规划的重要环节。它与园区项目规划共同构成规划的两个重点。具体来说，规划要对园区的组织管理模式、资金筹措机制、土地流转机制、科技研发与应用机制及园区经营机制和风险保障机制等做出安排。

7. 效益分析

园区的基础设施建设主要由政府财政投入，产业开发项目采取市场运作方式，由企业投资建设。无论资金投入源自何种渠道，带动农民增产增收、获得直接经济效益，都应作为首要任务。同时，间接效益要讲社会效益和生态效益。

（1）投资概算与资金来源

现代农业产业园的投资概算由固定资产投资和流动资产投资两部分组成。固定资产投资主要包括用于园区内部厂房、建筑物兴建和机器设备的采购等固定资产费用和其他费用，如土地征收费、勘察设计费、平整场地费、建筑工程费、公共基础设施费及人员培训费等。流动资金主要包括购买农业生产所需原材料、燃料、动力等费用，以及支付员工工资、设备维护和园区经营活动等费用，此外，还包括管理费用和不可预见的费用。

资金来源主要包括政策资金、社会融资、企业投资、招商引资及农民投资等。园区在规划建设过程中，应鼓励和吸引社会各界力量参与园区建设，拓宽资金渠道，充分发挥政府对园区投资兴建的带头作用，鼓励以各种生产要素加入园区建设中。同时，大力引进外资，加强国际科技合作，引进国外先进的生产设备和管理理念，以保证园区建设更好地进行。此外，为了减少财务风险，不应盲目扩大借入资金，而应使借入资金保持适当的比例，从而使项目运作始终保持有一定的偿付能力和良好的财务状况。

（2）效益分析与风险评估

国家现代农业产业园的效益主要涉及经济效益、社会效益和生态效益3个方面。经济效益主要是指园区及其辐射区产生的直接经济利益和间接经济利益；社会效益是指园区建设对促进就业、改善环境、增加财政税收等方面的社会影响力；生态效益是指在提高园区生产效率的同时，对减少周围环境污染、改善生态环境所带来的效益。

国家现代生态循环农业园区的主要风险包括经济政策风险、市场风险、科研技术风险、工程风险、财务风险、投资估算风险、环境风险及社会影响风险等。园区的风险评估是基于市场行情、时间衔接、技术安排和项目管理等各方面出现偏差时造成园区效益改变、威胁投资安全等现象出现所提前做出的预测与评估。国家现代生态循环农业园区的风险评估体系一般采用专家调查法、层次分析法、CIM法和蒙特卡洛模拟法等基本方法进行。由此可见，在规划建设时期应认真利用风险评估方法分析园区建设潜在的各项风险，针对具体风险因素提出相应的对策和建议，并在实施过程中采取切实可行的方法降低投资风险。

8. 其他专项规划

道路交通规划、园林景观规划、建筑工程规划、农业工程规划、服务设施及市政工程设施规划都是为园区产业规划和项目规划落地服务的。本节不做详细介绍。

四、现代生态循环农业园区规划成果

现代生态循环农业园区的基本规划程序是：委托规划→基础资料收集与分析→规划；设计阶段→方案优选→方案实施；规划成果为基础资料汇编→可行性分析→规划说明→规划图集。

现代生态循环农业园区总体规划的设计成果包括规划文本和规划图纸两部分。

1. 规划文本

现代生态循环农业园区规划文本（亦称规划说明、规划报告）应全面反映规划设计的内容，对规划的条件、理念、依据、建设方案、组织方案、运营方案、可能的效益等做全面的介绍和说明。在一般城乡规划中，要求总体规划的文本在格式上采用条文形式，文字要规范，语言要准确、肯定，而将一些技术上的论证和说明放在说明书中予以表达。现代生态循环农业园区规划成果可以借鉴城乡总体规划的要求，但也不必拘泥于形式。多数情况下，允许将文本和说明书融合在一起，用规划文本的形式予以表达。也有一些文本会附带专题研究报告，对委托方关注的一些内容进行更加深入的研究分析，成果作为规划文本的附件，附在规划文本和图册的后面。

（1）文本编制要求

现代生态循环农业园区的规划文本提纲一般包括以下内容：项目纲要、政策背景、规划依据，现状分析，园区建设的基本原则、指导思想、发展目标（战略目标和具体指标），园区定位、空间布局及功能分区，园区的产业规划，园区的建设内容，园区的组织架构与管理体制，园区的运营模式和机制，园区的投资估算和资金筹措，园区建设的效应分析，保障规划措施建议。

规划文本成果内容要前后关联，包括数据、内容前后一致；基本逻辑清楚，各个章节表达目的清晰明了。除了正文，还要注明规划编制双方单位、规划项目名称、编制成员等。

（2）规划文本框架

现代生态循环农业园区总体规划文本框架

第1章 规划背景

1.1 规划背景

1.2 主规划依据

1.3 政策规划阶段

1.4 规划范围期限

第2章 基础分析

2.1 区位与交通

2.2 自然资源条件

2.3 历史人文资源条件

2.4 社会经济条件

2.5 产业现状与趋势

2.6 案例借鉴与分析/SWOT分析

第3章 目标定位

3.1 基本原则

3.2 指导思想

3.3 功能定位

3.4 发展目标

3.5 空间结构/功能分区

3.6 平面布局/产业部署

第4章 分区建设内容

4.1 ××××区

包括建设地点、建设规模、建设内容。

4.2 ××××区

4.3 ………

第5章 基础设施工程系统规划

5.1 道路系统规划

5.2 农田水利规划

5.3 给排水规划

5.4 电力规划

5.5 供暖规划

5.6 通信规划

5.7 环卫工程规划

5.8 综合防灾规划

第6章 专项规划指引

6.1 农业科技服务

6.2 景观游憩系统

6.3 品牌农业培育

6.4 综合生态环境

6.5 招商引资

第7章 组织运营与投资估算

7.1 组织运营

包括组织机构、运营模式、管理机制。

7.2 投资估算

包括开发时序、投资估算。

7.3 综合效益分析

包括经济效益、社会效益、生态效益。

第8章 保障措施

8.1 政策保障

8.2 人才保障

8.3 资金保障

8.4 技术保障

2.规划图纸

现代生态循环农业园区规划设计图是规划设计成果的重要表达形式，包括现状分析图和规划图。现状分析图主要表达区位、交通布局、水资源、场地综合风貌、土地利用现状等内容；规划图包括空间结构、总平面、功能分区、鸟瞰图等空间布局图，土地利用、道路、水系、电力等专项规划图，景观、游憩、环保等专项规划图。

（1）图纸绘制要求

现代生态循环农业园区的规划图绘制比例一般为1：（5000~10 000），重点规划区域为1：（1000~3000），图纸印制提交，一般是A3幅面，彩色印刷，附在规划文本后面，或单独装订成册。规划设计图的格式有JPG、DWG等。各图的一般要求如下，但不限于此。

1）前期分析部分

A. 区位图表达园区在本级区域及上一级区域中的地理位置及所包含子区域情况，并可叠加道路、水系、铁路、机场等地理要素。

B. 现状图利用色块照片符号等方式，在一幅或多幅图上展现园区内土地利用现状、地形地貌、植被、基础设施现状、人工建筑等空间地物信息。

2）规划图部分

A. 功能分区图以色块方式勾勒不同功能区界线，并以文字、表格等形式说明。

B. 总平面图利用图形、图像、图片、符号、色块、线条等多种方式在一幅图上反映出各种规划要素等。

C. 分区平面图在总平面图基础上，按不同功能分区、分布方式显示规划要素，并配以说明文字、图片、意向图等。

D. 专项规划图编制道路、农田水利、林网以及观光、休闲、旅游、加工、流通等规划图件。

E. 效果图利用三维建模软件、图像处理软件等将总平面图鸟瞰，以三维等方式展现规划实施后的效果。

（2）规划图纸目录

现代生态循环农业园区规划设计标准流程：

现代生态循环农业园区的工作内容有景观概念规划设计、方案设计、初步设计、施工图设计、后期对接现场等。

现代生态循环农业园区工作根据项目情况可分为4个阶段进行。

第一阶段：概念设计

1.1 场地现况勘察，项目相关资料收集及整理（设计任务书、现状图、规划图、地质勘查报告等，搜集气象资料、水文地质资料、当地文化历史资料等，实地勘查、拍摄照片等）；

1.2 与业主沟通，确定主题构想（客户需求）；

1.3 场地空间概念分析，景观方案总平面图、景观轴线分析图、交通分析图及分区分析图等（展示、解析方案）；

1.4 主要景点透视意向（草图或意向图）；

1.5 概念设计说明；

1.6 根据项目情况议定设计周期；

1.7 阶段成果提交。

第二阶段：方案设计

2.1 概念设计的进一步完善，与业主再次沟通，使概念构思达成共识，并经专家评审最终确定的景观方案（客户需求分析）；

2.2 深化概念设计，完善设计概念、总平面图。增加竖向、设施、绿化、铺装等分析图，并进行效果图展示如总体鸟瞰、重要节点透视效果图等（深化方案）；

2.3 专项设计：竖向（总体地形变化处理方案，包括各主要剖面图、立面图），植物专项（植物绿化原则、框架、分区、特色、品种意向等），海绵城市专项（原则、雨水收集应用方式解析、LID设施分析等），铺装专项，设施服务等（根据项目需求添加）；

2.4 方案设计说明；

2.5 景观工程估算书；

2.6 根据项目情况议定设计周期；

2.7 阶段成果提交。

第三阶段：初步设计

依据方案成果与业主再次沟通，对现代生态循环农业园区方案达成共识，并进行深化方案的初步设计；

现代生态循环农业园区总体规划图册目录如下：

前期分析图

3.1 区位分析图

3.2 道路交通现状图

3.3 土地利用现状/规划图

3.4 用地现状综合风貌图

3.5 空间结构及产业布局图

3.6 功能分区图

3.7 总平面图

3.8 鸟瞰图

3.9 重点项目建设时序图

3.10 功能分区意向图

3.11 道路交通系统规划图

3.12 景观系统规划图

3.13 游憩系统规划图

3.14 灌溉系统规划图

3.15 排水系统规划图

第四阶段：施工配合

4.1 在施工图设计完成后,进行施工前设计交底;
4.2 对施工单位选送的主要铺面材料进行确定封样,以确保景观工程按图施工;
4.3 主要施工节点需到现场进行协调;
4.4 协助发包人解决工程施工中出现的技术性问题,解答施工单位有关设计方面问题,负责设计变更;
4.5 参加所有与园林有关的工程验收,签发竣工证明书。

第四节 现代生态循环农业园区综合评价系统

我国现代生态循环农业园区建设现状体现在两个方面,分别将农村作为主体建立示范园区、企业作为主体建立示范园区。基于农村和企业作为主体建立示范园区的基础上,使农业呈现多元化发展特点。其中观光农业、旅游农业等成为新的经济增长点。此外,积极应用现代化生产技术及设备,使农业生产更加规范化,在提高农业产品的产量和质量的基础上,有效提高农业资源的利用率,实现农业生态循环发展,获得良好的经济效益。

一、现代生态循环农业园区建设与评价标准化路径研究

1. 我国现代生态循环农业园区建设现状
(1) 建立以村为主体的示范园区基地

我国许多地区将农村作为主体,大力发展现代生态循环农业园区,并采用合作社等模式,不断强化农村在现代生态循环农业园区建设的主体地位,并在农村成立农业企业,使农业企业与农户建立合作关系,从而加快农村现代生态循环农业园区发展速度。这种模式的优势在于以乡土关系为依托,能够快速达成合作,土地连片使用,且通过合作社的形式保证了农民的最大权益。

(2) 建立以企业为主体的示范园区

许多企业参与到现代生态循环农业园区建设与发展中,将企业作为主体建立现代生态循环农业园区,企业与农户建立合作关系,并制定"一控两减三基本"的发展目标,采用种植和养殖结合的模式,将养殖产生的肥料使用在种植中,有助于实现发展目标。以企业为主体建立示范园区,可以将企业先进的管理经验和农业生产资源有机融合,还可以借助企业工业化手段,完善农产品深加工,提升农产品的附加值。

(3) 建立以县、市、省为主的生态循环

我国在各个地区建立的现代生态循环农业园区中,国家级园区超过100个,省级园区超过500个,生态农业示范点超过2000个。这些现代生态循环农业园区的特点是

规模大，吸纳企业多，能够对各类资源进行充分整合，同时还能够在水、电、道路、厂房等基础设施方面保持较高水准，能够实现规模效应。

2. 我国生态循环农业标准化园区建立路径

（1）转变传统农业生产模式，积极应用现代化理念建设生态循环

现代生态循环农业园区基于现代化理念，在园区构建经营体系、生产体系，并引入生态旅游项目，运用精细化管理模式，做好肥水管理工作，借助现代化技术对肥水进行无害化处理，实现资源再利用，同时农业与旅游业相互融合，促进农村第三产业快速发展，扩大农村经济发展规模。做好园区整体建设规划，合理安排农业生产空间布局，特别是要做好地下管网建设，要注重资源的二次利用。此外，还应重视对周边环境的保护，防止肥水对环境产生污染，发展健康绿色农业。在农业旅游中注重农产品品牌建设和宣传，提升农产品品牌价值和市场认可度。

（2）提高资源有效利用率，促进现代化生态循环农业快速发展

现代生态循环农业园区发展过程中，提高资源有效利用率，需要遵循4R原则，分别为减量化原则、再循环原则、再利用原则及可控化原则。遵循减量化原则时，园区内应加大节水技术的应用范围，使用低毒性农药，减少农药在农作物上的残留量。遵循再循环、再利用原则，要充分利用畜禽产生的粪便，同时对农作物产生的残渣进行无害化处理。其中，秸秆无害化处理是重要的组成部分，将秸秆农田单一的种植模式转变为与淡水养殖混合的种植模式，使农业产生的废物可以充分利用，有效提高资源的利用效率。遵循可控化原则时，基于农业生产内容建立农业生产体系、深加工体系、物流体系及销售体系，使生产出的农作物在较短的时间内由田间地头进入到用户的餐桌上，让消费者快速食用新鲜有营养的农作物。同时，要与高校和科研院所合作开展生态循环农业研究，提升科研成果的适用性和针对性，实现科研成果的快速落地和推广。此外，还要积极向国外学习现代农业生产模式，有效降低农业生产成本，提升农产品品质。

（3）建立标准化管理生态循环农业发展模式

在现代生态循环农业园区应用标准化管理模式，要求制定的管理内容必须简明扼要，使农户可以抓住标准化管理工作的要点，并能够在农业生产过程中严格按照流程进行标准化生产，有利于促进现代生态循环农业园区标准化发展。随着我国农民知识水平的提升和种养技能的完善，农户对通过标准化管理提升农业经济效益有强烈的内在需求。未来可以通过网络，以及图片、视频等方式向农户传授标准化管理工作方法，帮助农户提升标准化生产意识，提升农产品品质和农业生产标准化水平。

二、现代生态循环农业园区综合评价系统

系统评价是面向新开发的或改建的系统，根据预定的系统目标，用系统分析方法，从技术、经济、社会、生态等方面对系统设计的各种方案进行评审和选择，以确

定最优或次优或满意的系统方案。构建现代生态循环农业园区综合评价系统，是进行园区规划与设计非常重要的技术环节，科学评估和了解区域或项目所处的生态功能区的特质，是规划者、管理者制定相关政策的基础，对促进区域生态功能、维护区域生态安全、支撑规划、推进区域经济社会可持续发展具有重要的意义。在此基础上，综合运用频度分析法、专家咨询法和层次分析法，构建现代生态循环农业园区综合评价系统。

1. 现代生态循环农业园区综合评价步骤

评价步骤一般包括：①明确现代生态循环农业园区规划方案的目标体系和约束条件；②确定园区评价项目和相关的指标体系；③收集有关资料并制定评价方法；④园区综合评价系统的可行性研究；⑤园区技术经济评价；⑥综合评价。

以上是在系统评价的基础上制定的园区综合评价系统的一般程序。根据园区规划系统所处阶段来划分，现代生态循环农业园区系统评价又分为规划前评价、规划中评价、规划后评价及后续的园区发展跟踪评价等。

① 规划前评价：在园区的计划阶段的评价，这时由于园区建设体系还不明确，目标计划等尚处于摸索、探索、论证阶段，园区综合评价系统一般只能根据园区所在区域、当地经济状况、产业情况、技术条件、交通条件、政策等，参考已有资料或者用虚拟仿真的形式进行园区发展规划方案预判和评价，同时聘请相关专家学者进行方案论证，以此确定园区发展规划方向而进行评价。

② 规划中评价：在园区计划实施阶段进行的评价，重点论证、检验是否按照前期各论证基础上确定的方案计划实施，同时紧跟产业发展前沿，在原有方案的基础上，植入时代创新元素，与社会发展无缝衔接，如用计划协调技术对园区工程进度进行评价。

③ 规划后评价：在园区规划系统实施完成之后进行的评价，评价系统是否达到了前期园区规划制定的预期目标。因为园区规划已经全部完成，所以做出评价较为容易。对于园区规划系统有关社会、经济、生态、技术等因素的定性评价，需要通过相关专家学者共同参与，进行系统论证来进行。

④ 后续的园区发展跟踪评价：园区规划系统投入运行后对其他方面造成影响的评价，如园区秸秆还田再用对园区生态效益的影响等。

2. 系统评价方法

系统评价方法有以下4类。

① 专家评估，由专家根据本人的知识和经验直接判断来进行评价。常用的有特尔斐法、评分法、表决法和检查表法等。

② 技术经济评估，以价值的各种表现形式来计算系统的效益而达到评价的目的。如净现值法（NPV法）、利润指数法（PI法）、内部报酬率法（IRR法）和索别尔曼法等。

③ 模型评估，用数学模型在计算机上仿真来进行评价。如可采用系统动力学模型、投入产出模型、计量经济模型和经济控制论模型等数学模型。

④ 系统分析，对系统各个方面进行定量和定性的分析来进行评估。如成本效益分析、决策分析、风险分析、灵敏度分析、可行性分析和可靠性分析等。系统评估中最常采用的一种方法是相关树法。相关树可表示整个目标体系。通过相关树就能分析同一级水平中各个因素对上一级水平中有隶属关系因素的各自的相对重要性。逐级往上递推，直至零水平，即可知道每个因素对于完成总目标的相对重要性。重要性的数量评价一般是通过向专家征询意见而获得的。通常先由专家分别对各因素相对重要性给出估值，然后将这些估值平均。一级评估完毕即向上一级水平递推，每次递推都要考虑组合的相对重要性。

3. 现代生态循环农业园区综合评价系统

现代生态循环农业园区综合评价系统是在以上研究的基础上，针对现代生态循环农业园区制定的评价方法，利于园区的健康发展，建立园区规划综合评价系统，综合评价系统由"经济效益评价""社会效益评价""生态效益评价"3个评价系统构成。

① 经济效益评价系统：经济效益评价系统以科技子系统、资金子系统、人才子系统、产品规模子系统、生产经营子系统、产品子系统共7个子系统构成，用总量指标和单项指标综合评价园区的科技含量、成果转化、资金效益、生产规模、产品设计、市场经营、人才配置与流动、农民人均纯收入、农民人均纯收入增长率等。

② 社会效益评价系统：社会效益评价系统主要由定性指标组成，评价园区示范带动作用是否明显；是否带动该地区的农业走上品种优良化、产品优质化、生产规模化、结构布局合理化、服务社会化、经济特色化、市场化、高效化的发展道路；是否进一步增强农民"科技兴农"的意识；是否具有展示观赏功能，优秀的现代生态循环农业园区是农业发展的"窗口"；是否展示高新技术，进一步推动科技成果的转化；是否改善生态、生活环境，为居民提供观光、旅游场所，满足人们的精神生活需要；是否有助于缓解社会就业压力。

③ 生态效益评价系统：生态效益评价系统主要由农业高效和环境保护技术等技术性指标构成，评价园区是否推行了无公害作物种植、节水灌溉、生物防治等高新技术；是否建立了绿色食品、有机食品生产基地、其规模如何；绿化覆盖率是否得到了提高，能否达到国家或国际先进水平；是否有效控制了水土流失；是否对环境污染程度影响降低等等。

④ 持续更新发展评价：根据时代需求及产业发展做出及时的预判与接轨，园区区域主导产业和农业发展技术的不断创新会直接影响园区的发展方向，各指标直接会影响园区的核心区划分，产业布局、市场规划和现代化农业技术和技术装备的应用；社会环境的不断提升，影响园区旅游功能、服务功能等重要的农业相关产业的布局和开发，影响区域或项目的景观系统、基础设施的规划；相关规范和法律的确认体系保证

项目规划建设的安全性和科学规划设计；专家咨询和评估系统的不断更新发展应用能开阔设计人员和管理者的知识面和思路，为园区规划和设计的前瞻性和科学性、可持续性提供有力的保障。

三、我国现代生态循环农业园区标准化评价体系建立措施

1. 确定评价指标

确定评价指标过程中，必须确定评价指标的设定原则。在设定的原则中，应保证设定的原则具有科学性和实用性特点。科学性原则是现代生态循环农业园区确定评价指标的基础，遵循科学性原则，使评价指标更加全面，评价过程更加贴近现代生态循环农业园区真实发展状态。在确定评价指标和设定原则过程中，应根据现代生态循环农业园区所在的地区经济、环境等进行综合考虑，设定的评价指标有利于现代生态循环农业园区健康稳定地发展。此外，确定的评价指标应体现出动态和静态相互融合的特点，确定动态评价指标，可以根据现代生态循环农业园区对动态发展状态进行评价；确定静态评价指标，可以对影响现代生态循环农业园区发展的内外部因素进行评价。

2. 确定评价内容

现代生态循环农业园区在发展过程中，需要不断对园区进行规划，在规划中保证各项资源得以充分利用，并且园区发展有利于为社会创造更多的效益。基于园区规划、资源利用及社会效益，可以确定现代生态循环农业园区评价内容。确定园区评价内容后，可以使园区明确发展目标，同时园区会将提高资源利用率作为发展核心，围绕资源再生利用建设园区，有助于园区规划更加规范合理，园区资源利用率不断提升。在为社会创造更多效益的过程中，经济效益、生态贡献是效益的重要组成部分，将经济效益、生态贡献作为评价标准，不断扩大现代生态循环农业园区在当地经济建设的影响范围。

3. 构建评价指标体系

现代生态循环农业园区标准化评价指标体系建立过程中，将规划指标、资源指标及社会指标作为重要的组成部分。在规划指标中，重点对产业规划、资金保障以及队伍建设进行评价。产业规划是指景观论证的产业发展战略规划和产业类型与规模；资金保障是指生产设施设备年投入专项资金和资金管理办法；队伍建设是指组织架构和规章制度。在资源指标中，重点对资源循环再利用、资源减量化进行评价。资源循环再利用是指化肥有效利用系数、畜禽粪便资源化率及秸秆再利用率；资源减量化是指农药使用水平、农膜使用水平以及农业机械化使用强度。在社会指标中，重点对经济贡献、社会贡献及生态贡献进行评价。经济贡献是指单位面积农业 GDP 产值、农业总产值增长率；社会贡献是指人均年收入增长率、标准化培训人次数及科技推广指数；生态贡献是指土壤改良面积增长率和标准化生产基地年增加幅度。

4. 设置指标权重

由于现代生态循环农业园区涉及多种产业，在设置指标权重过程中，应具有综合性特点。现阶段，在设置指标权重过程中，会设立指标得分，将指标得分与指标权重建立关系，在指标得分不断变化过程中，指标权重也会发生变化。在现代化背景下，原有的指标权重进行优化和完善，将矩阵方差计算公式应用在指标权重计算中，使获得的指标权重更加科学，同时产生的指标权重为现代生态循环农业园区发展提供参考依据。

（1）指标权重排列顺序

对指标权重顺序进行排列时，一般会设定比较等级，按照等级的变化体现出指标权重的重要性。许多现代生态循环农业园区在设置指标权重时，会分别使用 B_i 和 B_j 表示。B_i 和 B_j 表示重要程度量化值，现代生态循环农业园区会将指标权重设定 4 个等级，量化值由高到低按照顺序排列。在一级量化值中，B_i 和 B_j 表示具有同等重要性；在四级量化值中，B_i 的重要性大于 B_j。多数现代生态循环农业园区采用一级指标对 B_i 和 B_j 的重要程度进行比较。

（2）指标权重计算

在计算指标权重过程中，以某现代生态循环农业园区为例，该园区使用 YAAHP1.0 软件，对规划指标、资源指标及社会指标进行计算。以规划指标为例，计算出的权重值为 0.5714，其中产业规划权重值为 0.3143，经过论证的产业发展战略规划权重值为 0.2095，产业类型与规模权重值为 0.1048；资金保障权重值为 0.1373，生产设施设备年投入专项资金权重值为 0.0915，资金管理办法权重值为 0.0458，队伍建设权重值为 0.1199，组织架构权重值为 0.0799，规章制度权重值为 0.0400。在资源指标中，该现代生态循环农业园区资源指标权重为 0.2857，资源循环再利用权重为 0.1905，化肥有效利用系数权重值为 0.0952，畜禽粪便资源化率权重值为 0.0476，秸秆综合再利用率为 0.0476，资源减量化权重值为 0.0952，农药使用水平权重值为 0.0381，农膜使用水平权重值为 0.0190，农药机械化使用强度权重值为 0.0381。

5. 措施与建议

（1）坚持标准化评价体系，促进现代生态循环农业园区可持续发展

现代生态循环农业园区应坚持标准化发展道路，在标准化发展过程中，建立标准化评价体系，通过评价优化和调整现代生态循环农业园区发展模式，使现代生态循环农业园区与外部市场建立联系，同时使园区不断提高产品质量，有效扩大园区在市场环境中的影响。现代生态循环农业园区应给予建立标准化评价体系足够的重视，通过标准化评价体系，可以获取园区真实的运行信息，根据信息掌握园区发展状态；利用标准化评价体系，可以不断积累园区发展经验，避免出现重复性问题，保证园区健康稳定的发展。

(2) 坚持标准化模式，加大生态循环园区管理力度

建立标准化生产模式，可以不断提高园区的生产质量，并且促进园区现代化发展。在标准化生产模式中，园区应严格贯彻和落实国家以及行业相关标准，使园区每个生产环节均能得到有效管理。例如，从农作物选种育种到农作物种植，种植期间应用的节水灌溉等技术，全部在标准化管理模式中，可以方便农户按照流程进行操作，操作过程不仅直观形象，降低生产操作难度，还能使农户统一思想，有助于现代生态循环农业园区更好地开展管理工作。

(3) 坚持标准化长效机制，加快生态循环园区经济发展

建立标准化长效机制，是加快生态循环园区经济发展的关键。在建立标准化长效机制期间，园区应积极培养本土化人员，建立适合本土发展的人才管理制度，以便机制中各项规定可以落实到实际工作中。此外，建立标准化宣传机制，宣传机制用于引导农户形成正确的发展意识，积极参与现代生态循环农业园区建立与发展，农户通过宣传获取标准化农业生产技术，使传统农户转变为职业农户，有助于提高现代生态循环农业园区现代化发展水平。现代生态循环农业园区应建立标准化奖励机制，各地方政府应加大政策及资金的投入力度，不断强化标准化长效机制的应用价值。

在现代生态循环农业园区发展过程中，应将现代化理念应用在园区建设与发展中，给予建立标准化评价体系足够的重视，确定评价指标、内容及权重，通过评价体系可以提高现代生态循环农业园区现代化发展水平。同时，各地方政府应给予足够的帮助，使现代生态循环农业园区不断扩大规模，积极融合多种经济生产模式，有效提高现代生态循环农业园区经济发展能力，带动更多的农户参与现代生态循环农业园区的建设与发展，为社会创造更多的效益。

第五节　现代生态循环农业园区规划发展中存在的主要问题及策略

近年来，随着我国经济的不断发展，自2006年中央"一号文件"明确提出要发展循环农业和2007—2016年的中央"一号文件"继续强调鼓励发展循环农业以来，循环农业作为一种新的农业经济形态和生态农业发展方向，越来越受到我国各级政府和社会的重视，我国各地也纷纷出现一些循环农业示范园区建设。现代生态循环农业园区是一类以生态农业、循环经济为发展理念，构建种养加循环发展产业链条的现代生态循环农业园区，注重生态技术和新能源的引进应用。在实际发展过程中，现代生态循环农业园区规划与设计、机遇与挑战并重，如何根据市场发展需求，将园区发展中存在的不利因素转为有利条件，是当前现代生态循环农业园区规划发展的重中之重。

一、现代生态循环农业园区规划与设计的问题

1. 缺乏科学的规划和设计管理

现代生态循环农业园区规划与设计主要目标是进行生态示范、拓展生态农业和观光农业的功能。但一些现代生态循环农业园区的规划设计缺乏对原始资料的充分调查、缺乏对主导产业和产品规模及产业运营模式的科学规划，没有充分发掘现代生态园丰富的资源内涵，导致园区规划核心功能区缺乏高效农业真正的技术含量，功能分区不够明确、资源利用率低。另外，园区建设盲目追求占地规模，资金利用率低，直接导致园区建设速度缓慢，导致生产能力的空间受到限制，严重地打击了投资者的信心。再者，由于认识上的差距，园区管理市场化、农业生产标准化没有受到重视，示范带动作用不明显，甚至起了反作用。

2. 定位不准确

现代生态循环农业园区开发既以现代农业生产为核心，就需要有合理的整体定位和科学规划，需要加大科技成果转化的投资力度，提升生产能力，开拓市场、并结合农村生态环境，形成休闲旅游业。现在，一些农业生态园中，只限于利用气候环境、绿化资源等，单方面提高了园区的旅游、餐饮功能，度假村规模越建越大，远远偏离了现代生态循环农业园区的主题。没有"培植精品"的观念，缺乏拳头产品和特色产品，整个生态园形象不明显，品牌不突出，最终导致生产能力上不去、旅游项目单调，缺乏吸引力，深度开发遇到政策和市场方面的限制。

3. 示范性作用不强

现代生态循环农业园区规划是采用生态可持续性模式进行园内农业的布局和生产，以生态农业高科技含量为基础，体现"生态文化"内涵，体现生产、产品、运营、观光、餐饮模式的和谐统一。一些园区以追求高新技术、设施生产等"精品农业"为主，热衷于引进国外设备和设施，脱离当地农业和农村经济发展现实需求，对区域发展的带动能力不强。也有生态园区开发单纯追求营利，没有采用标准生态农业的模式来规划和设计，产品市场、产品质量得不到保障，没有完全遵循有机农业的生产模式。这种缺乏文化内涵的生态园经济，其投资价值和发展潜力将大大降低，所以，也就很难起到相应的生态农业示范作用，同时也不具备通过有机农业来进行绿色食品生产的能力，很难实现经济、社会与生态效益三者的统一。

4. 基础条件薄弱

现代生态循环农业园区实质上是我国农业科技发展的一种模式或一个区域性农业科技创新基地，但其地位并未得到完全确认。在各级政府与科技部门扶持上仍缺乏明确、稳定的支持政策和支持途径；土地流转问题是多数农业科技园区建设中遇到的敏感问题，也是一个难题；对于产业发展需求，现代生态循环农业园区的科技人才和科技成果不足问题仍比较突出；各地区的园区发展不平衡问题比较突出，园区之间的项

目雷同、相互模仿的现象依然存在。以上原因导致的现代生态循环农业园区规划立项不认真或没有立项，缺少专家的科学论证。

二、现代生态循环农业园区规划设计与建设解决策略

1. 加强循环农业和现代标准化

生产管理知识的普及、家庭联产承包责任制的推行，使我国改革开放后的农业解放了生产力，农业生产关系调整得更加活跃，给农民带来了实惠，使他们走上了富裕之路。但从现代农业标准化的要求来看，土地承包带来分散的小规模生产和经营，缺乏对种植、养殖产前、产中、产后的统一规划和指导管理，无法保证产品严格按照规格化、标准化的要求生产。因此，新形势下的现代生态循环农业园区规划，必须符合国家产业政策，探索和创建新的生产体制，改变家庭联产承包责任制所形成的小农业生产方式，逐步发展适度规模经营，推广专业化与规模化经营相结合的农业生产新形式，尽快形成标准化、规模化、生态园区化的大农业格局，为顺利实现循环农业及生态园区的规划和建设，实现农业标准化生产打好体制基础。

2. 加强农业工程规划队伍的建设

2011年7月6日，中共中央组织部、人力资源和社会保障部发布《高技能人才队伍建设中长期规划（2010—2020年）》，这是中国第一个高技能人才队伍建设中长期规划。加强农业工程规划队伍的建设，一方面符合国家人才规划要求，另一方面对现有农业工程规划队伍的培训、技术整合及招聘高级人才起到促进作用。加强现有农业工程规划队伍的建设，对园区规划、建设有着非常重要的意义。

3. 完善现代生态循环农业园区建设项目的综合评价系统

现代生态循环农业园区建设项目的综合评价实际上是考验规划能否实施的关键环节，运用综合评价系统内的各项指标合理评价，实际上是对优化设计、优化资源配置、优化产业布局和产品规模等的定性和定量考察。现代生态循环农业园区的功能定位、产业定位、功能区布局及其他基础设施、景观、旅游等是否科学合理，产品市场化程度能否实现，投资回收期预判等在综合评价系统过程中会不会出现不合理、不完善的环节？在园区还没有正式投资建设的过程中发现问题、解决问题、优化规划方案，有利于更好地为现代生态循环农业园区建设项目节约前期成本，为后期投资贡献最大的综合评价系统。

4. 完善现代生态循环农业园区建设项目的投资风险管理

项目风险管理作为项目管理当中重要的组成部分，是处理因为不确定因素所产生的各种问题的一整套方案和方法。现代生态循环农业园区规划引入投资风险评价制度，是对投资者投资安全非常有效的项目前期管理制度。规划者和管理者及主管部门本着为人民服务、为农业生产做贡献的精神，完善并改进现代生态循环农业园区建设

项目的投资风险管理制度。

发展循环农业是实现农业清洁生产、农业资源可持续利用的有效手段，也是解决现代农业发展困境的必然选择。循环农业作为一种环境友好型农作方式，具有较好的社会效益、经济效益和生态效益。目前，我国循环农业发展遵循"5R"原则，即"减量化原则、再利用原则、再循环原则、资源的再生性原则、资源的替代性原则"，较好地推进了我国农业资源循环利用和现代农业的可持续发展。

我国具有良好的发展农业循环经济的基础，农业循环经济的内容十分丰富，包括农业生产部门内部的物质与能量的耦合、农业产业部门之间的物质与能量的耦合、农业与工业之间的物质与能量的耦合、城乡资源利用之间的物质与能量的耦合等具体发展模式。生态农业是农业循环经济发展的重要形式和载体，但要达到循环经济的目标和要求，还必须实现向产业化、规模化、标准化、市场化与功能多元化的更新改进。同时，必须采取相应的对策措施，包括加强农业循环经济的法制建设、生态农业产业化建设、农业废弃物处理与资源化利用及现代生态循环农业园区建设等内容。

第六节　现代生态循环农业园区管理与运营

美国学者弗雷德·R.戴维给战略管理下的定义是："制定、实施和评价使组织能够达到其目标的、跨功能决策的艺术与科学"。他认为，战略管理是致力于对市场营销、财务管理、生产作业、研究与开发及计算机信息系统进行综合的管理，以实现企业的成功发展。战略管理的目的在于为明天的经营创造并利用新的和不同于以往的机会。

战略管理的关键词不是战略，而是动态的管理，它是一种崭新的管理思想和管理方式。而制定战略和实施战略的关键，都在于对企业外部环境的变化进行分析，以及对企业的内部条件和素质进行审核，并以此为前提确定企业的战略目标，使三者之间达到动态平衡。战略管理的任务在于通过战略制定、战略实施和日常管理，在保持这种动态平衡的条件下，实现企业的战略目标。

由此，可以将战略管理定义为：企业确定其使命，根据特定的组织外部环境和内部条件制定企业的战略目标，为保证目标的正确落实和实现进行谋划，并依靠企业内部能力将这种谋划和决策付诸实施，以及在实施过程中进行控制的一个动态管理过程。

一、现代生态循环农业园区战略管理

现代生态循环农业园区战略管理是在充分占有信息基础上的一个系统的决策和实施过程，它必须遵循一定的逻辑体系，包含若干必要的环节，由此而形成一个完整的体系。

1. 确立现代生态循环农业园区战略指导思想

现代生态循环农业园区战略指导思想是在一定的社会经济条件下，在园区经营实践中不断演变而成的指导现代生态循环农业园区经营活动的一系列观念。它是战略研究的根本出发点，是现代生态循环农业园区开展经营活动的行动准则，指出了园区生存和发展的基本任务及园区应该达到的目的和行为规范。

现代生态循环农业园区的战略指导思想必须顺应当时的社会经济发展水平、国家政策、法律、法规及人与人、人与社会、人与园区、园区与社会之间的关系。这些制约因素对于所有现代生态循环农业园区是基本相同的，但在事实上，不同现代生态循环农业园区的战略指导思想是有差别的。即使是同一园区，在不同的经营时期，其战略指导思想也会发生变化。

2. 内外环境分析

由于现代生态循环农业园区组织是开放性体制，外界环境是园区组织的重要信息源，所以战略管理必须对环境进行更为细致的分析。通过大量的研究、调查来了解园区目前的状况，并在此基础上判断现代生态循环农业园区所面临的挑战与发展机遇。外部环境分析主要包括所有可能影响现代生态循环农业园区行为的现实与潜在的因素，如政治、经济、科技、文化和社会环境等，掌握其变化规律和发展趋势，以及这些变化将给现代生态循环农业园区带来更多的机会还是更多的威胁，重点是研究市场结构的性质和竞争对手的优势、劣势及其战略。内部条件分析要着重认识农业园自身所处的相对地位，具有哪些资源及战略能力，还需要了解与现代生态循环农业园区有关的利益相关者的利益。

3. 制定战略目标

现代生态循环农业园区应该根据战略指导思想，来确定其经营的战略方向和战略目标。一般而言，现代生态循环农业园区战略目标主要包含：①市场目标。主要是指现代生态循环农业园区所预期达到的市场地位，这就要求对顾客、对目标市场、对产品或服务、对销售渠道等做仔细的分析。②创新目标。主要是指在环境变化加剧、市场竞争激烈的社会里，现代生态循环农业园区实现技术创新、制度创新和管理创新的程度。③盈利目标。现代生态循环农业园区盈利目标的达成取决于其资源配置利用效率，包括人力资源、生产资源、资本资源的投入产出目标。④社会目标。现代生态循环农业园区社会目标反映其对社会的贡献程度，如环境保护、节约能源、解决就业、文化传承、科普教育等。

4. 战略方案的评价与选择

在未来的经营领域里，可以有多种途径和方法来达到战略目标，由此形成多个可能的战略方案，应对这些方案进行具体的论证，主要是技术上是否先进、经济上是否合理，从而做出综合性的评价，以此比较各个方案的优劣，选择一个比较满意的、切实可行的方案。日本战略学家伊丹敬之认为，优秀的战略是一种适应战略，它要求战略适应外部环境因素,包括技术、竞争者和顾客等。同时,现代生态循环农业园区战略也要适应园区的内部资源,如自然资源状况、经济社会发展水平、人力资本等。此外,现代生态循环农业园区的战略也要适应其组织结构。具体而言,主要从以下几个方面进行判断:①战略要实行差别化。要与竞争对手的战略有所不同。②战略要集中。现代生态循环农业园区资源分配要集中,要确保战略目标的实现。③制定战略要把握好时机。现代生态循环农业园区战略的制定应该紧跟国内外同类园区发展大趋势,并在有利的政策窗口期推出。④战略要能利用自身核心竞争力。现代生态循环农业园区要善于利用自身优势,并将其转化为核心竞争力。⑤战略要能够激发员工士气。⑥战略要能巧妙组合。现代生态循环农业园区战略应该能把园区各要素巧妙地组合起来,使各要素产生协同效果。

5. 战略实施

把战略制定阶段所确定的意图性战略转化为具体的组织行动,保障战略实现预定目标。新战略的实施常常要求一个组织在组织结构、经营过程、能力建设、资源配置、企业文化、激励制度、治理机制等方面做出相应的变化和采取相应的行动。在现代生态循环农业园区战略实施实践中,相应部门要向管理层提交园区经营战略报告,管理层根据内外形势及园区特点,确定现代生态循环农业园区战略并向各部门宣布,然后在各项工作中将战略目标进行分解、落实并监管督促基层人员执行。战略实施需要在"分析—决策—执行—反馈—再分析—再决策—再执行"的不断循环中达成。

6. 战略控制与评价

战略控制与评价是战略管理过程中一个重要环节。战略控制主要是指在企业经营战略的实施过程中,建立与健全控制系统,是为了将每一阶段、每一层次、每一方面的战略实施结果与预期目标进行比较,以便及时发现战略差距,分析产生偏差的原因,纠正偏差,从而保证全部战略方案的完成。战略评价就是通过评价企业的经营业绩,审视战略的科学性和有效性,它是战略管理的最后阶段,主要包括重新审视外部与内部因素,度量业绩,采取纠正措施。

战略实施的控制与战略实施的评价既有区别又有联系,要进行战略实施的控制就必须进行战略实施的评价,只有通过评价才能实现控制,评价本身是手段而不是目的,发现问题实现控制才是目的。战略控制着重于战略实施的过程,战略评价着重于对战略实施过程结果的评价。

二、现代生态循环农业园区运营

现代生态循环农业园区作为现代农业发展过程中出现的一种高级形态和生产组织方式,引起了农业组织方式、资源配置方式、农业经营方式和管理体制的变革与创新，其建设和发展对加速我国农业科技成果的转化与应用，推动农业技术进步，提升农业发展的质量和效益具有十分重要的现实意义。园区的建设，不仅要有较高的科技含量，注重先进性、适用性和可推广性相结合，而且要有一个结构合理的组织形式和协同、灵活、高效、完善的运行机制，以保证园区能够永续地生存和发展。

1. 运营目标

管理学有一句名言："无法衡量就无法管理"。现代生态循环农业园区的运营，处在一个不断变化的环境中，这就需要设定明确的、可以衡量的目标，引导园区朝着规范化、标准化、现代化的方向发展。运营目标是对运营活动预期所达到效果的设想，是运营活动目的性的反映与体现。运营目标的确定是现代生态循环农业园区运营的重要依据和逻辑起点，是决定园区未来发展导向的重要指南，同时也可以用来衡量现代生态循环农业园区的运营绩效、效率及运营质量。现代生态循环农业园区作为一种投入相对较高的产业模式，相应地也要求其在运营过程中所体现的产出水平、科技含量、社会经济职能等与投入水平保持一致，要能够通过良好的运营提高自身的市场竞争力。总体而言，创建一个经济效益好、社会服务功能强、生态价值高的现代生态循环农业园区，是现代生态循环农业园区运营所应达到的预期效果。因此，本节内容将从经济效益、社会效益和生态效益3个方面分别对现代生态循环农业园区的运营目标进行详细介绍。

（1）经济效益目标

传统意义上，经济效益是指经济活动中投入和产出的比例关系，是衡量经济活动所取得成效的一项综合指标。一方面，现代生态循环农业园区作为一个独立核算的经济组织，在市场导向下，以发展效益型农业为目标，通过科技成果产业化经营以获得最大的经济效益是其基本功能。经济效益这一目标的实现程度，反映了现代生态循环农业园区利用资源要素和市场创造价值的经营业绩，代表园区今后发展的速度和潜力。另一方面，现代生态循环农业园区运营作为一种开发投资行为，必然要以经济效益为中心，特别是对广大经济欠发达地区来说，通过对现代生态循环农业园区的合理运营，不仅可以增加经济收入，还可以改善当地的农业结构，促进地区的经济发展。经济效益最大化是现代生态循环农业园区得以生存和发展的根本，是现代生态循环农业园区先进性和竞争力的直接体现，经济效益难以保证的现代生态循环农业园区也很难发挥社会效益和生态效益。因此，现代生态循环农业园区的运营，首先要注重经济效益，要在获取经济效益的基础上，实现园区生存和发展的经济效益最大化，这是对现代生态循环农业园区这个经济组织运营的最基本要求。

在现代生态循环农业园区的运营过程中，要坚持以市场为导向，以效益为中心，本着"引进、集成、示范、推广、服务"的宗旨，按照市场经济规律进行运营，在对园区的组织管理和常规运营进行成本控制的同时，积极探索农业增效、农民增收的途径，通过增加投入、扩大规模、强化管理等措施，切实提高园区的经济效益。在充分发挥现代生态循环农业园区应有功能的前提下，不断提高自身的经济效益，从而更进一步地促进园区的可持续运营。

一般而言，经济效益的定量分析通常可以采用相对指标和绝对指标进行衡量。其中，用相对指标衡量，即经济效益等于劳动成果与劳动耗费和资源占用之比。

经济效益（M）= 劳动成果（W）/劳动耗费和资源占用（K）

该公式中，经济效益的计算结果值越大，说明经济效益越好。

用绝对指标衡量，即经济效益等于劳动成果与劳动耗费和资源占用之差。

经济效益（M）= 劳动成果（W）- 劳动耗费和资源占用（K）

该公式中，经济效益的计算结果为正数代表有盈余，且盈余数越大，经济效益越好。实际中，衡量现代生态循环农业园区运营经济效益目标的实现情况，常用的评价指标有园区总产值、园区年利润率、园区效益费用比、园区劳动生产率、园区土地生产率等。其中：

园区总产值指现代生态循环农业园区建设及正常运行期间每年的总产值。

园区年利润率指现代生态循环农业园区每年获得的利润总额与年生产创造总值的比值。这一指标可以用来衡量园区运营的获利水平。

园区效益费用比指园区正常运行年份中平均效益与年平均费用的比值。其中，效益指园区产品的销售收入；费用指园区支出，包括年运行费用和税金等。该指标可以反映园区资金的运用情况。

园区劳动生产率指农业园生产值与园劳动力人数之比。这一指标可以用来衡量园区内部员工的平均生产率。

园区土地生产率指农业园时间生产产值与所需要耗费的土地总面积之比。

园区生产者收入水平指园区内从事农业生产的劳动者平均收入的高低程度。

（2）社会效益目标

人类的任何组织都具有社会性。组织作为社会的一个器官，对社会有着直接的影响，必须随时关注本组织活动对社会的影响，承担应尽的社会责任，现代生态循环农业园区也不例外。现代生态循环农业园区不只是一个营利性的经济组织，它还必须随时关注本组织活动对社会的影响，承担应尽的社会责任。社会效益是指现代生态循环农业园区的建设、发展对社会所产生的间接效益和辅助效益。作为对现代农业新品种、新技术、新装备和新模式进行引进、集成、示范、推广、服务的新型组织方式，现代生态循环农业园区自产生以来，就受到了社会各界的广泛关注，也承担了一系列社会功能。

具体而言，现代生态循环农业园区承担的社会功能主要体现在以下几个方面：第一，一个成功的现代生态循环农业园区的建成和持续、稳健的运营，可以创造或提供大量的工作岗位，从而增加区域劳动力的容纳量，能够在一定程度上帮助解决园区及周边地区的农民就业问题。第二，现代生态循环农业园区的一项基本任务是要运用高新技术改造传统农业，通过农业高新技术示范项目，为科技成果转化创造一个全新的机制和环境，从而使充分"熟化"的科技成果迅速转化为现实生产力。因此，现代生态循环农业园区在持续运营过程中，要能够辐射带动周边地区农业科技进步与农业综合开发。第三，目前大多数现代生态循环农业园区将农业高新技术的培训作为园区的主要功能之一，在运营过程中依托园区高科技资源密集、人才密集和高效服务机制，大规模地开展形式多样的农业技能培训活动，通过培训，培养出新型农业人才，为提升区域农业整体素质和发展后劲提供人力资源保障。

根据上述现代生态循环农业园区承担的社会功能，对应设定现代生态循环农业园区的社会效益目标，并将其转化为可以量化的指标，用这些指标衡量现代生态循环农业园区为社会所做的贡献和所创造的效益，并据此评价现代生态循环农业园区的运营是否充分发挥了自己所承担的社会功能。这些衡量社会效益目标实现程度的指标具体包括：

带动劳动力就业人数，可以反映现代生态循环农业园区为社会提供就业机会的能力。具体指现代生态循环农业园区建设及正常运行期间，为周边地区创造的就业机会的个数，或带动劳动力就业的人数。

园区辐射推广面积，可以反映现代生态循环农业园区引导、示范和带动作用的强弱。具体指在每一年度现代生态循环农业园区周边地区引进该园区的新品种、新技术而进行生产经营的总面积。

年培训农民人数，可以反映依托现代生态循环农业园区在培养农业人才方面的贡献程度。具体指现代生态循环农业园区建设及正常运行期间，每年为周边地区培训农民的人数。

年带动农民致富人数，反映了园区的带动作用。具体指园区当年实际带动农民致富的人数，衡量标准是年纯收入大于当地农民人均年纯收入的园区辐射区域的农民数。

成果推广转化比重，具体指农业产业园本年度已推广的农业技术成果数量与本年度总的科研成果数量的比值。

2. 生态效益目标

现代生态循环农业园区不仅是园内企业生产运营的场所，它作为农业发展的"窗口"和"典范"，还应该是企业合作交流的商务区、吸引投资的温床、周边居民日常休闲观光的乐园，这一切都倚仗着园区内良好的自然生态环境而存在。因此，良好的生态环境是现代生态循环农业园区得以持续生存和发展的保障，也是现代生态循环农业园区运营成功与否的重要价值标准。

对于现代生态循环农业园区运营这一特定过程而言，生态效益是指园区在运营过程中对人类的生产、生活和环境条件产生的有益影响和有利效果。其生态效益主要体现在发挥洁、净、美、绿的特色，营造优美宜人的生态景观，改善自然环境，维护生态平衡，提高生活环境质量，充当都市的绿化隔离带，防治城市环境污染，以保持清新、宁静的生活环境，防止城市过度扩张等方面。

习近平总书记在党的十九大报告中指出，加快生态文明体制改革，建设美丽中国。这更突出了在经济发展已经进入新常态的时代背景下，要将生态和环保等理念融入农业园的运营过程中，要从过去"向规模要效益"转向"向生态要效益"。现代生态循环农业园区的持续运营，应当有益于园区及周边资源永续高效的利用及生态环境的改善，遵循"无害化、低排放、零破坏、生态好、可持续、环境优"的思路，通过合理高效地利用资源，达到提高效率、降低成本、保护环境的目的。通过带动种植农户运用生物防治、配方施肥等农业高新技术，在种植过程中减少农药、化肥等对环境造成的污染，维护生态平衡，推进农业可持续发展。在延长农业链的加工生产中，应以清洁生产为主，并对生产过程中所产生的环境污染做出明确限制，尽可能减少农产品加工业对环境的污染，并强制园区内的企业采取有效措施，减少企业产品残留对环境的影响，为现代生态循环农业园区所在地区建立长期稳定的生态安全格局。对加工产生的废弃物和排污物要进行无害化处理，有效防止对农产品、土壤、水体等的污染，以保证农产品质量安全、生态安全。与此同时，优化设计和配置名、优、特蔬菜花果等观光农业，加强现代生态循环农业园区内及园区周边环境的绿化、水源涵养林的营造，形成优美的园区景观生态格局，保护和改善区域生态环境。

现代生态循环农业园区运营过程中，生态效益目标的实现程度可以采用以下指标进行衡量：

① 园区绿化面积比率，指现代生态循环农业园区当前的绿化面积与园区总面积的比值。

② 园区无公害产品的比重，反映了园区农业环境保护情况和农产品安全性状况。具体指现代生态循环农业园区正常运营期间，每一年度的绿色无污染的农产品的产值占所有产品产值的比重。

③ 园区清洁能源利用状况，反映了现代生态循环农业园区建设及正常运营期间，利用洁净能源进行生产等的情况。

④ 园区循环经济发展水平，反映了现代生态循环农业园区建设及正常运营期间，循环经济的发展水平，包括园区的废弃物处理、污水排放情况等。

三、运行机制

机制是一个机体或组织内部各个组成部分相互制约、相互影响或者相互矛盾统一的关系。运行机制是系统所具有的，使系统整体保持正常运行所需要的各种功能的组合、联动和循环，是系统各要素之间相互联系的运行方式。现代生态循环农业园区的运行机制是指在现代生态循环农业园区的建设和发展过程中，推动、调节、制约现代生态循环农业园区内各要素正常运行，以实现现代生态循环农业园区发展目标的功能体系。现代生态循环农业园区建设与健康发展，离不开高效的园区运行机制。运行机制的完善，不仅可以解决现代生态循环农业园区中发展的问题，还可以提高园区运作效率，保障园区功能的发挥和运营目标的实现。通过现代生态循环农业园区的运营，能够整合园区内外的优势资源，为园区产业的发展搭建平台。为拓宽园区的服务范围，还要对整条农业产业链中的各个要素进行有机整合，从而使现代生态循环农业园区在市场竞争中保有核心竞争力，并不断缩短产品化过程，发挥现代生态循环农业园区的规模效应，并追求持续盈利的经济目标。

现代生态循环农业园区的运行机制可以分为投融资机制，管理机制，技术选择、对接与扩散机制，生产机制，示范带动机制，品牌营销机制和风险防御机制等。现代生态循环农业园区的健康发展，并不是各种机制简单的集成和堆砌，它是一个复杂的系统工程，各个不同的机制相互作用、相互影响，在持续的发展中逐渐统一和协调一致，并且运行机制一旦形成就具有相对的稳定性。

1. 投融资机制

现代生态循环农业园区的建设和持续运营发展离不开资本，而资本的积累需要借助良好的投融资机制来实现，如果缺乏资金，就会影响现代生态循环农业园区的规模建设、效益及先进技术的推广和发展。投融资机制是指现代生态循环农业园区在运营过程中投融资活动的组织形式、投融资方法和管理方式的总称。建立多层次、多渠道、多元化的投融资机制，是保障现代生态循环农业园区获得持续不断的物质基础和财力支持并得以发展的前提。

现代生态循环农业园区主要的融资渠道包括政府的政策性投资、社会主体投资、银行贷款和企业上市融资等，无论采取何种融资渠道，都应当遵循"谁投资、谁受益"的原则进行市场化运作。通过形成越来越宽的融资渠道，使融资数量越来越大，并逐渐形成稳定的融资机制，从而能更有力地促进现代生态循环农业园区的良性运营和发展。

其中，政府的政策性投资主要是各级政府作为投资主体，投入一定数量的启动和引导资金，这些资金主要用于现代生态循环农业园区的基础设施建设，重要的服务设施和具有间接效益、长远效益的项目上，并且往往是后续招商引资的"催化剂"。政府政策性投资的资金，主要是通过整合资金，整合土地整治、农业综合开发、退耕还林、农村交通、扶贫开发等涉农项目和社会事业发展项目，同时激活大量社会资本投入现代生态循环农业园区的建设和日常运营。也有在充分整合各类涉农资金的基础上，将用于贫困户发展的各类涉农资金入股到现代生态循环农业园区中，促使贫困户享受入股分红。这种融资机制一举多得，既可以使贫困户获得稳定的收益，又可以使现代生态循环农业园区的运营得到有效的资金支持。

现代生态循环农业园区的运营和发展，依靠政府的政策性投资是必要的，但是政府投入的财政支持是有限且较为分散的，单纯依靠政府投资，很难实现可持续发展，同时会给财政带来很大压力。因此，政府投资只能作为现代生态循环农业园区初期建设的引子和催化剂，按照市场经济法则，依靠社会法人、个人和外商投资才是筹资的根本之策。现代生态循环农业园区要通过优化园区内的投资环境，制定配套的优惠政策，以承包、租赁、拍卖、股份合作等形式，允许以土地、技术、管理入股，或赋予技术成果所有人以一定的产权权益来吸引社会各界主体投资。要加强招商引资，以优惠的政策和良好的基础配套设施，吸引各类企业及社会资金参与建设，推动园区的运营和发展。

现代生态循环农业园区的运营离不开金融体系的发展和变革，因此还应积极寻求银行的支持。目前，包括中国农业发展银行在内的多家银行都推出了用于满足建设运营各类现代生态循环农业园区等资金需求的贷款项目，但对注册资本、信用状况、管理经验、资质能力等方面做出了一定的限制和要求。

资本市场是中长期资金融通的场所，资本市场上的各种金融工具以不同的方式和期限为园区持续运营的投融资提供了多种渠道，有效减轻了政府的财政负担，同时也降低了银行贷款过高而带来的信贷风险，实现各类投资者风险共担、各资本要素有效配置。因此，农业企业进入资本市场，应成为现代生态循环农业园区融资的主要渠道。对园区内符合在资本市场上融资的农业高科技企业，要积极引导其在证券市场上发行股票或债券来融资，用股份制的形式加快和提升园区的后续建设运营。

2. 管理机制

制度经济学认为，经济发展的主要原因不是资金、人力和自然资源，而是组织制度，其本身决定着组织运转和发展状态，制度也是生产力。现代生态循环农业园区的管理机制，在一定意义上也是农业园区制度建构的产物，它是现代生态循环农业园区建设和运营管理组织形式与管理方法的总和。对于现代生态循环农业园区这一具有高级形态的特定生产组织而言，科学有效的管理机制是园区发展的组织保障。现代化的企业管理机制，强调对内部资源的整合和建立秩序，它贯穿于整个运营过程，是现代生态循环农业园区运行机制的核心。现代生态循环农业园区的运营有多种模式，不同运营模式下，涉及的参与现代生态循环农业园区建设的主体不同，其组织架构应根据现代生态循环农业园区的实际情况进行设定。但无论采取哪种运营模式，在管理过程中都应遵循"自主经营、自负盈亏、自我约束、自我发展"的原则，并逐步建立"产权明晰、责任明确、政企分开、管理科学"的现代企业制度，不断完善市场导向与技术创新有机结合的现代企业管理体制，保障现代生态循环农业园区的有序运营。

① 管委会制。是指对于开发型的现代生态循环农业园区，主要由农业园区管理委员会来全面管理园区的开发建设和日常运营过程。现代生态循环农业园区管理委员会可以是一级行政政府型管理模式，与当地政府合二为一；也可以是地方政府派出的管理机构。一般管理委员会都拥有征地、规划、项目审批和劳动人事等管理权限，负责园区的整体开发、建设和园区科技成果转化、科技企业孵化、培养科技人才、进行

新技术培训和指导工作,对农业产业结构调整起到示范带动和辐射作用。以经济为纽带,把周边农民组织起来参与种养业的生产,逐步形成以园区为龙头的企业化生产流通体系,促进和带动农村经济的发展。

② 公司制。对于通过农业公司和多方联合模式成立的现代生态循环农业园区,其运营过程中的管理机制多采用总公司制。即由开发总公司来全面管理和运营现代生态循环农业园区,若成立股份有限公司或有限责任公司,还应相应设立股东大会、董事会、监事会和总经理来规范运作。这种管理机制的优点是易于建立现代企业制度,直接进行企业化运作;缺点是总公司没有行政管理职能,在征地、规划、项目审批和劳动人事方面没有行政权,使园区在运营和发展过程中受到诸多方面的限制。

③ 承包管理制。科技承包型的现代生态循环农业园区,一般采用承包管理制。由政府或集体经济组织建好园区公共设施后,将园区生产经营的设施和项目,通过承包制或租赁制落实到农户和个人手中,实行以家庭经营为基础、统分结合的双层经营体制,农民既是投资主体,又是经营主体,而且采取统一规划建设、统一品种、统一技术、统一品牌、统一销售,实现生产、加工、销售等多个环节的衔接配套。这既有利于把农户和园区发展紧密结合起来,调动农民积极性,又有利于农民精心管理,专一经营。

3. 技术选择、对接与扩散机制

农业技术是现代农业生产力中最重要的支撑要素之一,现代生态循环农业园区代表性技术的选择和应用及对该技术持续地对接与扩散是现代生态循环农业园区运营发展的基础和关键,也是加速农业科技成果转化、推动农业产业化发展的必然选择。总体而言,现代生态循环农业园区应根据规划和建立伊始所设定的功能定位,选择相应的技术体系,不断地对所选择的农业技术体系进行研发、试验、应用、创新和示范,以此突出园区自身的特色。在此基础上,进一步充分发挥园区的辐射带动作用,引导园区周边地区的农民使用新的品种、新的设施工艺和新的农业技术,通过标准化、集约化的生产,使农业技术从潜在生产力向现实生产力转化,以促进整个农业区域经济的发展和推动现代农业的进步。

一般而言,现代生态循环农业园区选用的技术可分为农业高新技术、具有重大推广价值的实用技术和具有示范带动作用的适用技术这3大类。在选择技术时,应遵循复合多元原则、智能原则、高产优质原则和资源节约原则,根据现代生态循环农业园区在规划时所确定的功能定位及所在区域的特点,选择适宜的技术体系,并配套布置相应的具有良好品质、功能强、性价比高的设备。

现代生态循环农业园区的技术对接,是指农业技术对接主体之间,通过技术对接载体,将技术自上而下逐步传播的一系列过程。技术对接的主体包括专家、企业和农户3个部分。其中,专家是技术的创造者和提供者,企业和农户是技术的接受者和需求者。技术对接载体由政府、市场和中介组织3个部分组成,它们共同构成技术对接主体之间的纽带和桥梁。我国现代生态循环农业园区数量大、种类多,并且由于各地的自然、社会、经济条件存在差异,选择适合本地区发展特点的技术对接模式,是实现园区技术有效推广的前提。目前,我国农业技术对接机制的典型模式有"专家+农户""专家+农业中介组织+农户""专家+龙头企业+农户""专家+市场+农户"等。其中,"专家+农户"的技术对接模式属于政府和专家自上而下地向农民传播技术的过程。"专家+农业中介组织+农户"的技术对接模式,就是现代生态循环农业园区专家大院基于当地产业结构和资源特色,按照农业中介组织的委托和要求,引进、组装适宜的新技术与新品种,通过农业中介组织与农户进行技术对接。"专家+龙头企业+农户"的技术对接模式中,龙头企业以获得高质量的初级产品和生产原料为目标,在园区农业专家大院项目执行机构的指导下,与农户签订供销协议;专家根据企业需求进行技术研发、引进和指导,龙头企业为农户提供产前、产中、产后服务,农户负责农业生产并将农产品提供给企业进行深加工,最后统一将产品推向市场。"专家+市场+农户"的技术对接模式是现代生态循环农业园区管委会和专家大院在靠近科研院所和专家密集的地区投资修建大型的专业技术市场,专家将自己研发的农业新品种及各种新开发的实用技术拿到市场上来,周边农户通过市场向专家购买技术和种子,同时管委会建立严格的监管队伍和技术反馈机制。

现代生态循环农业园区的技术扩散是指一项新技术、新产品或者新的农业经营方式,借助一定的信息载体,通过不同的渠道,由一个人或少数人、少数地区向更多的人或是更大的范围进行传播、扩散,最终被人们普遍采用的过程。一般而言,现代生态循环农业园区的技术扩散是技术供给结构和园区龙头企业以技术转让和入股联营为纽带,通过技术宣传、典型示范、教育培训后续服务,与园区广大农民一道,构成农业高新技术扩散的3大主体。此外,以大学为依托的农业科技示范推广模式也是一种具体的技术扩散机制,它是农业大学服务于农业和农村经济发展的新型农业科技推广形式,是在政府的指导和推动下,以市场为导向,依托大学的科技、人才、信息等资源优势,以项目为纽带,整合、利用有关科研单位和基层农业科技推广资源,以农民科技示范户为切入点,以农村经济合作组织、涉农企业为结合点,开展农业先进适用新技术示范推广和农业高新科技成果转化的新型农业科技推广形式。

4. 生产机制

现代生态循环农业园区的生产机制是对其生产系统的设置和运行的各项工作的总称。园区生产的目标是高效率、低能耗,灵活、准时地生产高质量产品,提供满意服务,做到投入少、产出多,并最终取得最佳的经济效益、社会效益和生态效益。

现代生态循环农业园区的生产机制包括以下几个方面:

首先,从宏观上安排生产组织工作。这项工作主要是确定园区的生产内容,并据此实行劳动定额和劳动组织,设置生产管理系统等。通过安排生产组织工作,按照企业目标的要求,对生产过程中的各个组成环节进行时间和空间上的合理安排,设置技术上可行、经济上合算、物质技术条件和环境条件允许的生产系统，使现代生态循环农业园区的生产活动能高效顺利地进行。

其次，制定现代生态循环农业园区的生产计划。生产计划是进行生产管理的依据，是对园区的生产任务做出统筹安排，规定企业在一定时期内生产产品的品种、数量、质量和进度等指标，是园区在计划期内完成生产目标的行动纲领，也是园区编制其他计划的重要依据。现代生态循环农业园区的生产计划要以均衡生产为原则，依照能够对生产技术充分挖掘和资源的充分利用来制定。

最后，对现代生态循环农业园区生产过程中的各个环节进行有效控制，包括生产进度控制、设备维修、库存控制、质量控制、成本控制等。其中，为了保证现代生态循环农业园区产品的规定产量和按期交付产品，应当对生产进度进行控制；园区生产会采用各种各样的机器设备，为了避免设备出现故障耽误生产进度，应对园区内的各种设备、设施进行及时保养和维修，减少设备出现故障的可能；为了保障园区经营活动的正常进行，要通过规定合理的库存水平和采取有效的控制方式，使库存数量、成本和占用资金维持在最低限度；对质量进行控制，是要保证从现代生态循环农业园区生产出的每一个产品都符合质量标准的要求，这是现代生态循环农业园区能够受到市场选择和信赖的关键；此外，还要对生产过程中的材料费、库存品占用费、人工费等成本进行控制，从而保证园区运营过程中的利润。

5. 示范带动机制

现代生态循环农业园区要能够展示最新的农业科技成果、最先进的农业管理手段、最具活力的农业经营方式，预示未来农业的发展方向，这要求园区运营过程中要充分发挥示范带动作用。示范带动机制是现代生态循环农业园区通过引进、吸收、集成现代农业要素条件，创新现代农业经营管理体制机制，形成现代农业产业发展路径的示范样板，代表先进的农业生产力，具有强大的扩大效应。主要通过政策示范、产学研联动、产业链带动、信息服务引领，对周边形成带动作用，促进更广阔地区的农业生产关系进一步适应农业生产力的发展，促进现代农业产业体系的构建，增加农民收入，推进城镇化发展，使其成为农村发展乃至区域发展的增长极，辐射带动周边地区发展，实现区域经济的良性循环发展。

现代生态循环农业园区运营过程中的示范带动，主要依靠园区自身的技术特色、人才支撑、产业优势，通过政策导向、产学研合作、全产业链带动等，促进现代生态循环农业园区对周围区域农业发展、农民增收的引导带动作用。第一，在政策导向方面，现代生态循环农业园区的建设和运营，在资金的筹集与使用、土地流转与规模经营、人才引进与培养及生产组织模式等方面都进行了一些先行先试，为创建优良的投资环境和宽松的转入环境提供了良好的前提条件，这些政策导向都值得其他地区学习和推广。第二，在产学研联动方面，现代生态循环农业园区作为技术研发与成果转化的平台，与国内外科研机构、科技企业等诸多单位建立起了合作关系，吸引各类人才和科研团队入驻，通过产学研结合，最大限度地整合了各种资源，提升并强化了园区的引进、集成、运用、示范、推广新品种（新技术和新设施）的功能。第三，在全产业链带动方面，现代生态循环农业园区结合区域的发展现状和产业特征，突破传统的生产经营模式，通过对园区内外资源及要素的规划整合，大力发展园区特色产业，以产业为依托，加快发展质量农业、绿色农业、特色农业，实现产业结构的优化升级，使经济效益最大化。与此同时，加强生产、采收、包装、运输、销售的紧密衔接，形成完整紧密的产业链条，持续推动我国农业整体竞争力水平的提升。

6. 品牌营销机制

一个好的品牌可以产生"溢价"。品牌营销，是企业利用消费者对产品的需求，用产品的质量、文化及独特性的宣传来创造一个牌子在用户心中的价值认可，最终形成品牌效益的营销策略和过程。对现代生态循环农业园区而言，品牌意味着现代生态循环农业园区的代表性技术和所生产产品的口碑，在市场和消费者心中是质量的保证和安全的象征。品牌营销，就是要将园区生产的产品像工业产品那样进行加工和经营，在品牌的持续打造过程中，建立和保持与目标市场之间互利的交换关系，不断提升农产品的附加值，获得市场的信赖，并以此提升园区竞争力，获得满意的综合效益。现代生态循环农业园区运营过程中的品牌营销，应是层次化、多样化与品牌化经营策略三者的有机结合。

当今的社会是一个符号化的社会，对于品牌营销而言，具有差异化个性，能够深刻感染消费者内心的品牌核心价值就是现代生态循环农业园区的一个特殊符号，它能让消费者明确、清晰地识别并记住园区所生产产品的利益点与个性，是驱动消费者认同、喜欢一个品牌的主要力量。在品牌的塑造过程中，现代生态循环农业园区要根据自身的市场定位，首先保证产品的质量，获得消费者的认可。在此基础上，要依靠产品新颖的包装、独特的设计和富有象征力的名称，来进一步吸引更多的消费者。此外，在产品的销售过程和售后服务中，同样让消费者有良好的体验。通过上述要点，打造最能体现园区特色的产品品牌。

在塑造好了品牌，对外进行营销的过程中，要能够充分利用信息时代的资源，全方位、多角度地对外传播和推广园区品牌，达到宣传效果的最大化和最优化。应对目标市场的顾客综合运用广告、营业推广、公共关系、人员推销等方式进行推销。针对本地市场，可以依靠政府及相关部门的指导与扶持、结合本地新闻媒体进行宣传；也可以主动上门对机关团体、社会机构进行宣传，并给予一定的业务折扣、数量折扣、现金折扣、季节性折扣等优惠。一方面，可以减少资金负担和仓储费用；另一方面，也可以逐渐形成一批回头率高、消费稳定的团体。针对外地市场，可依托中心城市设立办事处拓展业务，并多渠道参加依托中心城市及邻近城市举办的交易会、展销会，为园区产品进行大力宣传。要能够充分利用如传统节假日、大型展销会等时机，积极进行品牌营销。

7. 风险防御机制

风险的基本含义是损失的不确定性，代表一种消极的不良后果。随着我国农业现代化进程的加快，农业集中生产和集中经营的步伐也在不断向前迈进，这使得推进农业发展的同时也将农业置于更大的不确定性中，所面临的风险也越来越大、越来越复杂。现代生态循环农业园区作为现代农业发展过程中出现的一种高级形态和生产组织方式，在运营过程中也将面临诸多风险。为规避园区运营过程中生产经营的风险、降低风险造成损失的程度、保证园区运营目标的实现，必须要明确现代生态循环农业园区运营中可能面临的风险种类并构建风险防御机制。

现代生态循环农业园区面临的风险包括制度风险、决策风险、生产风险、科研风险、信息风险和市场风险等。其中：①现代生态循环农业园区的运营是在国家的政治制度、金融制度、税收制度、土地制度等规范下进行的，这些制度可能会对园区的运营产生影响。比如，现行的各项制度可能存在一定程度的不完备性，并且会因为这种不完备性，造成预期收益与实际收益之间的偏离，偏离程度的大小与最终产生的风险大小呈正比例关系。当制度供给不足、制度变迁或制度创新的过程中会给现代生态循环农业园区的效益带来不确定的影响时，尤其是当这种影响朝着无益的方向变动或发展时，就会产生风险。②决策风险是指现代生态循环农业园区由于决策者自身能力水平及决策所依据的条件限制，或在未能将与决策有关的各项条件掌握完全而做出运营决策时，会导致园区在运营过程中与预期的效益产生偏差，出现风险。③生产风险是指在现代生态循环农业园区生产的各个环节中可能遇到的风险，包括农业机械设备或设施的购置、安装、调试，品种的培育、选择、培养，生产管理方式等方面产生的风险。④科研风险是指现代生态循环农业园区品种的选育、引进、改良、研制，各种生产方式及各个生产环节的技术研究、推介等，由于技术创新周期较长、受地域环境影响，以及技术转移推广、受农户经营规模的制约等产生的风险。⑤信息风险是指现代生态循环农业园区的运营必须要充分掌握社会、经济及其他各类信息。如果在信息采集不完备或各个部门之间存在着严重信息不对称的情况下进行园区运营，有可能会使运营结果与社会的实际发展需求出现偏差，无法获得预期收益，从而导致风险。⑥市场风险是指现代生态循环农业园区生产的产品是否能够被市场和消费者所接受，如产品是否适销对路、销售渠道是否畅通、资金是否能及时回笼等方面可能遇到的风险和挑战。

针对现代生态循环农业园区运营过程中可能出现的上述风险，应采取一系列风险管理措施、方法和手段来构建园区运营的风险防御机制，规避园区运营的风险，降低风险造成损失的程度，保障园区运营目标的实现。具体可以采取以下措施：①建立一套完整的现代生态循环农业园区运营风险评估体系，能够对风险进行全方位的识别、评估，使风险管理整个过程更加专业化、系统化和组织化。确立风险管理组织结构和组织关系，加强各部门间的联系，确保获得信息的完备性和可靠性。组织相关人员参与风险管理培训，提升风险识别、评估、防范和处理的技能。②制定现代生态循环农业园区运营风险责任制度。将园区运营过程中涉及的各项工作均责任到人，若其中某一环节出现因人为失误而导致的风险，则追究失误人的责任。此项管理对策有助于提高参与园区运营的工作人员的风险责任意识，避免出现因疏忽大意造成的风险。③加强农业保险扶持和推广力度。对于现代生态循环农业园区运营中面临的风险，特别是一些非人为性质的风险，可以通过农业保险来进行防范。

四、运营模式

对于现代生态循环农业园区这个组织而言，要追求经济效益、社会效益和生态效益的最大化，就要求在其管理过程中具有运营的理念。而模式可以看作是系统内部或系统之间各相关要素的组合方式及运作流程的范式。现代生态循环农业园区的运营模式可以理解为针对园区进行的具有特色的运营方式和运营特点的概括性描述。现代生态循环农业园区作为一个复杂系统，其构成要素众多，结构关系复杂，运行机制多样，易受周边环境影响，这些形成了园区运营模式的多样性。总体而言，现代生态循环农业园区的运营应结合地区特色，建立适合本地区的运营模式，实现现代生态循环农业园区与周边地区经济的协调及系统的可持续发展。现代生态循环农业园区经过多年的发展，逐渐形成若干具有代表性的运营模式，本书所介绍的运营模式主要是根据经营者性质的不同进行分类的，分别是政府主导型运营模式、企业主导型运营模式、"政府+企业"混合型运营模式、科研院校带动型运营模式。

1. 政府主导型运营模式

政府主导型的运营模式十分强调政府在整个运营过程中的重要作用，包括国家相关部门、省办、市办和县办等多个层次，投资主体是各级政府或者其职能部门。由政府主导运营的现代生态循环农业园区档次高、规模大、功能全，对国民经济和社会发展具有重大推动作用。因此，其运营的目标一方面着眼于区域经济的整体发展，要求现代生态循环农业园区的运营要能够提高农业水平、促进农民增收，并对现代农业建设和农业结构的调整起带动作用；另一方面也强调了园区运营的社会效益，要求园区的运营能够将农业技术向周边推广，形成辐射带动作用。在形式上，一般由政府牵头，组织有关职能部门组成领导小组，负责制订园区规划、确定运营目标、提供优惠政策和基础设施建设。在园区成立的管理委员会，作为政府的派出机构，具有较大的经济管理权限和相应的行政职能，对现代生态循环农业园区的存在、运营、发展具有一定的控制权，主要负责高科技企业入园审批、组织实施园区规划及政府授权的其他行政管理职能。在这种运营模式下，政府给现代生态循环农业园区的持续运营发展提供了大量的资金、土地、专业人员，并制定相关政策吸引企业、科技人员参与到园区进行深度建设。

政府主导型运营模式的特点：①投资主体相对单一，注重园区运营的公益性。②政府直接参与园区的建设和运营管理。③园区具有较强的引导和示范作用。政府对园区的建设、管理和运营通过园区管理委员会来进行，管理委员会实际上行使了政府的部分职能。因此，园区管理委员会权力集中，有利于运营体制的改革创新。

政府投资运营的现代生态循环农业园区，在实际运营过程中，由于政府承担农业技术推广职能，使得园区能够获得较好的社会效益；但由于存在政府管理体制复杂、市场敏感度较低、运行机制不灵活等缺点，使得现代生态循环农业园区存在办事人员效率不高、运营效果不够突出、缺乏激励机制、经济效益相对较难实现大幅度提高等问题，从而限制了园区的发展。政府应结合区域的实际情况，制定园区运营过程中的各项规章制度并强化运营监管和完善考评体系，使园区的建设和运营有章可循，以实现效益的最大化。

2. 企业主导型运营模式

企业主导型的运营模式是指在政府的引导下，由单个或多个企业组建成现代生态循环农业园区的建设和运营管理机构，以实现经济效益为主要目的，综合运用市场手段和企业经营理念来进行农业技术的研发和技术成果的转化，形成完整的运营体系。在这种运营模式下，政府并不设立派出机构，而是通过开发公司作为经济法人，进行现代生态循环农业园区的规划设计、基础设施建设、项目选择、招商引资、资金筹措，以及组织现代生态循环农业园区内的运营活动，并承担部分政府职能。投资企业按照"自主经营、自负盈亏、自我约束、自我发展"的原则，从园区的投资起点开始，就采用企业化的管理，对园区内的人才、财务、资源等进行生产要素分配，围绕利益最大化的目标对园区的技术选择、产品定位、市场方向等进行设计和布局，充分运用其资金、经营理念、信息、网络等资源优势进行园区的建设和运营，形成产业"扩散效应"，带动周边区域经济的快速发展，从而实现资本与资源的有效结合。一般而言，由企业主导运营的现代生态循环农业园区是由乡镇企业、民营企业、合资企业等不同性质的企业投资建立起来的，投资主体是企业，尤其是民营企业占较大比重。

企业主导型运营模式的特点：①采用公司制的运行方式。整个现代生态循环农业园区作为一个规范化的公司制合资企业，以法人为中心进行经营，组织农户进行生产，获取经济收益。②与市场密切接轨，追求经济效益，运行机制灵活。由于避免了行政干预，可以按照市场化规律和公司制管理的特点进行园区运营，企业会把自己最能盈利的项目作为发展的主要方向，并结合市场需求进行重点研发和技术创新，提高了运营的工作效率。

随着市场化进程的不断加快，企业主导型的运营模式由于其发展潜力和市场活力，已经成为越来越多现代生态循环农业园区的选择。但这种运营模式下，企业多以自身经济利益最大化为目标，运营过程中在社会效益和生态效益上往往难以获得持续的收益。

3. "政府＋企业"混合型运营模式

"政府＋企业"混合型管理模式是指在园区内既设立能够行使一定经济管理权限的园区管理委员会，又成立投资开发公司，由管理委员会负责政府行政管理职能，投资开发公司负责企业运作职能。其中，管理委员会是政府的派出机构，行使政府管理职权，用行政权力干预企业的经营活动，主要起监督、协调作用。现代生态循环农业园区在这种运营模式下，由政府选择一定合适的区域，编制园区总体规划和招商投资指南，投资园区的基础设施，设立园区管理委员会等管理机构，出台相关的管理办法和优惠政策，营造出良好的投资环境，为企业进园提供一个基础平台。园区建设项目经审定批准后，由企业投资进行自主经营。在现代生态循环农业园区运营中，将政府和企业"混合"之后，园区内产业发展可选择的余地会变得宽广起来，相应的技术、投资主体和融资渠道都会变得更宽，有利于充分带动各方面的积极性。

这种园区建设运营模式体现了"小政府、大企业"的原则，一方面有利于充分发挥政府的行政职能，另一方面也可以借助企业的雄厚资金、先进技术和管理运营经验来弥补现代生态循环农业园区本身的不足，充分发挥企业的经济职能。该模式符合市场化规则，能够充分发挥政府的宏观调控作用和企业的专业化经营能力，两者相互促进、互相配合，有利于现代生态循环农业园区的建设、管理和持续运营。

4. 科研院校带动型运营模式

科研院所和高等学校为了加快农业先进技术的创新开发，促进高科技成果转化而兴建现代生态循环农业园区，这些园区一般科技能力强、技术水平高，产业的科技先导性开发作用明显，有利于推动产业创新。科研院校带动型的运营模式一般可分为两种：一种是科研院校着眼于农业高新技术的研制和开发，推进产学研紧密结合设立的现代生态循环农业园区。其特点是集人才培养、知识创新、成果转化、技术咨询、企业孵化和产业化开发于一体。另一种是科研院校实施工程中心带动战略，以市场为导向，以农业技术成果的工程化、产业化为目标，通过新技术、新成果、新产品进行招商引资联合组建的现代生态循环农业园区。

但是在实际运营过程中，科研院校带动型的现代生态循环农业园区风险投资基金相对薄弱，很难到位，孵化器不够健全。因此，科研院校可适当借助政府的力量建立风险投资基金，来鼓励和支持园区的持续运营发展。

第七章 现代生态循环农业园区规划设计案例

第一节 生态循环理念在农业园区中的应用

面对我国循环农业发展的问题及相关对策，以某企业于2020年规划的"乡村振兴背景下的现代生态循环农业经济产业园"为例，对农业园区规划建设进行分析。

一、项目背景

项目位于暖温带半湿润季风气候区，自然资源丰富，土肥水美，盛产油料、药材、粮食、生猪等农副产品。项目区交通便利，且位于城市经济圈的中心服务地区，产业发展潜力巨大。

二、项目建设目标与原则

1. 建设目标

建设成为以循环经济为主体功能，示范培训和展示观光为辅助功能的综合性园区。

2. 规划原则

因地制宜、生态优先、循环利用、节能减排、产业优化、科技提升、项目带动、滚动开发。

三、功能区规划

将项目区分为现代农业循环示范区（核心区）、高效粮油生产区、设施果蔬种植区和生态循环养殖区4个部分。

1. 现代农业循环示范区（核心区）

核心区重点建设循环经济，发展种养加循环产业链，兼具观光展示和科技培训功能。核心区占地面积约3.2万亩，分为生态节能型新农村社区、绿色设施果蔬示范区、高标准粮油生产区、精品花卉苗木示范园、畜禽养殖示范区、循环经济展示区和农产品加工物流中心7个分区。

2. 高效粮油生产区

对项目区周围村庄的农田进行集中规划建设。依托良好的生态环境，利用先进的耕作技术，实现粮油的规模化、标准化、现代化耕种，提高项目区粮食作物生产单位的经济效益、社会效益、生态效益。

3. 设施果蔬种植区

大力发展果蔬产业，主要面向城市核心区区域市场，调整叶菜、茄果菜、瓜菜等产品内部结构，加强无公害绿色设施果蔬生产、加工、营销，延伸产业链。同时与项目区其他产业相互结合发展循环经济，建成省级绿色标准果蔬生产供应基地。

4. 生态循环养殖区

通过发展生态循环养殖，调整项目周边乡镇的农业产业结构，综合高效利用资源，提升农业效益，使农民持续增收。利用生态环保、最新畜牧饲养技术和循环经济原理，通过秸秆、家禽粪便、场地空间等各种资源的有效利用，降低项目区种植业和养殖业的生产成本，保护环境的同时，最大限度地提高经济效益和社会效益。

四、重点项目生态循环设计

1. 产业园区农业循环经济的主要模式

通过与园区禽畜养殖业及加工物流业结合，发展以清洁能源沼气为纽带的综合利用生态型模式。此模式具体表现形式为：养（猪、牛、鸡）—沼—种（粮油、蔬菜）等。模式原理：以土地为基础，以沼气应用为纽带，结合种植业与畜牧业，形成以农带畜、以畜促沼、以沼促农、种养结合的配套发展和生产良性循环的生态模式。

2. 养殖业禽畜粪便生态利用模式

（1）猪粪的综合利用模式

例如：核心区存栏生猪10万头，每天产生猪粪300 t。猪粪用于沼气生产，所生产沼气用于居民生活用气，沼气工程所产生的沼渣与沼液可加工为生物有机肥，资源再利用。

（2）牛粪的综合利用模式

核心区拥有黄牛万头，每天产生牛粪200 t，牛粪用于居民生活用气以及循环经济展示区的蚯蚓养殖与双孢菇种植，鲜蚯蚓可部分供应核心区内黄鳝养殖。通过蚯蚓养殖与双孢菇种植，可利用牛粪4058 t，生产的蚯蚓可满足16亩黄鳝养殖，剩余牛粪可用于堆肥，形成资源再利用。

（3）污水处理循环利用

通过针对不同产业产生的污水量建立污水处理池，污水进入处理池，经格栅、沉砂池、氧化塘、二次沉砂池等处理净化后，排入回用池的处理水，符合COD_{Cr}、BOD_5、SS和pH值标准，可用于：①果树生产加工场所的清洗；②种苗引繁中心育苗生产用水及新品种示范区的灌溉；③厂区绿化植物的浇灌；④洗手间的冲洗等。

对于回收池中多余的处理水，还可通过渠道排放，用于农田灌溉，形成"工业废水—污水处理—二次利用"的循环利用链条。

生态循环理念下的现代生态循环农业园区建设规划是对规划区域农业生态系统进行整治协调的复杂工作。此章节讲解的现代农业循环经济产业园的规划建设，解决了农业在发展过程中产业链关联性不强、资源利用不合理、农业面源污染严重、禽畜粪乱堆乱放等问题，促进了园区顺利和高效的建设与发展，使得园区对所在地区的循环农业产业经济发展更具有示范、指导和推动意义。

"畜禽粪污—沼气—电—热—有机肥—农作物—饲料—养殖"绿色农业生态循环模式，走出了一条以企业为核心、产业为纽带、生态环境保护为目标的绿色发展之路。园区规划遵循"减量化、再利用、再循环、资源的再生性、资源的替代性"的发展理念，以现代化的沼气工程为纽带，将种植、养殖、加工、旅游与新农村居住有机结合，构建种养加循环发展、一二三产联动互促的能源链条、产业链条。各种循环种养模式的引入，使园区内的动植物资源得以充分利用。同时，新技术的应用，又为园区实现循环发展提供了有效的技术保障，达到了社会效益、经济效益和生态效益的统一。

第二节　朝日绿源生态循环园区设计

朝日绿源农场位于山东莱阳市沐浴店镇，于2006年由日本朝日啤酒株式会社（73%）、住友化学（17%）、伊藤忠商事（10%）3家公司合资组建。农场租地1500亩，涉及5个村（吴家疃、大明、南汪、小店、中汪）、1000户农民，每亩每年租金1000元（这个租金是很高的，当时每亩地租金580元），租期20年，以蔬菜、水果、牛奶等高附加值农产品的生产、加工、销售为经营目的。

朝日绿源农场引进日本先进农业技术，以约100公顷的农场为基地，有效地利用循环农业、节能设备、IT，实施构筑从农作物的栽培到物流、销售一条龙的食品系统及严密的质量管理，向中国国内提供安全、安心、美味的高附加值的农产品。农场旨在通过与山东省政府协作，以为提高中国的饮食生活做贡献，实现全新型农业经营示范模式，培育下一代农业领域领导技术人才，为解决中国农业问题助一臂之力作为目标。

建园之初，朝日绿源农场提出："盈利不是我们的目的，我们是要建立一个'生态循环农业示范项目'，生产出安全、安心和高质量的产品是我们目前唯一的目标。这是中国第一家由外商独资经营的农场，也是中国第一家遵循'循环型农业生产'的农场，朝日准备赌一把。"

一、朝日绿源农场的循环农业模式

朝日绿源采取循环农业模式，投资 300 万元建造了堆肥工厂，利用农场奶牛的粪便和有机物生产有机肥料来改善土壤，从而提高农作物质量，探索循环型农业生产。此外，朝日绿源还花 200 万元引进了风力发电和太阳能发电设备，为办公楼和农场解决能源问题。

1. 养地

养地已经成为日本、德国等欧美国家发展生态农业和有机农业的第一步，"土地的健康与否"直接决定农产品品质。

早在 2003 年 12 月，朝日集团开始组织农业专家进行可行性调研，先后对山东淄博、潍坊、胶南、章丘和莱阳等市 8 块农地的水及土壤等环境指标进行了两年多严格考察。

朝日集团派驻农场的监事孙英豪说，检测结果表明，8 块农地的土壤均不符合种植绿色果蔬的要求，最终选择的莱阳沐浴店镇也不行，所以朝日绿源农场光养地就养了很多年。

朝日绿源农场希望通过养地实现土壤的健康和有机质的提升，从而影响农产品的品质。表面上看，日本人租种的土地产量仅是当地人的一半，是在犯傻，但实际上日本人的种地方式却代表了世界农业未来的康庄大道，有许多内容值得我们咀嚼。不施化肥、不打农药、不用除草剂的种地方式，其实是农业返璞归真的希望所在。现代农业最重要的特征是施用化肥和农药，目的是为了增产，但是，这种做法的长期后果和恶果已日益显现。年复一年大量使用化肥固然在短期内可以让农产品增产，但是其副作用是使土壤板结、盐碱化，导致土壤退化，而且种植出来的农产品质量也在下降。因为施用尿素等氮肥，使农作物硝酸盐含量增加，品质变差。

同样，大量施用农药固然能消灭害虫和杂草，提高农作物的产量。但是，杀虫剂的广泛使用却使得生态逐步恶化。因为害虫和杂草会对农药越来越产生耐药性，从而导致道高一尺、魔高一丈的恶性循环，不仅增加用药量和农业生产的成本，而且也会杀灭其他对人类和环境有益的昆虫和杂草，对生物多样性造成破坏。从这个意义上说，日本人不施化肥、不打农药、不用除草剂的种地方式并非在糟蹋土地，而是在保护土地。朝日绿源副总前岛启二认为，他们的种地观念是按照日本的古训，"种植之前先做土，做土之前先育人"。虽然莱阳土地肥沃，但化肥和农药的长期使用和渗透，已使得土地退化。他们要做的是，在前几年投入大量的精力和财力以恢复土壤。

当然，日本人不施化肥、不打农药、不用除草剂的种地方式可能会赔本。但这只是吃小亏，从长远来看他们是在占大便宜。其一，他们可以通过不同农产品的收益来弥补亏本。例如，朝日绿源的牛奶每升定价 22 元，是国内牛奶价格的 1.5 倍；他们生产的草莓每公斤定价120元。5年前，朝日绿源的草莓就在上海上市，刷新了草莓价格纪录。

2. 敬畏之心

干农业不仅仅是从农业资源里获取，而是要对农业资源怀有敬畏之心，如此才有可能尊重自然规律。日本人对"农业或自然的敬畏之心"的确值得称赞。比如，朝日绿源农场的管理制度极为严苛，工作人员不得用手触摸奶牛，不得对牛大声喊叫；如果某天死了一头母牛，职工要集体默哀；生产后的母牛要喂食日本味噌汤（以鲷鱼、红白萝卜、鱼骨、味噌等材料制作而成的一道日本料理），以促进食欲。

3. 定位高端市场

朝日绿源的绿色产品瞄准的是中国和海外的高端市场，而且将来也会被低端市场所接受，原因在于，他们的产品是绿色无公害食品。他们不撒化肥，全用牛粪堆肥；去草不施除草剂，而是手拔锄除；农药极少打，偶尔用，也需由专家指导；土壤定时检测，确保养分均衡。这种种植模式符合自然和生态的规律。

这样的产品不仅高质、安全，为所有人接受，而且由于保护了环境，维护了生物多样性，也最终能获得永续发展。这样的种植理念和方式无论从经济效益，还是社会效益及长远发展来看，都是丢了芝麻捡到西瓜，而且符合中国传统的天人合一理念，也是遵循自然规律的生存方式。

4. 循环农业模式

日本的循环农业模式核心是建立生态价值链，日本农业通过生态价值链的建立维护各个生态链上的关系与平衡。

无独有偶，朝日绿源农场的最大特色就是践行了日本的循环农业。简单来说，就是基本上不施用任何农药、化肥，种植的玉米秸秆可以作为奶牛的饲料，而奶牛的粪便又可用作农作物的肥料，以改善土壤质量，从而提高农产品的品质。整个生产过程几乎不使用农药、化肥。

采用与自然之道相符合的方式种植庄稼并非只是日本人的智慧，在中国早就有农民实施并取得了成功。如果绿色种植理念和行动能推广开，受益的不仅是当今国人，而且还会泽被千秋万代。

除了以上4点值得借鉴之外，日本农业的现代化经营理念和精准的市场定位也值得中国农业经营者学习。

二、朝日绿源农场循环农业模式失败原因分析

朝日绿源农场是一个设计完整的商业模式及经营模式，但为什么朝日绿源农场会失败呢？原因有以下3个。

1. 产品定价高于消费水平

朝日绿源农场的养地成本都被核算到产品的定价中,该农场希望借助高定价减少农场的亏损和投入。

比如,朝日绿源农场的产品问世之初走进一线城市的商超。其主要产品的价位是:草莓320元/斤、甜玉米8元/个、牛奶每升价格超过20元,以上价格均是国内最高端产品价格的数倍。如此高的价格,导致朝日绿源农场的销售量跟不上,甚至产品到了无人问津的窘地。

2. 规模化之殇

生态农业的产量一直处于低产,这是行业不争的事实。

由于中国土地的分散性,再加上中国"农药"为主的农业等等诸多原因导致朝日绿源农场在规模上一直上不去。

"朝日绿源农场一直处于亏损,很大程度上是因为没有突破规模瓶颈。"朝日绿源监事孙英豪称,农业是典型的规模经济,以现有的生产规模来看,企业很难实现盈利;若要扩大种植规模,土地又是首要的制约因素。

3. 形态单一

从朝日绿源农场的计划书中看到,朝日绿源农场以养奶牛为主,农产品种植为辅。或者说,该农场的农产品基本上是内部消化的。众所周知,农产品溢价之一来自加工端。但朝日绿源农场在加工端还未涉及,而是以"高价农产品"走向市场。因此,由于产业形态单一,从而导致朝日绿源农场单纯溢价部分的短路。

三、朝日绿源农场循环农业模式学习借鉴

我们能从朝日绿源农场失败的案例中学到什么呢?

1. 找到溢价点

循环农业模式只是相对于传统农业模式而言,但循环农业的关键还是靠产业衍生,尤其是循环农业产业的溢价点。比如,采用鸭稻共生模式,该产业利益点除了生产出优质的水稻,同时依靠鸭子还能衍生出新的利益链条,从而降低生态农业带来的低产效应,以其他产业提升亩产效益。

2. 规模效应

从总体上来说,农业利润来自规模效益,依靠规模来降低成本,提高农业整体利润。其实,提高农业利润的方法有两个:①降低投入成本;②提高亩产效益。无论智慧农业还是循环农业模式,都是如此。所以,循环农业模式的关键是建立产业价值链,从而实现降低投入成本,提高亩产效益。

朝日绿源农场的失败原因很多,但核心的原因是循环农业商业价值链的缺失,因此,我们想要借助循环农业模式,一定要在商业价值和产业价值上发力。

第三节　凤凰生态循环农业示范园方案设计

一、项目概况

2015年，中央"一号文件"《关于全面深化农村改革加快推进农业现代化的若干意见》坚持农业基础地位不动摇；全国农业工作会议也指出，扎实深化农村改革，加快发展现代农业，发展集农业生产、农业观光休闲度假、参与体验于一体的休闲农业，提高农民生活质量、改善农村生态环境，实现民富村美。农业部《全国休闲农业发展"十二五"规划》；某市农业委员会《XX市美丽乡村建设规划纲要》提出，XX市XX区加大财政资金扶持，提高金融服务保障前提，创新产业发展模式，培育新型技术人才，大力发展特色效益农业。加快滨河休闲农业示范带建设，以点带面、以农为本，突出滨江休闲农业示范带建设，打造渝东南乃至全市和武陵山区有知名度和影响力的农业品牌。基于以上政策支持，实现政府引领＋公司牵头＋农户参与＋互助共赢的循环生态模式。

当前农业发展，在政府的支持下正处于一个前所未有的时期，农业市场活力十足。现代农业发展需要转变农业发展方式，积极开发农业多种功能，需要有特色、创新、生态，且能保证农民切身利益的项目来拉动和引导。

二、政策分析

1. 响应政府政策

凤凰生态循环农业示范园，力图打造成为区域山区特色养殖龙头企业，同时响应政府的号召，坚持"民办、民管、民受益"的原则，因地制宜，分类指导，与农民专业合作社、农户建立更加紧密的利益联结关系。深化"龙头企业＋基地＋农户"模式，加快发展见效快、附加值高、产业链完善的特色效益农业。推进相关扶贫项目建设，把扶贫攻坚工作抓紧、抓实、抓细、抓好，帮助贫困群众发展产业、脱贫越线、增收致富。凤凰生态循环农业示范园扶持一批有劳动能力、可以通过生产实现脱贫的贫困人口，加大产业培育扶持力度。促进区域农业增效，以农民增收为目标，坚持积极发展、逐步规范、提质增效的原则。凤凰生态循环农业示范园是生产技术先进、生态环境优良、带动经济发展的现代化项目。

2. 具体措施

凤凰生态循环农业示范园是集生态养殖示范、生态种植示范、农旅结合于一体，具有高水平的经济效益、生态效益和社会效益的综合园区。

凤凰生态循环农业示范园扶持政策：

① 充分发挥此区农村合作经济组织联合会作用。

② 鼓励农民专业合作社积极参加区外农产品展示展销，推广本地产品。
③ 建立合作经济人才培训机制。
④ 农业养殖业：聘请当地农民，优先为贫困户提供培训机会，培育高技术人才，从而增加贫困户人员的就业率和经济收入，提高贫困户的经济水平。
⑤ 农家乐观光餐饮住宿：对凤凰生态循环农业示范园内原有农户的房舍进行改善，同时也利用先天的优质环境，打造农村休闲旅游业，增加收入来源，提高当地农民的经济收入。

三、区位基础及资源分析

1. 区位自然条件分析

项目所在地位于山东省某山区，交通便利，风景宜人。项目地块西面有一条乡间道路，连接国道，其往北可直达市区。

场地气候属暖温带季风气候类型。降水集中，雨热同季，春秋短暂，冬夏较长。年平均气温 11～14 ℃，场地光照资源充足，光照时数年均 2290～2890 小时，热量条件可满足农作物一年两作的需要。年平均降水量一般在 550～950 毫米之间。降水季节分布很不均衡，全年降水量有 60%～70% 集中于夏季，易形成涝灾，冬、春及晚秋易发生旱象，对农业生产影响最大。

2. 场地高程分析

凤凰生态循环农业示范园海拔较高，约为 49～569 m，最高点和最低点的高程相差 75 米左右。基地东西两面均为高山，东面临溪沟处有较平坦开阔的梯田，基地整体地形呈南高北低，地势起伏多变，水、梯田一应俱全。示范园东西两面的山体坡度较大，基地原为耕作田地，地貌已经被划分成大小不等相对较平缓的开阔田地，在溪沟边及东西两面山体上，因地形原因坡度较大，局部甚至是陡峭斜坡，山上适宜布置一些游览步道和种植果树；对于不同坡度条件的区域采取不同的开发和打造方式，有利于打造整体空间的多样性和趣味性，提高土地利用率。示范园地形主要是东、东南和西、西南，呈两两相望的走势。基地的溪水流向是由南向北流动，溪水两旁是梯田，梯田顺应溪水走势南高北低，有利于水体的汇集。在排水设计时根据坡向作指导设计。

3. 现状分析

项目地形大致呈两边高中间低的趋势，视线良好，观景点较多，环境优美，可塑造性强，可挖掘点多。

现状水源：本项目内有一个天然泉眼和一条自南向北流的天然水源，水质优良，雨季短期水流量较大，形成洪峰。泉眼水源基本可满足园区内用水，现已从山泉中分流了一条灌溉沟渠分布在园区内。

现状道路：基地西面边缘现有一条村级公路，宽约 3.5 米，是园区外来车辆的主要交通要道。这条乡村道连接国道，通往城区。园区内部多为乡间小路，内部交通需重新组织。

现有植被：区内植被现状主要分为麦田、旱地、林地及灌木林等。麦田多为梯田，其所占面积比重较大；东面山上多为灌木林，西面两山头现有茂密树林，树林内乔木生长良好，具有很好的景观效果。

4. SWOT 分析

优势（Strength）：项目区内环境本底优良，可打造较为丰富的景观；本项目自然环境好，资源丰富；距离城区近，市场潜力大。

劣势（Weakness）：项目区内现有农业产业结构单一，传统农业所占比重大；基础设施条件差，配套设施不完善。

机遇（Opportunity）：国家政策的支撑、当地政府对乡村发展的重视、现代化农业发展的引导、农业发展越来越多元化、乡村旅游渐成为趋势、推动农业发展。

挑战（Threat）：如何把项目内各个产业串联起来，形成一个完整的产业链；怎样利用好当地资源，形成良好的生态循环；怎样把景观与经济效益结合起来。

5. 市场分析

（1）养殖业市场分析

园区的养殖业主要分为岩蛙、黄粉虫、蚯蚓、蛇、牛和林下鸡等。目前人们多追求健康、高品质、新奇的绿色食品，园区养殖的岩蛙、黄粉虫、牛、鸡等，都是绿色、生态、有机的品种，满足人们的需求，势必会有良好的经济效益。

① 食用价值：养殖业为人民生活提供肉、乳、蛋、禽等丰富产品，具有很好的食用价值。例如，园区的岩蛙，蛙肉中含有高蛋白、葡萄糖、氨基酸、铁、钙、磷和多种维生素；走山鸡含有人体所必需的多种氨基酸，营养价值丰富；黄粉虫和蚯蚓不仅可以用作饲料，而且也具有很好的食用价值。

② 药用和保健价值：园区的养殖品种不仅有丰富的食用价值，还有较高的药用和保健价值。例如，黄粉虫特有的甲壳素和抗菌肽，有降低血压、提高免疫、预防癌症、抗衰老等功效，可用于医药产品和保健产品；养殖的眼镜蛇是一种剧毒蛇，但也是主要的药用蛇之一；蚯蚓体内含有地龙素、地龙解毒素、黄嘌呤、抗组织胺、维生素 B 等多种药用成分。

③ 其他价值：园区的养殖品种不仅有食用价值和药用、保健价值，还有其他价值。例如，蚯蚓和黄粉虫是鸡、蛙等动物的最佳饲料；利用蚯蚓可以处理鸡、牛粪便；蚯蚓粪还可以加工成绿色有机肥料，是园林绿化、养花种草的高级肥料，在国内蚓粪价格每吨 1000 元左右。

（2）种植业市场分析

园区的种植业主要分为果木种植、经济作物种植、苗木种植。果木多选用经济价值高和当地特有优良品种，如樱桃、苹果、山楂、寿桃、柿子、枣、石榴等；经济作物种植为小麦、玉米、花生、地瓜；苗木种植为高杆石楠、桂花、紫叶稠李。由于人民生活水平提高，对绿色生态食品的需求日益增大，有机生态食品在市场上需大于供，市场前景良好，园区种植的绿色生态水果、经济作物及绿化苗木可以供应市场需求，带来较高的经济价值。

① 健康价值：人们由于本能和对科学的认知，开始越来越关心健康，注重食品安全，注意保护生态环境。特别是对没有污染、没有公害的农产品倍加青睐。园区生产的绿色有机农产品，如樱桃、桃、李等水果保持了较好的发展势头，具有很高的健康价值。

② 品牌价值：地理标志效应的存在是一种客观现象，而绿色农产品的地理标志效应更加明显，对农产品消费者的产品品质评价和购买行为的影响很大。将示范园打造成为山东地区绿色农产品龙头企业，发挥品牌效应，提高园区的知名度，带动园区种植业的发展。

③ 其他价值：园区的种植业是养殖业和观光农家乐的纽带，种植业的发展促进园区的宣传，吸引了人气，带动了养殖业和观光农家乐的发展。

（3）观光农家乐市场分析

凤凰生态循环农业示范园农家乐旅游策划以"吃、住、游、玩、购"的丰富功能延伸，带动园区养殖业和种植业相关产业的发展为主，带动水果等农产品的生产和销售，促进了农产品加工制作等产业的发展，带来可观的经济效益。园区与周边农户互助结合形成农家乐，也为当地农户创源开流，提高了当地农民收入。

① 农业附加值：提高农业附加值，带动农村第三产业发展。乡村旅游植根于农村，与农业生产息息相关，园区的樱桃、李子、桃等水果和五谷等农产品直接面对消费者，产品可以跳过流通环节直接到达消费者手中，适时解决了当地农业产业化中购销体制不畅等难题。旅游需求还直接增加了园区养殖农产品的需求量，提高了农业附加值，推动了农村产业结构调整，为发展园区农业产业化经营提供了一个良好的平台。

② 其他价值：改善当地农村环境，提高农民生活水平。大部分游客对乡村旅游目的地的餐饮、住宿的卫生状况、接待服务水平、旅游接待地居民态度等方面十分关注，尤其是对卫生与安全的要求更高。这必然促使乡村旅游景区发展的同时加大基础设施投入，改善人居环境、健全农村社会化服务体系，如给排水建设、美化洁化、道路改善、住房改造、卫生厕所建设、生活垃圾处理等生活细节的处理，从而使当地居民客观上享受到现代化生活。

（4）市场分析

走绿色农业产业化、精细化、品牌化发展之路是推进农产品产业腾飞发展、可持续发展的最佳选择和成功之路，对"产、加、销"各个环节具有强大的吸引和带动功能，必将推动新农村建设和全面建设小康社会的发展步伐。园区的养殖业、种植业、观光农家乐是一条循环经济产业，市场广阔，具有很高的经济价值。

① 立足本地市场、做大本地常规产业：园区旅游资源丰富，游客量大，市场规模较大，是养殖业、种植业和观光农家乐发展的基础。市场路途短，交通便利，运费成本低，需求量大，可带来可观的经济收益。种植梨、桃、李等果树及高杆石楠、桂花和桂花绿化苗木；养殖走山鸡、黄粉虫、蚯蚓等中低端品种，其中低端市场需求类型为连锁超市、道路绿化、养殖基地、大排档、农家乐等。

② 深挖近程市场、做强近程高端产业：园区具有一定的区位优势，周边公路成网络式交织。近程市场交通发达、便捷，同时市民对健康、绿色农产品需求量增大，并且由于都市人对乡村自然的向往，养殖业、种植业和观光农家乐都有着广阔的市场潜力。省内城市是他们的重要市场，开拓近程高端市场，将使得示范园产业的腾飞胜券在握。园区种植樱桃、枣、柿子等果树及高杆石楠、桂花绿化苗木，养殖了牛、蛇等高端品种，其高端市场需求类型为园林绿化、知名餐馆、星级酒店等。

③ 大力发展齐鲁市场、面向全国市场

省内市场是凤凰生态循环农业示范园产业发展的延伸市场，面向全国市场是生态农业示范园发展的最终目标。交通便利，是发展山东及全国产业的前提条件。在立足周边之后，应当跳出低消费区域，直接面向国内重点市场，进行开拓性促销。园区种植樱桃、苹果等果树及高杆石楠、桂花绿化苗木，养殖岩蛙等精品品种，其精品市场需求类型为精品盆景、主题餐馆、星级酒店。

四、设计原则

凤凰生态循环农业示范园设计原则为：现代模式 + 农民互助 + 种养结合 + 综合示范，按"适度规模、标准化生产、产业化经营、促进循环经济发展"的总体思路，进行规划设计和施工建设。建设现代农业创新与示范基地，开展养殖、种植新品种、新技术，特别是循环农业技术创新与示范；示范园是秉持经济效益与社会效益并重的综合园区，力图既能给企业带来经济效益，又能极大地促进当地地方农业经济发展，带动当地农户共同致富；示范园实行"严格保护、统一管理、合理开发、永续利用"的发展方针，使生态园成为集生态养殖示范、生态种植示范、瓜果采摘、农旅结合式休闲度假于一体，具有高水平的经济效益、生态效益和社会效益的综合园区，同时依托现代农业设施、现代生产过程及优美生态环境，融入我国悠久的农耕文化，为周边市民提供休闲和体验服务的同时，带动生态农业的发展。

1. 生态优先原则

生态是凤凰生态循环农业示范园的重要前提，应结合园区的地形地貌，在园区体现生态保护、生态种养、生态景观、生态能源、生态循环。创造恬静、适宜的生态是自然的生产生活环境的基本原则，是提高园区景观环境质量的基本依据。

2. 高效益原则

从效益原则出发,规划中将考虑最佳投入和综合效益最高的项目组合。生态养殖示范、生态种植示范、乡村旅游业3方面的项目将有机联系,互相促进,创造出比单独经营更大的经济效益,同时获得生态效益和社会效益。

3. 参与性原则

直接参与体验、自娱自乐已成为当前旅游的重要方面。规划中将强调旅游项目的参与性、娱乐性和知识性的紧密结合,使城市游客广泛参与园区生产、生活的方方面面,通过亲自动手获得乐趣和相关的知识,多层面地体验农产品饲养、种植、采摘、加工、收获,农业高新技术的操作及农村生活的情趣等,使游客享受到源于乡村又高于乡村的文化氛围。

4. 突出特色的原则

特色是农业示范园发展的生命之所在,愈有特色其竞争力和发展潜力就愈强,因而,规划设计要根据园区的实际,明确资源情况,选准突破口,使整个园区的特色更加鲜明。设计中考虑本园区应该突出的是:人工生态景观(特别是大面积的植物景观)、生态技术生产(利用生态技术进行新优动植物、水产品的养殖)、滨水景观(人工生态和自然生态相结合的景观),使本园区成为山东岩蛙养殖的示范园区。

5. 文化渲染的原则

通常我们谈及农业,首先想到的是其生产功能,很少考虑其中的文化内涵,以及诗情画意的文化渲染。事实上,农业的发展是与文化的进步紧密相连的,其中的花卉种植和欣赏,更受到历史文化的熏陶。所以,在园区的规划设计中,应深入挖掘出项目内在的文化资源,并加以开发,提升园区的文化品位,以实现景观资源的高水平利用。

五、发展方向与策略

1. 园区发展方向

凤凰生态循环农业示范园以高效农业、设施农业、立体农业、观光农业等经营项目,形成了生产、加工、旅游等多元化的产业发展格局。结合现代生态循环农业的发展模式,园区规划了养殖业、种植业、乡村旅游业三大产业类型。特色养殖与种植是园区内重点经济项目,乡村旅游则是让消费者实地品尝园区特产,为示范园积累口碑,提升人气,园区内3大产业之间相辅相成,互益成长,形成科学生态的循环农业。

2. 生态循环农业

① 对畜牧业污染问题的根治可起到很好的示范作用。

② 可推广成产业,吸引农民工返乡创业。

③ 成为当地科学循环、立体养殖的示范基地。

3. 园区定位

以循环农业＋高效农业＋设施农业＋立体农业＋观光农业为基础，构建农业品牌示范园＋农业推广生态园＋农业休闲观光园。拟建设的现代生态农业示范园将以岩蛙、黄粉虫、蚯蚓养殖与绿色水果、精品苗木种植的种养结合的开发模式，实现科学生态的循环农业。按照产业培植与环境保护同步发展、经济效益与生态效益并重的发展思路，科学实施"沼气生态工程"，使园区的物质、能量、生态良性循环，成为全省资源节约型、环境友好型循环农业的示范。

（1）农业品牌示范

依托以岩蛙、黄粉虫特色养殖为主，蔬果采摘、乡村旅游为辅，打造一个可休闲、娱乐、品尝时鲜的绿色品牌农业园区。

（2）农业生态推广

保护园区内生态，科学地把养殖与种植混合形成3种循环农业，提高经济产值，减少农业园区对生态环境的破坏与影响。

（3）农业休闲观光

园区内自然环境优越，有独特的山泉资源，在山泉的滋润下，园区的蔬果更加甘甜。还可体验鲁家文化、采摘蔬果、DIY制作果蔬等，带动乡村游的发展，为园区主体经济、养殖销售带来人气。

六、重点建设项目设计

1. 生态循环养殖产业

（1）岩蛙、黄粉虫等

岩蛙因为肉味鲜美，生长速度快而且个大，是主要的食用蛙类之一。中医认为，岩蛙的肉味甘咸平，入肺胃肾经，有健脾消积、滋补强壮的功效，可用它来缓解消化不良、食少虚弱等症状。项目地养殖了一些营养价值高的岩蛙，岩蛙对水源要求较高，项目地的山泉水正好满足了养殖岩蛙的水源条件；岩蛙池周围建造了一个养殖黄粉虫的饵料池，方便喂食岩蛙；同时，饲料池旁边建造了一个过滤池，为岩蛙创造一个更好的养殖环境。

黄粉虫体内含有较高的蛋白质、脂肪、糖类等营养物，汁多体软，生命力强，极易饲养，故被很多动物园及各地养殖场选作上好饲料。同时，黄粉虫亦可作为人类的食物，长期食用，可以很好地补充高蛋白等营养元素。

蚯蚓又名曲蟮，中药名地龙。是一种软体多汁、蛋白质含量达70%的软体动物。蚯蚓以废物和农副产品下脚料为食，不与其他动物争饲料，可改良土壤、改土造肥、处理垃圾、净化城市、改善卫生。蚯蚓还是猪、鸡、鱼及各种动物和名贵珍稀水产品的最佳蛋白饲料来源。蚯蚓在人工养殖条件下，一年四季都可以繁殖，它的饲料主要是黄粉虫的粪便。它能作为畜、禽、鱼类等养殖的蛋白质饲料，可以利用蚯蚓改良土壤，培育地力。

园区主要养殖大王蛇和眼镜蛇，大王蛇是肉食鲜美的蛇类，很多地区的美食爱好者都喜欢吃。大王蛇不但肉可以吃，皮还可以制作工艺品并有药用价值。眼镜蛇是一种剧毒蛇，是主要的食用、药用蛇之一。人工饲养眼镜蛇，主要目的有3个：一是采毒；二是饲养一段时间后，赚取季节差价；三是观赏之用。蛇的市场求大于供，价格呈现年年递增的趋势。

（2）田野牧牛

养牛场主要养殖肉牛和一小部分奶牛，肉牛和奶牛的销售渠道主要是市场，其次是在农家乐进行加工食用。

项目特色：DIY制作加工内循环产品＋科普加工知识。

加工种类：DIY手工牛乳皂、牛轧糖、酸奶等。养牛场旁边建造了一个手工DTY作坊，可以通过手工加工制作成一系列的副产品。游客可以亲自参与制作产品，DIY手工牛乳皂、牛轧糖、酸奶等，特别是针对儿童，可以感受从认知到学习的体验过程，寓教于乐。

2. 生态循环种植产业

（1）果木类花海大道

春季开花的植物：樱桃、苹果、梨树等。其花期在3—4月，果期为6—10月，且花色艳丽清香，为早春重要的观花树种。盛开时花繁艳丽，满树烂漫，如云似霞，极为壮观。可大片栽植造成"花海"景观，亦可三五成丛点缀于绿地形成锦团，也可孤植，形成"万绿丛中一点红"之画意，还可作小路行道树或制作盆景。

夏季开花的植物：紫薇。紫薇花期在6—9月，果期11月。树姿优美，树干光滑干净，花色艳丽，开花时正当夏秋少花季节，花期长，故有"百日红"之称，又有"盛夏绿遮眼，此花红满堂"的赞语，是观花、观干、观根的盆景良材。根、皮、叶、花皆可入药。

秋季观叶的植物：红枫。红枫幼时紫红色，成熟时黄棕色，果核球形。果熟期10月。红枫是一种非常美丽的观叶树种，其叶形优美，红色鲜艳持久，枝序整齐，层次分明，错落有致，树姿美观，观赏价值非常高。

冬季开花的植物：蜡梅。蜡梅花期在11月—翌年3月，果期7—8月，蜡梅花在霜雪寒天傲然开放，花黄似蜡，浓香扑鼻，是冬季观赏的主要花木。蜡梅不仅是观赏花木，其花含有芳樟醇、龙脑、桉叶素等多种芳香物，是制作高级花茶的香花之一，由它提炼而成的高级香料，在国际市场上1000克相当于5000克黄金的价格。

（2）生态果园

李园：李树花期在3—4月，果实7—8月成熟，果实饱满圆润，形态美艳，口味甘甜，是人们最喜欢的水果之一。在春季，游客可以欣赏李花。李花有一种静美；在夏季，可以品尝李果，一个个紫红的果子悬挂在枝头，等着你去采摘，咬上一口，满齿留香。

梨园：梨树花期在4月，采果期7—9月。翠冠梨是沙梨的一种，果实近圆形，果形指数0.96，黄绿色，果肉雪白色、肉质细嫩、松脆多汁、化渣，石细胞极少，味浓甜；黄金梨果肉细嫩而多汁，含糖量可达14.7%度，味清甜，有香气；黄金梨是梨的一种，因其鲜嫩多汁，酸甜适口，所以又有"天然矿泉水"之称。

石榴园：石榴树花期在5—6月，果期9月上中旬。石榴花大、色艳，花期长，石榴果实色泽艳丽。由于其既能赏花，又可食果，因而深受人们喜爱，用石榴制作的盆景更是备受青睐。中国人视石榴为吉祥物，以为它是多子多福的象征。

桃园：桃树花期在3—4月，果实8—9月成熟。桃子素有"寿桃"和"仙桃"的美称，有着生育、吉祥、长寿的民俗象征意义，因其肉质鲜美，又被称为"天下第一果"。

樱桃园：品种主要有黑珍珠樱桃、紫红樱桃。樱桃树花期在2月中下旬，果期4月中下旬。果实可以作为水果食用，外表色泽鲜艳、晶莹美丽，果实富含糖、蛋白质、维生素及钙、铁、磷、钾等多种元素。

枣园：品种主要有猪腰枣、罗江贵妃枣。枣树花期在5月，果实8月上中旬成熟。枣的果实味甜，含有丰富的维生素C，自古有"一日三颗枣，百岁不显老"之说。

柿子园：品种主要有罗田甜柿、巧克力黑肉柿。其花期在5—6月，果期9—10月。甜柿是世界唯一自然脱涩的甜柿品种，秋天成熟后，不需加工，可直接食用，其特点是个大色艳，身圆底方，皮薄肉厚，甜脆可口。

（3）林下种植和林下养殖

林下种植，优势比较明显，树林可以很好地控制光照的时间和强度，减少强光对果的危害，提高果实的品质，减轻老化，延长采收期，起到自然保鲜的作用。在水果采收后再进行养鸡，防止鸡对花果的伤害。在提高了果树、鸡附加值的同时，也建立起鸡与果林互促互利的良性循环。

（4）瓜果长廊蔬果花卉种类

丝瓜、瓠瓜、八月瓜、葡萄、猕猴桃、草莓、紫藤、金银花等。瓜果长廊是一个种满蔬果花卉的长廊，力图打造一个绿意盎然，惬意且充满趣味性的地方，让游客穿梭其中，观赏一些奇特的瓜果花卉，还可以采摘瓜果，采摘后可到农家乐加工食用，感受一种被绿色自然包围的感觉。

（5）农旅结合

农家体验：项目特色为清新、自然、淳朴的文化体验。项目地与周边农家住户合作，利用农民房屋，将其改造为农家乐。游客可以在周末带上孩子到这里走走，呼吸新鲜空气，品尝特色农家菜。游客不仅可以到田间认识各种蔬菜，而且能感受采摘蔬果的乐趣，还可以现磨制作豆浆，感受农耕文化，体验淳朴自然的乡村生活，让小朋友在大自然里游戏享受童真。

餐饮住宿：项目特色为休闲、度假、生态。游客来到这里，可以暂时告别拥堵的道路、喧嚣的城市生活，享受乡村的质朴和宁静，清新的空气、碧绿的树木、清脆的鸟鸣、绿油油的草地让人有融入自然、回归自然的感觉。

休闲娱乐规划了4个项目：登山步道、魔幻树屋、休闲垂钓、趣味溪流。

① 登山步道：项目特色为休闲健身、体验丛林冒险。在绿野仙踪区域内，规划了一条登山步道，游客可以在这里享受穿梭在森林里，听鸟声，观植物，体验野外登山的乐趣。登山步道在用材上也选用了贴近大自然的木材。

② 魔幻树屋：项目特色为魔幻趣味、俯瞰园内风景、喝茶聊天。在绿野仙踪区域内，沿着登山步道设计了几个魔幻树屋，给园内增添几分魔幻趣味，游客可以爬上魔幻树屋，鸟瞰园内风景，也可以在树屋内乘凉休憩、聊天喝茶。

③ 休闲垂钓：项目特色为休闲钓鱼、赏荷花。在此区域规划的是一个垂钓区，养殖草鱼、白鲢等鱼类；还在池塘里种植了荷花，游客在钓鱼的同时，还可以赏荷花。游客们可以自带渔具或者租借渔具垂钓，钓到的鱼归顾客所有，可以在农家乐进行加工食用，也可自行带走，平时还可以举行一些钓鱼比赛。

④ 趣味溪流：项目特色为戏水歇凉、抓螃蟹、抓蝌蚪。项目地有一条沿着山谷流下来的溪流，清澈见底，游客们可以在趣味溪流里玩耍、戏水、抓螃蟹、抓蝌蚪等，一些摄影爱好者也可以在这里进行摄影创作。

参考文献

［1］2011—2015年中国农业生态化市场供需预测及投资前景评估报告［R/OL］.［2022-05-01］. https://www.docin.com/P-1583929523.html.

［2］常文韬,袁敏,闫佩.农业废弃物资源化利用技术示范与减排效应分析［M］.天津:天津大学出版社,2017.

［3］邓玉林.中国农业生态化建设成就与发展［J］.农业环境保护,1997（1）:32-34.

［4］杭小帅,王伟,张镭,等.红色粘土对模拟及畜禽养殖废水中磷的去除［J］.环境科学学报,2012,32（6）:1399-1405.

［5］李纯.农业生态［M］.北京:化学工业出版社,2009.

［6］李文君,蓝梅,彭先佳.UV/H_2O_2联合氧化法去除畜禽养殖废水中抗生素［J］.环境污染与防治,2011,33（4）:25-28,32.

［7］刘思华.绿色低碳循环农业［M］.北京:中国环境出版社,2016.

［8］刘涛.记农业生态化的内涵和产业尺度［J］.农业现代化研究,2002（23）:38-40.

［9］刘维平.资源循环概论［M］.北京:化学工业出版社,2016.

［10］吕卫光.上海生态农业典型模式研究［M］.上海:上海科学技术出版社,2018.

［11］欧阳超,尚晓,王欣泽,等.电化学氧化法去除养猪废水中氨氮的研究［J］.水处理技术,2010,36（6）:111-115.

［12］于鹄鹏,钱运华,胡涛.新型凹土吸附剂去除养殖废水中氨氮的研究［J］.广州化工,2009,37（7）:140-141,161.

英文篇

Chapter 1 Overview of the Development of Ecological Agriculture

Section 1 Overview of ecological agriculture

I. The concept and characteristics of ecological agriculture development

The concept of "ecological agriculture" was first proposed by William Albrecht, an American soil scientist, in 1970. In 1981, M. Worthingter, a British agriculturist, defined ecological agriculture, pointing out that ecological agriculture is a kind of small-scale agriculture with economic vitality, low ecological input, self-recommendation, and can be accepted by the society aesthetically, ethically and environmentally. At present, the definition of ecological agriculture in China is as follows. To protect and improve ecological environment as the goal, to make modern science & technology and engineering management as the basic means, to make ecology and economic principles as the theoretical basis, it is a modern agriculture with intensive management and ecological benefit established on the basis of traditional agricultural technology experience.

In 2011, the report "2011–2015 Forecast and Investment Pprospects Assessment on the Chinese Ecological Agriculture Market Supply and Demand" was released by Chinese Industry Research Report website. In this report, the concept of ecological agriculture was made more completely, pointing out that ecological agriculture is an important model of agricultural development, on the premise of protection and improvement of ecological environment and in accordance with the law of ecological economics, ecology. We will use modern science and technology and systematic methods to develop intensive production and management. Comparing the definition of ecological agriculture at home and abroad, we can see as the dominant model of modern agriculture, ecological agriculture has the following basic characteristics: first, ecological agriculture is modern agriculture, with the target to achieve higher social benefits and ecological benefits realizing the rational utilization of resources, which can increase the benefits and wealth, and effectively improve the agricultural ecological environment. Secondly, ecological agriculture can increase agricultural output value and improve production efficiency. Through the combination of nature, society and economy in rural areas, the ecological system structure can be reformed and adjusted, and the waste can be effectively utilized, and the application amount of pesticides and fertilizers can be reduced, so as to effectively reduce the cost of agricultural production. The development of ecological agriculture can not only prevent pollution, protect the ecological environment, maintain ecological balance, but also enhance the safety of agricultural and sideline products, changing agriculture from conventional development to sustainable development, making the development of agriculture full of momentum.

The characteristics of ecological agriculture are mainly reflected in the following aspects.

1) Sustainability. The basic feature of ecological agriculture is sustainability. The so-called sustainability refers to the mutual coordination between human development and natural ecological environment capability, which is reflected in four aspects: ecological sustainability, economic sustainability, technological sustainability and social development sustainability. ① Ecological sustainability of ecological agriculture can make agricultural production adapt to the natural environment on the basis of existing agricultural resources, so that natural resources can be used sustainably. ② The economic sustainability of ecological agriculture can make agriculture, forestry, fishing, animal husbandry, processing industry and other industries get a good coordination, so as to improve production efficiency. ③ The technological sutainability of ecological agriculture can produce more organic products without causing harm to the natural environment. ④ The social sustainability of ecological agriculture can transform the extensive development model of human beings, improve food safety, and realize the all-round development of human beings with being people-oriented.

2) Intensification. The word "intensive" is mainly embodied in three aspects: capital intensification, technology intensification and labor intensification, and ecological agriculture just has these three kinds of intensification. The capital intensification of ecological agriculture is mainly reflected in the industrialization management, and the capital flows into agricultural production. The technology intensification of ecological agriculture is mainly manifested in changing agricultural production factors, perfecting agricultural production methods, improving production efficiency and industrial management level. The labor intensification of ecological agriculture is most obviously manifested in China's abundant labor resources. By promoting the production and management of ecological agriculture, workers can improve their technical level and labor efficiency.

3) High efficiency. Ecological agriculture efficiency mainly displays as two aspects: the economic and social benefits. First, as to the economic benefits, ecological agriculture has a broad market space for development, organic products have obvious characteristic of high quality, high yield, ecology, safety, and as "organic" consumers continue to increase, the demand for organic agricultural products become more and more urgent. In addition, organic products have a price advantage over other traditional agricultural produces, and because of their safety characteristics, people are more inclined to organic produces. Second, ecological agriculture is a kind of green technology revolution, which improves the supply structure of agricultural produces, and meet the structural overproduction problems in agricultural produce market, so as to achieve sustainable development.

II. The basic principles of ecological agriculture development

1. Principle of agricultural biosymbiosis

The principle of crop symbiosis is that two different species of organisms live together for mutual benefit. In agricultural production, we should properly arrange the spatial and temporal distribution of biological populations and their complementary and mutually beneficial relationships, make full use of light, heat, time, space and other conditions, establish a multi-level configuration to make a variety of organisms coexistenting , and develop efficient and sustainable agriculture by adopting the combination of three-dimensional planting, three-dimensional breeding or three-dimensional planting and breeding. The principle of crop symbiosis is mainly applied in three aspects: multidimensional land use, that is, to combine the agricultural organisms, which uses land vertically, horizontally, temporally and spatially to fully exploit the production potential of natural resources of the land. The production model mainly includes breeding in different water level, interplanting different species, three-dimensional planting and so on. For a long time, the developing countries focus on the one-sided pursuit of the rapid development of agricultural economy, the use of fertilizers and pesticides is becoming more and more widely in the agricultural production process, making the toxic ingredients of food, vegetables, fruits and other agricultural and sideline products becoming more and more serious. Thus influenced the food safety, and people more and more desire for green ecological environmental protection food. The development of ecological agriculture will change this situation fundamentally. Moreover, with the increase of rural living garbage output and types, we cannot rely on their own given environmental ability to solve the problem of waste. The lack of the governments' unified management of the rural living garbage disposal and regulation made the rural living garbage collection, transportation and processing in a disordered state, and the living garbage can not be timely disposed. Secondly, low government investment leads to weak sanitation infrastructure and lack of necessary garbage containers, transportation machinery and treatment facilities. Moreover, rural residents have a weak awareness of environmental protection. They are not aware of the harm of garbage and do not know how to classify and collect it, not to mention harmless treatment. These problems lead to the existence of the serious phenomenon of dumping and piling at will, thus it not only caused visual pollution, and with the wind and rain, it caused soil and water pollution. In the long term it will bring a serious threat to health and life safety for the people. Especially the rural wastes contain more organic compounds, such as pesticide and fertilizer, which are more harmful than that in the city. The contamination of water and food in some areas will have serious consequences. Therefore, to strengthen the protection of agricultural ecological environment as soon as possible and vigorously promote the development of ecological agriculture are bound to become the inevitable requirement for the sustainable development of modern agriculture. It can also be said that promoting ecological agriculture will improve the living environment of rural residents, influence and change their way of life, and better improve people's quality of life.

2. The principle of multifunctional agriculture

In the late 1980s, the Japanese government first proposed agricultural multifunction in "rice culture". In the early 1990s, *Agenda 21* was formed and adopted at the United Nations Conference on Environment and Development, marking the formal adoption of the concept of agricultural multifunction. *The 1996 Rome Declaration and Plan of Action* explicitly identified the recognition and use of agricultural multifunction; At the end of 1990s, when the European Union was exploring a new path of agricultural development, it put forward the "European Agricultural Model" based on the core theory of agricultural multifunction. In 1999, the United Nations Food and Agriculture Organization (FAO) emphasized that agriculture has a variety of functions in the international conference, and so did Japan in the *Basic Law of Food Agriculture and Rural Areas*. That is to say, in addition to economic functions, it also has political, social and ecological functions. The sixth Hong Kong Ministerial Conference held by the World Trade Organization (WTO) negotiated a new round of multilateral trade in agricultural products, and then the issue of agricultural multifunction became the focus of academic debate.

The agricultural multifunction means that in addition to the basic function of providing agricultural and sideline produces, agricultural industry can also provide political, social, environmental and other functions for human beings. And each function is manifested as a variety of sub-functions, and each function is interrelated, mutually restricted and mutually

Chapter 1　Overview of the Development of Ecological Agriculture

promoted to form a multifunctional organic system characteristics. At present, non-economic functions such as improving the ecological environment, protecting biodiversity, ensuring national security and realizing sustainable, stable and healthy development of rural areas have become important contents of agricultural multifunction. The theory of agricultural multifunction is put forward under the background of the constant changing agricultural functions in the process of economic development. Agricultural production activity is the basis of human survival and development, because it provides agricultural produces that other industries can not give. With the change of the output efficiency in the agricultural activities, the status and role of agriculture in the development of national economy are constantly changing, and the continuous improvement of social and economic development stage leads to the increasing diversity of agricultural functions. According to the stage theory of social and economic development, the main function of agriculture is to provide agricultural produces for people in the age of agricultural economy. In industrial economy times, it is to improve the quality of agricultural produces. And in the post-industrial economy era, it is to improve the survival and development environment. From the policy implication of agricultural multifunction theory, commercial output and non-commercial output are easy to become joint products due to their technical connection. The change of non-commodity output supply is affected by the change of supply quantity of agricultural products caused by the increase of free trade of agricultural produces, which indirectly affects the welfare level of the countries importing and exporting agricultural produces. On the contrary, the welfare effect of the countries that import and export agricultural produces is diametrically opposite. In view of this, the theory of agricultural multifunction is the product of the continuous transformation of agricultural industry functions, and the theory of agricultural multifunction provides theoretical support for agriculture to seek the rationalization of protection.

3. New public management theory

New public management movement was put forward in the 1980s, being a theory of new public administration and the management pattern, which once thrived in Britain and the United States and other western countries. In recent years, it is one of the main guiding ideology for the western administrative reform, and its theoretical basis is on the theory and method of modern economics, the private management, for example from "rational man" assumption the basis of performance management was gained. According to the theory of public choice and transaction cost, the government should be market-oriented or customer-oriented to improve the efficiency and quality of service. According to the cost-benefit analysis method, the government performance target can be defined, measured and evaluated. Private enterprise performance management, management by objectives, organization development, human resources development and other management methods should be mastered. In addition, on the basis of the theory of public choice and transaction costs and the theory of new managerialism, the theory of new public management began to develop in different directions. Friedman and Hayek put forward the "theory of small government", pointing out that the main activity of the government is to provide services that the market cannot do well, that is, the public goods and services provided are non-exclusive, and the scope of jurisdiction should be reduced. Hammer and Ciampi developed the theory of "process reengineering", which mainly aimed at the bureaucratic system and emphasized the reconstruction and transcendence of the bureaucratic system. Aiming at the needs and satisfaction of customers, they established a new process organization structure, which greatly improved the cost, quality and service of the organization. Starting from government performance, Holzer regarded performance evaluation as a method to improve performance management. He designed a set of specific performance evaluation process and emphasized that citizens should be widely involved in the performance evaluation process, because only such evaluation results and information can have greater significance for government policies and project management. Holzer also studied the theory of "Responsive Government TQM", which mainly focuses on customer, continuous improvement, and empowerment and collaboration as the basis for TQM. Osborne and Gabler put forward the theory of "reinventing government", hoping to shape the government into a catalytic, competitive, mission-driven, results-oriented, customer-driven, enterprising, forward-looking, decentralized and market-oriented government.

III. The advantages of ecological agriculture

Agriculture is the foundation of the development of our national economy. The development of agriculture is related to the national economy and people's livelihood. Agricultural ecological model is the only way to solve the problems of rural economic development and environment. As an important agricultural base in Weifang city in our country, in recent years, it also has such problems as backward mode of production, serious loss of resources, shortage of cultivated land and fresh water resources, serious environmental pollution, shortage of in-depth processing of agricultural produces, lack of agricultural competition, and the serious contradictions between agricultural development and the ecological environment. Compared with the traditional agricultural industrialization model, ecological agriculture has the following advantages: production and management advantages. First, high-benefit ecological agriculture has provided a sustainable way for the modernization of Chinese traditional agriculture. Ecological agriculture has changed the single management idea of our people who only focus on large farming and agriculture of the land itself. They have expanded the development idea to a broader background of paying attention to the harmonious coexistence of man-land-man, and at the same time, it is also in line with the long-term desire of farmers to get rid of poverty and get rich. Therefore, based on "favorable conditions, favorable geographical conditions, and favorable people" and new ecological agriculture, it will become one of the mainstream directions of agricultural modernization from traditional agriculture to higher depth development. Second, the development of eco-agricultural tourism becomes the inevitable choice for social funds to seek new investment fields under the country's macroeconomic adjustment, and will also become a new economic growth point. In recent years, urban real estate and automobiles have far exceeded the current economic capacity of the public, and at the same time, they lack the support of corresponding financial policies, which eventually leads to slow development. Eco-agricultural tourism is directly favored by the national investment policy by virtue of its agricultural characteristics.

IV. The significance of ecological agriculture development

From the perspective of the development of our agriculture, ecological agriculture industry has a more profound significance, which can improve China's rural social productive forces, promote the continuous development of socialized production, optimize rural industrial structure, improve the comprehensive competitiveness of China's agriculture, increase farmers' income level, and further our opening to the outside world.

Developing ecological agriculture industry can further improve our rural social productivity. Through the practice of reform and opening up, China's rural social productivity had certain development, but at the same time we should realize that the present small-business pattern of our agriculture resulted in the high marginal cost to absorb new technology, and the underpowered urge to adopt new technology to develop efficient agriculture. At the same time, it is also difficult to spread the use of large machinery and new agricultural technologies because small-scale operations make land too fragmented. How to realize the leap from traditional agriculture to modern agriculture on the basis of stable household contract management is a major issue facing agricultural development under the new situation. As a kind of organizational form of mass production, ecological agriculture industrialization, with domestic and international market as the guidance, focuses to improve the agricultural comprehensive benefit. Based on leading enterprises and based on the production base of agricultural and sideline products, it can realize the regional layout, specialized production, enterprise management and socialized service, integration management of each industry or product of the agriculture. Through the establishment of professional production consortium or large-scale agricultural production base, small farmers in the past scattered through industrial management organization can explore a new way to improve the overall scale efficiency of agriculture on the basis of small-scale family management in China, without changing the original way of family contract management. This kind of organization form and mode of operation does not change the farmers' land relations, but it can improve the input mechanism and mode of production. And it can make a lot of capital and advanced technology for industry and commerce, agronomic measures, such as modern equipment organically integrate into the agricultural production and operation, which can promote agricultural labor productivity, improve the farmers' quality, and speed up the transitional pace of the traditional agriculture to modern agriculture.

Developing ecological agriculture industry can optimize rural economic structure and drive agricultural structure to realize strategic adjustment. After entering the new century, the agricultural structure of China began to adjust from the adaptation stage to the strategic adjustment. With the acceleration of the world economic integration process, Chinese agriculture has been integrated into the world agriculture, and become an important component of the global agricultural structure. Therefore, the adjustment of Chinese agricultural structure cannot be limited to China's economic development. We must fully consider the division of labor and cooperation of global agriculture and the world economy. The adjustment of agricultural structure is no longer a problem what should be planted more or less, but a comprehensive adjustment process related with the agricultural product variety, quality, regional distribution and the latter processing, and it is also a process to speed up agricultural scientific and technological progress, improve the quality of the laborers, transform the mode of agricultural growth, and promote agriculture deeply. This kind of structural adjustment can not be carried out by the way of administrative orders in the past, nor can it rely solely on the spontaneous regulation of the market. It must be adjusted and optimized in every level and every link of the agricultural and rural economic structure through the development of agricultural industrialization. By promoting the industrialization of ecological agriculture, for one thing, we can accelerate the pace of agricultural structure adjustment, accelerate the concentration of dominant industries to dominant regions, accelerate the expansion of leading industries, and accelerate the optimization and upgrading of variety structure under the drive of leading enterprises. For another, it can accelerate the pace of rural economic restructuring, promote the two-way flow of production factors between urban and rural areas, increase the proportion of the second and third industries in the rural economy, and promote the process of rural industrialization by vigorously developing the processing and marketing industry of agricultural products. By perfecting coupling mechanism and large-scale production and operation, it will enable the farmers, in accordance with market demand, avoid scattered farmers' disadvantages brought by the spontaneous adjustment structure of management. At the same time, it will realize scale economic benefit, and achieve the new industrial structure adjustment of "the government regulating the market, the market leading the enterprises, the enterprises guiding the farmers". In addition, the in-depth promotion of agricultural industrialization can promote the construction of small towns, where the processing, marketing and related services of agricultural products will be carried out, thus it will expand the scale of towns, enhance the ability to absorb rural surplus labor, accelerate the pace of farmers to cities and towns, and drive the adjustment of rural employment structure.

Developing ecological agriculture is beneficial for improving China's agricultural comprehensive competitiveness. Market economy is a highly competitive economy, but also the competition of other elements such as market, product, price, technology and profits. Due to the special historical reasons, our closed, self-contained, semi-self-contained natural economy has experienced quite a long period. Only when the Third Plenary Session of Party's 11th Central Committee enacted the reform and opening policy in the eighties of the last century, did our country gradually begin to develop to a planned commodity economy. And the 14th National Congress of our Party further confirmed the development direction of the socialist market economy. Nowadays, the competition in the international market is not only the competition of product quality, but also the competition of economic strength and scale between the two sides of trade. The farmers are the micro-foundations of long-term unchangeable agricultural production in our country and the micro-subjects of our agriculture participation in the international competition. The country's more than 220 million farmers, with small scale of operation, low level of organization, weak economic strength, are difficult to adapt to the international and domestic agricultural market competition after China's accession to the WTO. Decentralized family management can not ensure product quality, and there is no competitive strength with the international consortium. Under such circumstances, the development of industrialization management is more conducive to dealing with the challenge of economic globalization and accession to WTO, which makes our agriculture integrate more quickly into the world agriculture. Through the development of agricultural industrialization, we can have some large-scale and strong market main entities, and they organize the dispersed farmers, according to the requirements of the international market, standardization, large-scale production. We should as soon as possible expand the agricultural production scale which has a comparative advantage, improve the level of deep processing, and create a number of brand-name products and brand enterprises with strong competitiveness. In addition, to implement the strategy of agricultural internationalization, it is necessary to improve the level of agricultural industrialization management, rely on the industrialization management organization to introduce foreign capital, technology, equipment, management experience, and further develop and strengthen our international competitiveness.

The development of ecological agricultural industry can improve the comparative profits of agriculture and greatly increase farmers' income. At present, the outstanding contradiction of our national economy development is the slow growth of farmers' income, and the larger gap between urban and rural income, which is also the key problem of "San Nong" (three words related to agriculture, "agriculture, countryside and farmers"). There are many reasons that restrict the growth of farmers' income, such as the difficulty of selling agricultural products, backward marketing methods, low degree of marketization, short production chain of agricultural products, insufficient in-depth processing, low added value, narrow income channels for farmers, and high dependence on land. Due to the fact that to some extent agriculture is a weak industry with high social benefits but low economic benefits, in the role of the law of comparative benefit, the capital, technology, talents and other factors of production will inevitably flow to the non-agricultural industries with high profits. So the agriculture is in the plight of serious insufficient investment and powerless drive, which is the difficulty in the development of agriculture in the market economy. At the same time, due to the increase of the urban unemployed population, the employment position is limited, the surplus rural labor force can not be transferred to the cities and towns. Deepening the industrialization of agriculture can not only help farmers solve the problem of selling agricultural products, but also ensure a good price and reduce the cost of marketing farmers. In addition, by developing the processing and marketing of agricultural products, the first, second and third industries in rural areas will be integrated, so as to open up new employment fields for farmers and drive them to move to the second and third industries. Through scale management and multi-level processing, agricultural products can achieve multiple added-value, improve agricultural added value and comprehensive benefits, and broaden the channels to increase income for farmers.

Section 2 Theory of ecological agriculture development

I. The background and development of ecological agriculture

Under the background of the emergence of ecological economy, compared with a series of problems brought by the development of petroleum agriculture, ecological agriculture has attracted much attention and concern from its emergence to development. Ecological agriculture can be traced back to Europe in the 1930s and developed in Britain, Switzerland and Japan in the 1930s and 1940s. Since A. Howard, a British agronomist, put forward the concept of organic agriculture in the early 1930s and organized experiments and promoted accordingly, organic agriculture has been widely developed in Britain. Rodale (J. I. Rodale) is the earliest practitioner in the United States. He founded the first organic farm in 1942. In 1974, he established the Rodale Institute on the basis of expanding the farm and past research, and became a famous institute engaged in organic agriculture research in the United States and the world. But the ecological agriculture at that time was mostly confined to the self-enclosed biological circular production model and developed very slowly.

By the 1960s, many farms in Europe had switched to ecological farming. Since the 1970s, with the development of industrialization in some developed countries (especially the United States, Europe and Japan), environmental pollution has become increasingly serious, endangering human life and health. In order to ensure the quality of human life and the healthy development of economy, various thoughts of alternative agriculture have been set off to protect the agricultural ecological environment. In 1971, W. Abrecht, a professor at the University of Missouri, put forward the idea of ecological agriculture. The United Nations Conference on Human Environment held in 1972 put forward the ecoligical strategy of social economic development. In 1974, American ecologist G. E. Hutchinson and others published a collection of papers called *Biosphere*, which put forward the view of agroecological economy. In 1981 the British environmental scientist Washington (M. K. Washington) published a book *Agriculture and Environment*, regarding the planting industry, animal husbandry, breeding, natural resource conservation, family small products processing and sales, environmental beautification as an agricultural ecological economic system, which further demonstrated the macro strategic significance for the development of ecological agriculture. During this period, France, Germany, the Netherlands and other developed countries in Western Europe also started organic farming movements, and the International Federation of Organic Farming

Movements (IFOAM) was established in France. The focus of the ecological agriculture in Japan is to reduce agricultural thermal alkalinization, reduce the agricultural non-point source pollution (pesticides, fertilizers) , and improve the quality and safety of agricultural products. The Philippines is an early and fast developing country in ecological agriculture. Maya Farm is the most influential example of ecological agriculture in the world. In this period, the development of ecological agriculture has attracted extensive attention from all over the world. Both developed and developing countries generally believe that ecological agriculture is one of the important ways of sustainable agricultural development.

Since the 1990s, ecological agriculture has developed greatly in various countries. Constructing the ecological agriculture and taking the road of sustainable development has become the common choice of agricultural development of all countries in the world. Especially since the 21st century, sustainable development strategy has become a global common action, and ecological agriculture, as a kind of important practice patterns of sustainable agricultural development, has entered a vigorous development in the new period. There has been a qualitative change in the level, besides in the scale and speed, transforming from a single, dispersed, spontaneous folk activities towards the global production movement advocated consciously by the government.

II. The connotation and characteristics of ecological agriculture

1. The connotation of ecological agriculture

In order to understand ecological agriculture correctly, we should first have a brief understanding of ecological agriculture. There are different interpretations of the meaning or concept of ecological agriculture. For example, M. Worthington, a British agronomist, defines ecological agriculture as a small agriculture that is ecologically self-sustaining, low inputting, economically vital and acceptable in terms of environment, ethics and aesthetics.

Ecological agriculture is modern agriculture established according to the principles of ecology and economics, using the achievements of modern science and technology, modern management methods, and effective experience of traditional agriculture, which can obtain higher economic, ecological and social benefits. It requires the development of grains and a variety of economic crops, the development of field planting and forestry, animal husbandry, sideline production and fishery, the development of combining agriculture with the second and the third industry. Using the essence of traditional agriculture and modern scientific and technological achievements, and using human ecological design, it coordinates the contradiction between the development and environment, between resource utilization and protection, forming two better circle of ecological and economic, being the unification of economic, ecological and social benefits.

Ecological agriculture refers to the agricultural development model of intensive management under the premise of protecting and improving the agricultural ecological environment, following the laws of ecology and ecological economy, using system engineering methods and modern science and technology. Ecological agriculture is a comprehensive agricultural production system based on the principles of ecology and ecological economy, using systems engineering and modern scientific and technological methods. Ecological agriculture is to plan, organize and carry out agricultural production according to local conditions in a certain area on the basis of ecological theory.

Ecological agriculture is a system, a larger complex system integrating ecological system and agricultural production system. It is not a simple addition or mechanical accumulation of various elements, but an organic whole of mutual connection, interaction, mutual restriction and mutual promotion. It is instructed by ecology theory, using ecology principles of coexistence within species in the agricultural ecosystem, material cycle, energy multi-level. It uses agricultural natural resources reasonably and efficiently, well protects the ecological environment applies the agricultural high-tech, absorbs the essence of traditional agriculture, uses modern management methods, and organizes scientifically the agricultural production. Thus, it would achieve ecological self-maintenance and dynamic balance, sustainable development of agriculture, and achieve the efficient unity of ecological benefits, economic benefits and social benefits.

2. The characteristics of ecological agriculture

① Comprehensiveness. Ecological agriculture emphasizes the integrated function of the agricultural ecosystem, taking macro-agriculture as a starting point, in accordance with the principles of "integration, coordination, circulation and regeneration". It plans comprehensively planning, adjusts and optimizes the agricultural structure, makes agriculture, forestry, animal husbandry, sideline production and fishery industries closely combined with the rural first, second and third industry, makes the different industries mutually support, brings out the best in each other, and enhances the comprehensive production capacity.

② Diversity. For the difference of natural conditions, resource base, level of economic and social development, ecological agriculture fully absorbs the essence of traditional agriculture, applies modern science and technology, and provides the equipment of agricultural production in a variety of ecological models, ecological engineering and colorful technology type. Thus, it can foster the regional advantage and circumvent weaknesses, and at the same time, according to the social demand combined with local reality, we will develop industries with distinctive features.

③ High efficiency. Through material circulation, comprehensive utilization of multi-level energy and serialized deep processing, ecological agriculture realizes economic added-value, implements waste resource utilization, reduces agricultural costs, improves benefits, creates agricultural internal employment opportunities for a large number of rural surplus labor, and protects farmers' enthusiasm in agriculture.

④ Sustainability development of ecological agriculture can protect and improve the ecological environment, prevent and control pollution, maintain the ecological balance, improve the security of agricultural products, change the regular development of agriculture to sustainable development, combine closely environmental construction and economic development, and improve the stability and sustainability of ecological systemthe with meeting people's growing demand for agricultural products as best as possible at the same time. We will strengthen the sustainability of agricultural development.

III. The basic principles of ecological agriculture

Ecological agriculture is a branch of ecology applied to agriculture. The discipline or science that deals with ecological agriculture is agroecology. The concept of agroecology is made up of three stems, from the Latin agrarius (field); the Greek oikos (home or housework), logos (science). According to Haeckel's definition of ecology above, agroecology can be defined as the science of living things in environments shaped by humans for the production of certain crops. American scientist Miller summed up the three laws of agricultural ecology are as follows. The first law is that any action we take is not isolated, and any invasion of nature has numerous effects, many of which are unpredictable. This law was put forward by G. Hardin and can be called the principle of multiple effects. The second law is that everything is connected and integrated with everything else. This law is also known as the interrelation principle. The third law says that nothing we produce should interfere in any way with the natural biogeochemical cycles on Earth. This law may be called the No-disturbing principle. The basic principle of agroecology is to imitate the biological production of energy flow, material circulation and information transmission of the natural ecosystem to establish human social organization. With natural energy flow as the main, it reduces artificial additional energy as far as possible, to seek the maximum comprehensive benefits with as little consumption as possible.

According to the basic principles of agricultural ecology, ecological agriculture runs in "integration, coordination, circulation and regeneration". Its principle is specifically reflected in the following aspects.

① The principle of mutual restriction between organisms. Following this principle to organize production according to food chain reasonably, we can dig out the resource potential and raise benefits.

② The principle of co-evolution between organisms and environment. According to this principle, in accordance with the conditions of the land, we should reasonably lay out, reasonably arrange the rotation of stubble, plant three-dimensionally, and combine the planting and breeding to obtain significant benefits.

③ The principle of multi-stage energy utilization and material recycling and regeneration. According to this principle, the organic waste can be recycled, the photosynthate can be reproliferate, the pollution can be reduced and the fertilizer source can be replenished, and the comprehensive benefit can be improved.

④ The principle of structural stability and functional coordination. We should follow the principle of giving full play to the advantages of biological symbiosis, take advantage of biological mutually seeking advantages and avoiding disadvantages, and use the principle of biological interdependence.

⑤ The principle of the unity of ecological benefits and economic benefits. We should follow the principle of rational allocation of resources, the principle of full utilization of labor resources, the principle of rationalization of agricultural structure and economic structure, the principle of specialization and socialization.

Ecological agriculture theory is an important theoretical basis of green and low-carbon circular agriculture. According to the above basic understanding of the theory of ecological agriculture, the basic principle of ecological agriculture is the symbiosis of species, material cycle, multi-level utilization of energy, and the realization of ecological harmonious cycle. Practice has proved that the ecological agriculture is a successful model to realize environment protectionand better resourses usages, an effective means to solve the contradiction among China's rural population, economic development, resources, and environment, under the existing conditions to balance the economic, ecological and social benefits. The development of green, low-carbon and circular agriculture is also inseparable from the development principle of ecological agriculture. It is also necessary to realize material circulation, multi-level utilization of energy and ecological harmonious cycle. The development model and goal of ecological agriculture are also important for the development of green, low-carbon and circular agriculture. Therefore, the basic principles and development ideas of ecological agriculture are the important theoretical basis for the development of green and low-carbon circular agriculture.

Section 3 The development and enlightenment of ecological agriculture at home and abroad

I. The main models of foreign ecological agriculture

From the middle and late 20th century, developed countries and districts such as Europe and the United States explored and practiced the ecological agriculture model in many parts of the world, and formed a good system, which should be learned from. The agricultural model with high consumption and high input should be replaced gradually, and the agricultural production model with harmonious coexistence with the ecological environment should be established.

1. Environment-friendly ecological agriculture development model

① Basic characteristics of the model: The development model of ecological agriculture, which takes environmental protection as the main goal, has been successfully reflected in the agricultural development model of the United States. Starting in the 1970s, the traditional modern petroleum agriculture characterized by high cost, being highly specialized, highly chemical intensiveness, high production, high pollution faced unprecedented challenges, in this case, the United States accordingly introduced such operation models as biological agriculture, renewable agriculture, green agriculture, in an attempt to solve the increasing contradiction between agricultural development and resources environment. But they all fell short of expectations. Therefore, low input and high efficiency ecological agriculture development model was introduced. The main goal of the ecological agriculture is to protect the agricultural ecological environment, put it into practice through laws and regulations, strive to reduce costs, reduce the use of chemical fertilizers, pesticides and other external chemical reagents as far as possible, based on the natural production characteristics of agriculture and the management of agricultural internal resources to protect and improve the ecological environment. The effect of this model is good, but the production efficiency is relatively low, mainly including the rational rotation mode, the integrated management mode of planting and animal husbandry, the integrated control mode of pests and diseases, and the management mode of using the organic fertilizer on the farm. Then they put forward the high efficiency ecological agriculture development model, paying attention to the scientific management of each link of agricultural production, strengthening the agricultural ecological principles, and the key is to rely on scientific and technological progress to promote the improvement of the production efficiency, to reduce the pollution of chemical reagents to the environment, to protect the ecological environment, and to create good material conditions. At present, the developed technologies used in agricultural production include: mechanization technology, chemical technology, biotechnology, information technology and so on. The government has also made great efforts in macro-control, policies and regulations to strengthen education management. This model is mainly aimed at some countries with more people and less land and scarce natural resources, in the face of many environmental problems in agricultural development.

② Main cases of this model: the typical successful development model is Japan. Since the 1980s, Japan has formally put forward the "maintenance and cultivation of green resources", emphasizing the rational utilization of resources and effective protection of the environment. The development path of environment-friendly agriculture was chosen. Under the guidance of national land planning and policy support, the measures of sustainable utilization of resources, environmental protection and productivity should be closely combined, and adhered to the ecological development. The main development direction is ecological agriculture and precision agriculture. This model requires reducing the input of farm external conditions such as fertilizers and pesticides to protect the environment, prevent damage to land resources, and improve farmland fertility. The use of biological breeding technology to cultivate crops that can be suitably planted in areas such as salinization and desertification lands, expand the area of cultivated land and improve the shortage of resources. They paid attention to the high efficiency of agricultural production and emphasized the combination between the proportion of agriculture, forestry, animal husbandry, fishery industry structure and the regional agricultural characteristics, thus improved efficiency. To evaluate and calculate the benefits of agricultural resources, they emphasized the role of forests in species diversity and air purification, and strengthened the protection of green resources. At the same time, the Japanese government attaches great importance to environmental protection and actively cooperated with the development of ecological agriculture in terms of legislation, policies and credit publicity, thus formed a good atmosphere for development. In technical measures and product quality and safety, a large number of policies and measures have been implemented, increasing the practical and feasible supervision from the management. And they formulated the "Ecological Agriculture Development Agricultural Law".

2. Comprehensive ecological agriculture development model

① Basic characteristics of the model. Comprehensive ecological agriculture development model refers to the management model which can meet the needs of human survival without destroying the natural agricultural environment. As the largest producer and user of chemical fertilizers and pesticides in the world, Germany has maintained high yield in agricultural production, while bringing serious damage to the natural environment, causing product quality problems and overproduction problems. In this case, Germany put forward the comprehensive ecological agriculture model, which mainly includes implementing comprehensive agriculture without destroying the natural environment and balancing with the development law of ecological system, comprehensive plant protection and water resources protection, comprehensive management of agriculture, etc.

② Main cases of this model. The model is represented by the Netherlands, which is a typical country with more people and less land. Therefore, the Netherlands attaches great importance to the utilization and development of its own resources and advantages to develop high-efficiency agriculture. In recent years, the Netherland government attaches great importance to the promotion and implementation of the agricultural strategy of ecological agriculture development, which combines with its own characteristics and advantages to improve the efficiency of agriculture. This is represented by the development of flower industry. Agriculture in the Netherlands is composed of planting industry and animal husbandry, among which planting industry is divided into field planting industry and horticulture industry. In terms of field planting, horticulture industry and animal husbandry, the main structural part of Dutch agriculture is animal husbandry, accounting for more than 55%, field planting industry accounting for 10%, and the rest is horticulture industry. The structure of Dutch agriculture shows obvious specialization pattern. Dutch agriculture is divided into three main production zones: horticultural production along the west coast, dairy production in the middle, and intensive livestock production in the east and south. Most farmers practice specialized production mode. This has greatly improved agricultural labor productivity and management level, meanwhile enhanced the market competitiveness of agricultural products, becoming a major exporter of agricultural products.

3. Intensive ecological agriculture development model

① Basic characteristics of the model. Relevant laws, regulations and policies are formulated to ensure the implementation of comprehensive agriculture, such as waste discharge regulations, fertilizer use regulations, environmental pollution punishment regulations, etc. At the same time, such measures are applied as the combination of agriculture and animal husbandry, planting rotation, renewable resources development and other reward and punishment systems. At the same time, German puts emphasis on training farmers and young agricultural workers. There are relevant laws that to establish or inherit agricultural enterprises he must receive training, and obtain the corresponding qualifications. There are strict regulations on the education of farmers, and a new program for agricultural education has been developed, emphasizing the comprehensive renewal of farmers' professional knowledge and skills. The government has also set up a number of part-time universities in rural areas to improve the quality of farmers.

② The main case of this model. France mainly adopts the intensive and modern agricultural development model, but it has a great impact on the environment and the negative effect is increasingly serious. Therefore, France adopted environmental protection system and farmland fallow system. In Israel, much of the land belongs to the arid or semi-arid areas, so they vigorously develop water-saving agriculture, ecological agriculture development to change the original mode of extensive cultivation, adopt the intensive farming methods and the efficient irrigation technology, depending on modern science and technology to realize self-sufficiency. Meanwhile, national policy gives strong support to promote the development of the company and farmers cooperation mode. South Korea has adopted a pro-environment development model, with huge government incentives. India has also developed a low-cost, energy-efficient agricultural development model and promulgated relevant laws and regulations to implement the development of ecological agriculture. Mexico also attaches great importance to the application of agricultural science and technology to the comprehensive development of rural areas.

Comprehensively speaking, in the selection of ecological agriculture model, developed countries, on the basis of modern production system, pay attention to comprehensive plan from its own national conditions, the ecological element and energy, while developing countries focus on growth and development of agriculture, but in order to avoid the disadvantages of the western way, we should pay attention to the ecological environment when focusing on the development.

II. The main models of domestic ecological agriculture

1. Contractual agricultural development model

Since 2000, Chongqing's ecological agriculture developed step by step. Chongqing Academy of Agricultural Sciences and other research units in view of the problems of ornamental vegetables, adopted "test + demonstration + popularization" scheme, which cultivated ornamental vegetables more than 80 kinds, and they include four kinds as the potted plants, garden-cultivation plants, entertainment plants and apreciation plants, etc. , the maintenance instructions of them are put forward to improve the adaptability. In the process of popularization, Chongqing optimized the structure of ecological agriculture and improved the production and cultivation technology of vegetables, which greatly increased the income of farmers and resulted in obvious social and economic benefits. In Chongqing in 2008, through the government support, they introduced 76 kinds of ornamental fruits and vegetables , built new fruits and vegetables nursery gardens of 0.33 hectares, and built test and seedling breeding base of nearly 1.7 hectares. To make good use of the information of the ornamental vegetables and to form a consulting system, they set up the information network platform of chongqing ornamental fruit and vegetable, perfecting the information database. Through careful screening again, Chongqing introduced 18 kinds of ornamental vegetables suitable for local cultivation, and built 2 large-scale demonstration bases, which created huge economic, social and ecological benefits. The ornamental vegetables Chongqing has developed are mainly leafy, fruity and flowery vegetables. Leafy vegetables are ornamental plants that have tastefully shaped and brightly colored leaves, such as purple cabbage, mint, asparagus, parsley, caraway, kale, colorful spinach, red beets, silver cabbage, cayenne pepper, colored leaf lettuce, chicory, etc. Fruity vegetables refer to the size, shape and color of the fruits are unique and novel, attracting customers. For example, in color, there are ornamental eggplant, colorful pepper, bright red pepper, etc. , and in shape, there are saucer-shaped pumpkin, multi-horn loofah, pumpkin and bitter gourd. The main types of flowery vegetables include bulbs-eating lily, bud-eating daylilies and caragana plants, safflower, and they introduced from abroad colorful cauliflower and emerald-tower cauliflower. Nowadays, Chongqing ornamental vegetables are mostly used for urban home life, urban landscaping, rural landscaping, festival fashion gifts, modern potted scenery and so on. When Chongqing introduced varieties from abroad, it continuously improved varieties, hybriding the foreign vegetables with local ones to cultivate their own brands. Through a wide range of collecting wild resources and high quality germplasm resources, after a long trial and error, domestication, monitoring to the resistance and adaptability, ornamental evaluation, they select the best quality germplasm resources, using modern advanced science and technology. The essence is to transplant the exogenous genes of plants with excellent germplasm into the original ones to optimize their comprehensive adaptability. The main measures to develop eco-agriculture in Chongqing are as follows: First, to strengthen the popularizaiton, improve the willingness of farmers to participate, and expand the scale of industry; Second, by optimizing the product structure, improve the product cultivation technology, improve farmers' income; Third, the government provides support in various aspects to promote the development of ecological agriculture, such as introducing new varieties, building information platforms and establishing demonstration bases.

2. Brand-driving agricultural development model

It is to implement agricultural brand strategy and build a safe brand. In order to further improve the industrial benefits of Weixian radish (former name of Weifang city), it is necessary to continue to develop storage and fresh-keeping technology, improve the level of deep processing, try to prolong the supply period of radish, and give full play to the role of storing when harvesting and replenishing when lacking. In addition, to gradually attach great importance to the post-processing work, we should do well the radish packaging, which can be guided correctly by the government to let the various processing enterprises introduce new technology at home and abroad, update the new equipment to strengthen the processing innovation of weifang radish. More kinds of radish series related with health and beauty should be developed, such as dried radish, preserved radish, radish pickles and so on, to improve the added value of radish products as far as possible, to find out more development space to create economic benefits. In Hanting district, they established the "110-type" scientific and technological information service network platform about agricultural products, in order to obtain more production and marketing information. First of all, they should further consolidate the information network technology, and give full play to the function of the marketing association, by setting up information center, contacting with large vegetables wholesale market in the country, doing well about the information collection, analysis, release and so on, thus in the shortest possible time, the actual situation of the real vegetables at home and abroad market will be reflected . The government should encourage each industry association to establish and improve information communication equipment, develop the production and marketing website of Weixian radish, realize sales through "online order" , broaden sales channels, establish distribution service center, effectively shorten the middle link of radish sales, and improve the economic benefits of Weixian radish. Traditionally Weixian radish was mainly planted and sold by the farmers on their own, mostly in the form of retail, which greatly limited the scale. With the increasing demand of Weixian radish by the large wholesale business and retail business, small scattered planting is very difficult to to meet the market demand. As a result, Hanting government carries on the appropriate guidance, according to the masses voluntary principle, and they established the independent organization of the radish planting, processing and sales organization. In the planting-concentrated area, the Weixian Radish Association was established to better organize radish planting, carry out technical training and education for farmers, do a good job of quality supervision for each farmer's planting, strengthen product upgrading and packaging, and timely contact buyers for sales. This solves all kinds of technical difficulties encountered by farmers in the production process, and improves the economic benefits of farmers. The scientific research units in Hanting also cooperated with the Radish Association to devote themselves to the development and cultivation of ecological radishes, which further improved the quality of radish produces. At the same time, more than 30 radish sales outlets have been established in Hanting urban area to meet the demand of the local market. In order to meet the demand of the foreign market, sales supermarkets have been set up in Jinan, Qingdao and Beijing, respectively, which greatly promote the continuous production. Brand is able to help the product quickly occupy the market and steadily expand the market share. In recent years, Chinese agricultural products market is increasing competitive. Only by taking the brand route, giving full play to the advantages of Weixian radish's famous specialty, and establishing and improving the industrial management model as far as possible, can we obtain greater benefits. Therefore, Weixian radish should further strengthen the publicity, give full play to its brand advantages. Since Weixian radish still has some problems, such as obvious differences in product quality, Hanting government emphasizes to have its own brand route, encourages trademark registration, and makes strict testing regulations on the certification of green radishes and the quality of processing packing cases. For the brand radish that has completed trademark registration, strict testing standards are put forward on the production base, management facilities or other packaging quality, not only to ensure the high quality of radish, but also to give play to its advantage in price. From a long-term perspective, Hanting district should lead farmers to further develop the radish sales market, establish a brand image, give full play to the effect of famous and excellent specialties, increase the scale of operation, and improve the competitiveness of Weixian radish.

III. The enlightenment from the develpment of ecological agriculture at home and abroad

1. The government provides policy, legal and financial support

As an important subject of ecological agriculture, the government plays a key role in promoting the development of eco-agriculture, and its functions are mainly manifested in regulation and management functions. First, it formulates and implements relevant policies, laws and regulations, focusing on agricultural tourism experience, planning, ecology, safety, etc. , such as Singapore's agricultural science park planning, (note: In this book, "park" means modern ecological circular agricultural park) and the Citizen Agricultural Park Law issued by Germany. The second is to provide financial support for the planning and construction of ecological agricultural areas. Compared with the policy delay effect, the financial support is an immediate effect. A large amount of capital input alleviates the pressure of ecological agricultural operators to a large extent and provides impetus for the sustainable development of the industry. Financial support is manifested in many aspects. Germany, Netherlands and other European countries provide a large amount of financial subsidies for tourism agriculture every year, and also take out special funds for agricultural scientific research to indirectly promote the development of agriculture. The Singapore government is directly involved in the construction of the agricultural science park; Chongqing government provides support to its ecological agriculture from many aspects.

2. Non-governmental organizations provide coordination

In the ecological agriculture of developed countries, although the government plays a leading role, at the same time, industry associations and social groups also play a role that can not be ignored. The main function of NGOs is to supervise and provide services to the third party, and realize the industry self-discipline and standardized management of ecological agriculture through supervision. The industry association is worth mentioning; Many countries with mature ecological agriculture have set up different forms of industry associations, and the non-governmental organizations achieved a lot in the protection of ecological agriculture in the process of long-term development. It is the responsibility of the industry association to formulate industry norms and manage the enterprises' behavior, according to the characteristic of local ecological agriculture. Due to the fact that its starting point is from inside the industry, the trade association can effectively promote the harmonious development of ecological agriculture. Germany and the Netherlands have set up corresponding trade associations, and other European countries, such as France and Italy, also have non-governmental organizations such as the French Network of Farmers and the National Association of Agriculture and Tourism. Weifang city has made great efforts in promoting ecological agriculture association. By referring to the forms and basic contents of agricultural associations at home and abroad, several ecological agriculture association organizations have been established, which provides opportunities for further internationalization. These organizations play the service function of the association and play a great role in maintaining the healthy and orderly development of ecological agriculture industrialization in Weifang city.

3. Paying attention to the characteristics of sightseeing agricultural products

Tourism agriculture is a product with distinct regional characteristics, but it is also a leisure and entertainment product in tourism market, which has a greater substitution. The ornamental vegetable industry adheres to the concept of sustainable development and must highlight the green ecological characteristics of the natural environment. At present, the popularity of ornamental vegetables in most areas is prominent, and most products are almost the same, which do not reflect the local characteristics. Therefore, it requires in-depth exploration of tourism products and in-depth investigation and analysis of tourists' psychological needs. In addition, on the basis of segmented consumption markets, many countries develop various types and grades of ornamental vegetable products. The industrial level covers outskirt and suburbs, and the prices are from high to low with obvious hierarchy. For example, in the outskirts of the city, the park-type popular culture ornamental vegetable demonstration is adopted. The suburbs adopt outstanding rural tourism products with high quality, distinctive characteristics. "Localism" refers to the rural culture which is different from the urban culture. The stronger the "localism" is, the more it can attract tourists with great differences in living environment, and the greater the difference, the stronger the attraction to tourists. Many countries in the world integrate local cultural characteristics into the development of tourism agriculture and design unique cultural tourism themes. Such examples are numerous. For example, Germany's tourism agriculture attaches great importance to natural and cultural characteristics and has distinct characteristics. The Dutch fruit and vegetable float show is very characteristic of the country. These sightseeing agricultural projects developed by relying on domestic resources have endowed the activities with rich rural cultural connotations and greatly improved the taste of ornamental vegetable products. From this point of view, tourism agriculture is not only a means to promote the development of local economy, but also a platform to promote the local culture. From the perspective of traditional culture inheritance, it is the inheritance and protection of national culture and folk culture. Taking regional culture as tourism products has a strong attraction for overseas tourists, and operators can get considerable income every year through cultural activity planning.

4. Strengthening the promotion and publicity of sightseeing agriculture

Tourism agriculture is different from the single first industry and third industry, but a combination of rural culture and natural landscape experience activities. This needs the rural community residents to have a common consciousness, needs them to create a social atmosphere conducive to the development of tourism agriculture , so it needs more promotion and publicity. The community members should be encouraged to actively take part in, making the sightseeing agriculture as the leading economic power, to develop agricultural cooperatives for the development of tourism and to strengthen the communication of community residents and tourists. Otherwise, even with the local culture and local characteristics, the local sightseeing agriculture would not be well developed, sightseeing agriculture ecosystem and the desired effect of experience would not be achieved. The low participation and enthusiasm of community residents will also lead to the inability to expand the scale of industry and the formation of large-scale effect. The main approach of many countries is to organize the local farmers to form a sightseeing agricultural cooperatives or associations, which not only can improve industry development, but also can provide the operator with a range of services, to enhance the enthusiasm of operators to participate in, ensure that community residents in participation of industry building, gradually they would form industrial cluster, deepen the daily communication between the operators, rural residents and tourists. The participation and recognition of rural community residents to sightseeing agriculture is conducive to promoting urban-rural integration, driving rural social and economic development, reducing friction and conflict between each other, integrating local agricultural tourism resources, and creating a civilized and harmonious sightseeing agriculture community environment. In addition, in order to increase the participation rate and promote the recognition of sightseeing agriculture, some countries reduce the control of agricultural land. For example, Chongqing city strengthens the promotion of ecological agriculture, so that farmers can understand its benefits, enhance their participation enthusiasm, and promote the rapid development of ecological agriculture.

Chapter 2 Theory of Ecological Circular Agriculture

Section 1 Circular agriculture

I. The origin and development of circular agriculture

With the birth and development of industrial civilization, especially with the continuous innovation of modern scientific and technological means, human activities have an increasing impact on the biosphere, which is related to the stability and prosperity of the whole biosphere. Due to people's irrational behavior under the lead of unscientific outlook on development, such as deforestation, overgrazing, pollution and population expansion, social conflict, etc., they caused direct or indirect threat on the whole biosphere including humanity, thus force people to reflect on their own problems and the protection problems for scarce resources and environment, seeking to coordinate the relationship between people and the biosphere, then the idea of sustainable development is raised. Starting from the concept of sustainable development, all the countries around the world have put forward a series of development models and strategies to transform traditional economy, trying to coordinate the relationship between economy and environment through these strategic plans and concrete measures, so as to maximize the overall benefit. Circular economy is one of the many strategic ideas. As a new economic development model with the concept of sustainable development, circular economy is an inevitable strategic choice for the combination of economic development and environmental protection in the 21st century.

The idea of circular economy originated in the 1960s from the "spaceship theory" put forward by Boulding an American economist. Boulding criticizes the "open loop" paradigm of "resources-product-emission" in the traditional industrial economies. At about the same time, in 1962, American biologist Rachel Carson published a book called *Silent Spring*, in which she accused chemical pesticides such as "insecticides" of destroying food chains and biological chains. In 1972 the Club of Rome advocated "zero growth" in its "Limits

Chapter 2 Theory of Ecological Circular Agriculture

to Growth" report. In 1992, the United Nations World Leaders Conference on Environment and Development issued the Rio Declaration and Agenda 21, and the concept of sustainable development gained popular support. In 2002, the World Conference on Environment and Development decided to promote cleaner production worldwide and formulate an action plan. Under the above background, the idea of circular economy arises at the historic moment and circular economy develops from it.

With the continuous development of industrialization and urbanization, traditional agricultural economic growth is facing more and more prominent problems of resources constraints particularly land and water resources. At the same time, such problems arose as the worsening agricultural ecological environment, the increasing agricultural waste, the expanding non-point source pollution, the reducing safety in agricultural production and farm produce. This model of agricultural production turns resources into products, and meanwhile caused wastes and pollution, which has affected agricultural ecological security, people's life and health safety and the farmers' income level. Therefore, in order to get rid of the above predicament, under the background of the emergence of circular economy, circular agriculture emerged at the historic moment.

Circular agriculture is one of the continuing models of sustainable development, it is the result of applying the idea of circular economy to agricultural system, and it is also the new trend of agricultural development at home and abroad. In the world, many countries have carried out a lot of fruitful practices on circular agriculture. However, due to differences in national conditions, natural resources, geographical conditions, climate and other aspects, they have adopted different agricultural recycling methods for production, forming some circular agriculture with their own characteristics, such as the material-reused recycling agriculture model of Aidong District in Japan, the "green energy agriculture" of Germany, the precision agriculture of the United States and the water-saving agriculture of Israel with the reduction model, and the "rotation ecological agriculture model" of Sweden. The Chinese government attaches great importance to the development of circular agriculture. In December 2007, the Ministry of Agriculture has designated 10 cities (autonomous prefectures) as circular agriculture demonstration cities: Handan city in Hebei province, Jincheng city in Shanxi province, Fuxin city in Liaoning province, Ji'an city in Jiangxi province, Zibo city in Shandong province, Luoyang city in Henan province, Enshi autonomous prefecture in Hubei province, Changde city in Hunan province, Guilin city in Guangxi autonomous region and Tianshui city in Gansu province. Handan city, Changde city, Fuxin city, Zibo city, Ji'an city have compiled circular agriculture development plans. Among them, Handan city has formulated the "34567" circular agriculture development goal, Jincheng vigorously to carry out the circular agriculture construction with the rural biogas, straw gasification as the core. Fuxin city from 2005 formulated "Fuxin City Circular Economy Implementation Plan" and "Fuxin City Circular Economy Plan", and in October 2007 issued "Fuxin City's Agricultural Circular Economy Management Measures". Some circular agriculture models have been formed, such as the "grain-pig-biogas-fertilizer" ecological model in Henan and other places. The planting industry provides feed for the breeding industry and the breeding industry provides fertilizer for the planting industry, making full use of the circulation of materials and energy in the ecosystem, greatly improving the utilization rate of materials and reducing environmental pollution. Thus using the growth characteristics of various organisms in the natural ecosystem to form a three-dimensional complex ecological model of biological ecological groups, and so on.

In a word, circular agriculture is developing in full swing all over the world. It will become one of the important models of modern agriculture and an important part of sustainable development, with broad prospects and far-reaching significance.

II. The characteristics of circular agriculture

Circular agriculture system should have two characteristics as follows.

1) It has the three obvious characteristics of general circular economy. ① Reduction, that is to reduce the quantity as far as possible in the agricultural production and consumption process, to save the agricultural resources, to reduce the emission of pollutants, and to protect the ecological environment; ② Reuse, this is to try to improve the utilization efficiency of agricultural and sideline products and resources, to reduce production pollution or achieve zero emissions; ③ Recycling agricultural products can be turned into renewable resources after the completion of the use function, for reuse.

2) It has its own characteristics of agricultural production. That is, agro-ecological food chain cycle, each subject of the material and energy cycle in the agricultural production is mutually complemented, symbiotically coexisted and strongly co-benefited; Cleaner production, that is to specially emphasize product safety, use high and new technology to control and minimize the application of chemical fertilizers and pesticides, or the application of organic fertilizer and biological control; Protection of land, water resources and agricultural ecological environment is to make sustainable use of agricultural resources without pollution, and promote a virtuous circle of agricultural ecological environment; Green consumption, it means that people not only meet the needs of life on green products, but also do not waste resources and do not pollute the environment, when the agricultural main & by-products were consumed, the waste were recycled, thus the wastes turned into treasure, returning to nature.

III. The ideas and strategies of the development of circular agriculture

1. The development ideas

Under the guidance of the theory of recycle economy, according to the material energy cycle and ecological food chain principle, we should give full play to the energy conversion, material circulation, added value, and information transfer function of the agricultural ecological economic system, promote the transformation of energy and material in the food chain cycle, and effectively utilize resources of various organic wastes (crop straw, livestock and poultry dung, processing residues, etc.) , to achieve the best production, maximum benefit, the most moderate consumption and the least waste. It is necessary to make full use of biological resources such as plants, animals and microorganisms effectively, and build a recycling system of "plant production, animal transformation and microbial restoration" , so as to realize the harmonious development between man and nature. In the process of agricultural production, we should grasp the "four modernizations" . ① Cleanization of agricultural production includes clean input (clean raw materials, clean energy) , clean output (clean agricultural products that do not harm human health and the ecological environment) and clean production (the use of non‑toxic and harmless fertilizers, pesticides, etc.) . ② Gradient utilization of resources within the industry, that is to reasonably arrange the production mode within the industry, optimize the production space structure, as far as possible to reduce the waste of water, fertilizer, soil, medicine and other resources, and to improve the efficiency of resource utilization. ③ The utilization and recycling of wastes among industries, that is to rationally arrange the time and space structure of agricultural industries, to establishment a mutually beneficial relationship between related industries, to reduce production costs, to improve economic benefits, and to improve the ecological environment. ④ Rational consumption of agricultural products, that is to guide consumers to choose their own consumption level objectively, product categories and brands, to provide accurate demand information to agricultural production operator, to direct product range adjustment, organization and management mode and technology improvement, to promote agricultural resources allocation efficiency, and to avoid structural, foam waste of resources.

2. The development concept

To develop circular agriculture, we should set up three development concept .

① The organic unified concept of economic benefits, social benefits and ecological benefits. To develop circular agriculture, we should change the concepts of emphasizing production over environment, emphasizing economy over ecology, and emphasizing quantity over quality. We should not only pay attention to supply in quantity, but also ensure safety in quality. We should not only pay attention to the improvement of production efficiency, but also pay attention to the construction of social benefits and ecological environment, so as to realize the organic unity of economic benefits, social benefits and ecological benefits.

② The concept of multi-level recycling of resources. Traditional agricultural production activities are generally manifested as a singlelinear growth mode of "resource-product-waste". The more the output is, the more resources will be consumed, the more waste will be discharged, and the more serious the ecological damage and environmental pollution will be. Circular agriculture makes the chain extension as the main line, promotes single-process agricultural growth pattern to the comprehensively recyling model of "resources-product-renewable resources". At the same time, circular agriculture through the optimization of the design and management of the agricultural ecological economic system, realizes the efficient use of the natural resources such as heat and other renewable resources in the agriculture system, minimizes pollutant emissions, conserves energy and reduces emissionson, and promotes low-carbon agriculture.

③ The concept of relying on high and new agricultural technology. Development of circular agriculture must rely on scientific and technological innovation, strengthen the science and technology research on the circular agricultural, promote the agricultural new and high technology of resources circular utilization and ecological environment protection, improve the technology content of agriculture, and realize the tranformation from the single agricultural technology system focusing on production growth to the circular agricultural technology system of resources recycling and high energy efficiency.

3. The development strategy

The development strategy of circular agriculture mainly includes the following aspects.

1) Optimizing the agricultural cycle structure. To optimize the agricultural cycle structure the following work should be done: ① Optimization of industrial chain cycle structure. It is mainly to promote clean production and pollution prevention and control in the whole process of agricultural products from where they are produced to the table, so as to achieve source control, intermediate treatment and terminal circulation, so as to minimize pollutant discharge. ② Optimization of internal circulation structure of agricultural industry. It is mainly the exchange of matter and energy within the agricultural industry, and minimization of waste emission. For example, three-dimensional cultivation, three-dimensional breeding and ecological processing of agricultural products are typical circular agriculture development models. ③ Optimization of external circulation structure of agricultural industry. It is mainly the mutual exchange of wastes between agricultural industries, so that the waste can be used as resources. For example, fish farming in rice field with the combination of planting and

breeding provides a good growing environment for fish, which can eat weeds and pests. And the fish wastes will fertilize the field, thus reduce the amount of fertilizer and pesticide used in rice, protect the ecological environment, increase economic benefits, and realize the closed cycle of material flow and energy flow.

2) Doing well in the industrial chain of circular agriculture. The industrial chain of circular agriculture is an organic integrity formed by ecological planting industry, ecological animal husbandry, ecological forestry, ecological fishery and ecological processing industry through waste exchange, recycling, element coupling and other ways, which are interdependent and synergistic. The relationship between the various industrial sectors is mutually dependent and mutually restricted in quality, and it is an organism composed in a certain proportion in quantity. For example, ecological industry chain of sugar should be established on the framework of sugarcane field planting system, sugar processing, alcohol brewing system, the paper industry system, cogeneration system, and comprehensive environmental treatment system, through such steps as re-vitalization, optimization, upgrading and expansion construction. The product output for the system and its resources would get optimal allocation, and the wastes would be efficiently used. These systems are connected by the exchange of intermediate products and wastes, thus forming a relatively complete and closed ecological industry network.

3) Building a circular agricultural technology system. The development of circular agriculture should be supported by high and new technologies, and the economic benefits, social benefits and ecological benefits should be organically integrated, so that material and energy can be fully utilized, and an agricultural technology system suitable for the development of circular agriculture should be established. Circular agriculture technology includes clean production technology, low carbon technology, ecological technology and so on.

Section 2 The theoretical basis of circular agriculture

Circular agriculture refers to applying the idea of sustainable development and circular economy theory and ecological engineering methods, combining with ecology, ecological economics, ecological technology principle and its basic rules, basing on the protection of the agricultural ecological environment and making full use of high and new technology, adjusting and optimizing the internal structure and the industrial structure of agricultural ecosystem, improving agricultural ecosystem multistage circulation of matter and energy use, strictly controling the external input of harmful substances and the production of agricultural waste to reduce environmental pollution as far as possible, taking the agricultural production into real economic activity in the agricultural ecosystem circulation, establishing dynamic equilibrium mechanism between agricultural economic growth and the ecosystem environment quality improvement, and realizing the virtuous circle of ecology and the sustainable development of agriculture.

I. The theory of circular economy

Circular Economy is short for Closing Materials Circular Economy. It is generally believed that the "Spaceship Theory" proposed by the American economist K. E. Boulding in 1962 is the rudiment of circular economy thought. In 1990, British environmental economists Perth and Turner formally used the term "circular economy" for the first time in their book "Natural Resources and Environmental Economics", after which the development model of circular economy has been widely valued by the international society. To summarize, circular economy is based on the efficient utilization and recycling of resources as the core, based on the principle of "reducing, reusing and recycling" (3R principle for short). The Reducing Principle requires to reduce the quantity of materials entering the production and consumption process. This principle is conducive to avoiding the traditional development mode of pollution first and treatment later. The purpose of the Reusing Principle is to prolong the time intensity of products and services and reduce the generation of waste in production and consumption. This principle can prevent items from becoming wastes earlier. The Recycling Principle calls for an item to be turned back into a usable resource after its use.

The core concept of circular economy is to transform the traditional "one-way single-loop" linear economy of "resources-product-pollution emission" into a comprehensive feedback economy and circular economy model combined "muiti-way multi-loop" and "multi-loop recycling" of "resources-product-renewable resources-product-renewable resources" The traditional extensive economic growth model with high consumption, high pollution, high input and low efficiency will be transformed into an intensive economic growth model with low consumption, low emission and high efficiency. At the macro level, circular economy requires the adjustment of industrial structure and layout, the concept of circular economy runs through all fields and links of social and economic development, and it is to establish and improve the whole society's resource recycling system. At the micro level, it is required to save energy and reduce consumption, improve the efficiency of resource utilization, realize reduction, and recycle the waste generated in the production process. At the same time, according to the resource conditions and industrial layout, the production chain should be extended and expanded to promote the symbiotic coupling between industries.

II. The principle of subtance circulation and energy flow in ecosystems

Substance and energy are the basic power of all life movement, energy flow is the power of substance flow, substance flow is the carrier of energy flow. In order to survive and develop, living organisms and ecosystems not only continuously input energy, but also continuously complete the substance cycle. Energy and substance entering the ecosystem are not static, but are continuously absorbed, fixed, transformed and recycled, forming an energy flow chain among all components of the ecosystem, that is "environment-producer-consumer-decomposer", maintaining the life of the whole ecosystem. The natural system relies on the food chain and food web to realize substance flow and energy flow and maintain the stability of the ecosystem. Agro-ecosystem relies on artificial inputs and assistant power to maintain normal production functions and system operation. In the ecological system, energy flow is unidirectional and gradually decayed during the process of transformation. The amount of effective energy decreases step by step and tends to be converted into inefficient heat energy. The sunlight fixed by plants is gradually consumed along the food chain and finally leaves the ecological system. Some of the stored energy in the ecosystem can also form a reverse feedback energy flow, but the energy can only be used once, and the so-called reuse refers to the unused part. Substance flow is not a one-way flow, but a cycle, the process of material from simple inorganic state to the complex organic state then to simple inorganic state regeneration process, meanwhile it is the process of biological fixation, transformation and dispersion, during which can not be used only once, but can be reused. The substance in the process of flow is changed in form rather than disappear, It can circulate in the system forever and never become waste.

The existence and development of any ecosystem is the result of the simultaneous action of energy flow and substance flow. If either of them is blocked, the continuation and existence of the ecosystem will be endangered. Many substances involved in the cycle of the ecosystem, especially the indispensable nutrients for the growth of some organisms, are not only the substance basis for sustaining life activities, but also the carrier of energy. Organic substances are synthesized with solar energy as the power and transferred step by step along the food chain. In each transfer process, there is loss of substances and dissipation of energy, but the lost substances will return to the environment and eventually decompose into simple inorganic materials, which will be absorbed and utilized by plants, while the dissipation energy will not be reused. But relative to the ecosystem, because the sunlight energy is the main energy, being infinite, but the substance is limited, the distribution is very uneven. Therefore, if the agroecosystem is properly regulated, the substances can be renewed in the system, and the energy efficiency can be continuously improved.

III. The principl of niche and biological complementarity

Niche refers to the characteristics of comprehensive adaptation of an organism to the environment when it completes its normal life cycle. It is the function and position of an organism in the species and ecosystem. Niche is closely related to the utilization of resources and interspecific competition in the community of organisms. The theory of ecological niche shows that there are no two species with exactly the same niche in the same habitat, and different or similar species must carry out some kind of spatial, temporal, nutritional or age niche differentiation and separation, so that direct competition can be reduced and species tend to complement each other. Communities composed of multiple species can use environmental resources more efficiently, maintain higher productivity, and have higher stability than communities composed of single species. In agricultural production, human beings rationally arrange the species composition of agricultural organisms from various aspects such as distribution, morphology, behavior, age, nutrition, time and space, so as to obtain high niche efficiency and fully improve the resource utilization rate and agroecosystem productivity. With the deepening of the concept of ecology, the concept of ecological niche has been transferred from the pure natural ecosystem to the complex ecosystem of social, economic and nature, and the concept of ecological niche has been further expanded. It is not limited to the simple planting system or breeding system, but even extended to the whole agricultural economic system.

In the process of long-term evolution, a variety of biological populations in the ecosystem have formed unique adaptability to the environmental conditions. There are interdependent and mutually restrictive relationships between species, and this relationship is extremely complex. For one thing, we can use the biological variety and the interpromoting relationship in the ecological system, to form a reasonable and efficient compound ecosystem, where we can, in limited space and time, accommodate more species, produce more products, make full use of resources and maintain the stability of the system, such as the widely-used methods of three-dimensional planting, mixed breeding, crop rotation, and the use of bees and insect-borne to pollinate crops, etc. For another, the interrestricition relationship between various biological populations can be used to effectively control diseases, insects and weeds. At present, the emerging biological control of diseases, pests and weeds, as well as biological pesticides, fungicides, biological herbicides and other biological pesticide technologies have shown broad development prospects.

IV. The principle of systematic project and integral effect

According to the principle of system theory and system project, any system is composed of several closely related sub-systems. By optimizing the structure of the whole system and using the interaction and feedback mechanism of each group, the overall function of the system can be greater than the sum of the functions of each sub-system. Agricultural ecosystem is a complex network composed of biological and environmental systems, made up of many different levels of system, and the levels of the system are closely connected. The link is realized through the substance circulation, energy conversion and value transfer and information transmission. Reasonable structure can improve the efficiency of overall system function and can increase the productivity and stability of entire agroecosystems. The famous ecologist Ma Shijun once highly summarized the basic principles of ecology into eight words: "integration, coordination, circulation and regeneration", among which "integration and coordination" points out the rational and coordinated horizontal relations of the ecosystem, while "circulation and regeneration" contains the characteristics of sustainable operation of the ecosystem.

The principle of overall effect of agroecosystem is to give full consideration to the interaction relationship inside and outside the system, the overall operation law and the overall effect of the system, to use the system project method, comprehensive planning, rational organization of agricultural production, and to maximize the overall function through ecological optimization design and regulation of the system. We should realize the harmonious coexistence of ecosystem of species; realize the coordination and adaptation between species and environment; realize the coordinated development of the ecosystem structure and function; realize the coordination of different ecological process; build a harmonious circulation mechanism; make the system productivity and resources environment continue to be value-added and updating; meet the long-term needs of human society, and achieve the virtuous cycle of ecological and economic system.

V. The principle of agricultural location and regional differentiation

In the early 19th century, T. H. Von Thiine, a German economist, explored the phenomenon of agricultural zonation caused by different land prices according to the relationship between capitalist agriculture and market, and established the agricultural location theory, which is a theory that quantitatively studies natural and social phenomena from the spatial or regional aspects. Economists have further developed this theory, expanding it from natural location to economic location, market location and ecological economic location. Combined with comparative advantage theory, this theory has effectively promoted the development of regional, large-scale and specialized agricultural production. Agricultural production is obviously restricted by natural factors, and agricultural development must be adapted to local conditions, and give full play to regional advantages, economic advantages, market advantages and scientific and technological advantages. Comparative advantages are the basic principle of regional division, and also an important theoretical basis for agricultural structural adjustment.

Due to the similarities and differences of geography, terrain, climate, land, social economy, humanities and other elements, the phenomenon of convergence and separation exists in each region. And agricultural production is a unity of natural and artificial environment with all kinds of agricultural organisms, and its regional differentiation features are remarkable. The law of regional differentiation of agriculture includes the differences of natural geography, human geography and biogeography, which result in the differences of agricultural production and ecological economy types. Although with the continuous development of social economy, agricultural development rules are the same, that is from traditional, self-sufficient, extensive agriculture to modern, commercial, intensive one, the regional nature of agriculture, diversity will still exist for a long time. Our country is vast in territory, natural and social economic conditions are extraordinarily complex, so to develop circular agriculture we must make the species and varieties adapt to the local conditions, and make the structure reasonable and coordinate with each other. According to the regional environment, the pattern of distinctive circular agriculture should be constructed. We should consider the reasonable use of all resources including human resources, land resources, biological resources and other natural resources, etc. In accordance with the natural ecological rule and economic rule, we would carry on the comprehensive planning and overall consideration, adjust measures to local conditions, and constantly optimize its structure. And we can fully improve the sun and water utilization and realize the virtuous circulation within the system to make economic benefit, ecological benefit and social benefit increase synchronously.

VI. The principle of sustainable agricultural development

The essence of sustainable development is to ensure that the development of contemporary people should not endanger the development ability and opportunity of future generations, to realize the optimal efficiency and fair allocation of resources, and to realize the harmony and co-evolution between man and nature. Since the emergence of sustainable agriculture in the 1980s, countries around the world have continuously deepened their exploration in theory and practice. Although they have different understandings and practices, the overall development goal is the same, which is to ensure the sustainability of agricultural resources and environment, sustainability of the economy, and the sustainability of society. Resources and environment sustainability mainly refers to the rational use of resources and their sustainable use, while preventing environmental degradation, especially to ensure the sustainable use of agricultural non-renewable resources, including fertilizers, pesticides, machinery, hydropower and other resources. Economic sustainability refers to making the economic benefits of agricultural production and its products maintain good and stable in the market competitiveness, which directly affect the production to maintain and develop, especially under the condition of market economy. Whether a kind of production pattern or a technical measures can be promoted and last mainly depends on its economic benefits, the product's competitiveness in domestic and foreign markets. Economic feasibility is the key factor that determines its sustainability. Social sustainability refers to the coordination of agricultural production with the national economy overall development, and agricultural products can meet the needs of the people's living standards. We should not only ensure product supply adequate, keep the agricultural market prosperous and stable, especially the effective supply of grain and meat and eggs products, but also ensure the product quality and reasonable price, which can meet different levels of consumption demand for high quality agricultural products and meet the needs of overall social and economic development. Social sustainability directly affects social stability and the overall situation of people living and working in peace and contentment.

The three objectives of agricultural sustainability are complementary and inseparable. It is one-sided and detached from reality to neglect any aspect and regard sustainability only as the sustainability of ecological environment. On the basis of rational utilization of resources and protection of the ecological environment, efforts should be made to increase output, meeting the growing material needs of human beings. Meanwhile, it is to promote the development of rural economy, improve farmers' income and social civilization.

Section 3 Overview of the development of ecological circular agriculture

I. The concept of ecological circular agriculture

Ecological circular agriculture is a new type of agriculture that can achieve good economic, ecological and social benefits by following the relevant principles of ecology and economic cycle theory, using modern science and technology and modern management tools, and combining with the indispensable, effective and remarkable comprehensive experience in traditional agriculture. It is not just limited to a certain mode of production, output and economic benefit, but on the basis of ecology, it is a renewable development model of agricultural production by changing the production mode, development mode and efficiency mode. At the same time, the economic benefit, social benefit and ecological benefit are highly unified turning the concept of "beautiful environment, safe products, scientific production and sustainable development" into reality.

Since the 1980s, ecological agriculture has taken shape in some countries with more developed agriculture in the world. Since its emergence, this new agricultural development model has received widespread attention and developed rapidly. In order to accelerate the ecological agriculture development speed and construction speed, we should pay attention to the summary and promotion experience and practice of ecological agriculture, such as reasonable crop rotation, organic fertilizer (green manure) instead of fertilizer and the utilization efficiency, which are the measures familiar to basic production personnel, such as farmers and are willing to be accepted. In addition, we should step up the research and promotion of advanced new technologies in ecological agriculture, such as the use of new photolysis film production, biological drugs, organic fertilizers, rational utilization of straw, water and fertilizer saving irrigation, etc. Only through the effective combination of these two aspects can the development effect of ecological agriculture become better.

Since the invention and use of pesticides, chemical fertilizers, all living things in the agricultural production are more or less threatened, and the biosphere suffered all kinds of harmful chemicals constantly. So did the human being, and we often hear someone was poisoned by consumption of contaminated crops. All kinds of chemicals polluted land, which lead to hundreds of thousands of "death land", "stink earth". Ecological circular agriculture arises at the historic moment, in order to explore how to rationally develop the land, produce safe agricultural products, safeguard human health of body and mind, reduce pollution, protect the ecological environment, effectively guarantee the agricultural sustainable development. In the ecological cycle of agriculture model, it generally does not allow a large amount of chemical fertilizers, pesticides and other chemicals like traditional agriculture. But by using the production environment, artificial design, scientific control and other ways, we should produce safe products of high quality and high efficiency. In breeding industry, various kinds of additives are not heavily used which can get more products to promote livestock growth, but we emphasize the utilization of wastes (mainly the feces, urine, waste water etc.) In the production, and comprehensive utilization of wastes generated in the planting and breeding to form circular and ecological use patterns.

II. The characteristics of ecological circulation agriculture

1. Comprehensive coordination

Ecological circular agriculture emphatically mobilizes and exerts overall function of agricultural ecosystem, with large-scale planting, breeding, in accordance with the overall consideration, overall coordination, recycling, the principle of sustainable development, makes a comprehensive coordinated planning, adjusts and optimizes the agricultural structure to the greatest extent, coordinates the involved agriculture with the other industry comprehensively, make the various industries support and cooperate with each other, so as to achieve the purpose of improving the comprehensive production capacity.

2. Diverse models

To develop ecological circular agriculture, we should take into account the actual situation of agricultural production in our country, because our country is a large country, and the differences of the natural conditions, resources foundation, economic and social development level are very big. The ecological circular agriculture fully absorbed the essence of traditional agriculture, combined with modern science and technology, applied many ecological model, ecological engineering construction and practical technology to the agricultural production. Through continuous and in-depth analysis and comparison, we can find the advantages of agricultural production in different regions, transform backward production modes, and integrate different and diverse development models to highlight the diversity of development.

3. High benefits

Ecological circular agriculture combines substance circulation, multi-level multi-polar utilization of energy and deep processing of agricultural products, which can recycle the wastes, reduce the production costs, improve product quality, improve efficiency, and increase the value of agricultural economy. At the same time, farmers will be liberated from the backward, inefficient and high-consumption agricultural production which is "to work at sunrise and stop at sunset" and "facing the lands with their backs to the sky", creating new ways of labor and jobs for them, making them willing to accept and actively work, so as to stabilize the development foundation of "farmers" in the link of "agriculture, rural areas and farmers".

4. Sustainability of development

The scientific development of ecological circular agriculture can effectively protect the environment, prevent harmful materials such as fertilizers and pesticides in agricultural production to pollute the land, crops, livestock and agricultural products, maintain the ecological balance and improve the quality of the agricultural products, safeguard the security, and closely combine the environmental protection, agricultural production and economic development. In ensuring the supply of agricultural products and meeting people's consumption needs, it can make modern agriculture in our country develop in the more stable, more harmonious and more sustainable direction.

III. The comparison between ecological circular agriculture and traditional agriculture

Traditional agricultural economic structure is one-way linear, that is, in the agricultural production, it is to use the resources land and varieties, produce the primary agricultural products, produce a large amount of waste which is not reused. The resources consumption is higher, the resource utilization ratio is low, and the waste emissions are more and not reused, which caused great wastes. In the whole process, the producer (mainly farmers) often only pursue the value of higher agricultural production, and the one-sided pursuit of economic efficiency. The production mode and the resources use mode are more extensive, which virtually increases the damage to the ecological environment, causes the serious waste of resources, and the output value is not high, and even leads to loss of money sometimes. But ecological circular agriculture emphasizes the rational use of resources, ecological environment protection, rational pursuit of economic benefits and the sustainable development of agriculture, which is a circular multithreading development, creating a circular development models of "reasonably using resources, producing the safe agricultural products, comprehensively utilizing the waste into agricultural production. Through the comprehensive utilization of wastes, only a small amount of pollutants are discharged, so that the utilization rate of resources is doubled, which not only ensures the ecological balance, but also develops agricultural production, and makes the development of agriculture into a sustainable "fast lane". The difference between ecological circulation agriculture and traditional agriculture is mainly manifested in four aspects.

Ecological circular agriculture emphasies the application of the ecological, recycling, environmental protection to the production practice, adhere to reducing the excessive use of resources and waste recycling, adhere to the scientific method in reducing waste and recycling of seemingly abandoned material, control pollution problems from the endogenes, try their best to realize recycling.

Ecological circular agriculture not only pays attention to economic benefits in the production mode, but also actively transforms some industries with high input and high energy consumption, and uses scientific methods to maximize and optimize the utilization of resources. Besides, new technologies, new products and new development models are also in an important position.

In terms of industrial development, traditional agriculture is usually limited to the fixed development of a certain industry. The planting industry is nothing more than grain, oil and cotton, while the breeding industry is nothing more than pigs, cattle and chickens. Ecological circular agriculture advocates the active adjustment of agricultural industrial structure. From the perspective of ecological circulation, it not only makes an industry bigger, stronger and better, but also effectively combines different industries on this basis to form industrialization, so that related industries can realize circular and sustainable development within the industrial chain.

Traditional agricultural focused on the ourtput and economic benefits of the products, which is a firm "truth" in thousands of years. In the society with underdeveloped agricultural science and technology, infierce market competition, and simple consumer demand, this "truth" really should be pursued, but the emergence of ecological circular agriculture altered the pattern. Agricultural production should not only pay attention to product output and economic benefits, Instead, we should learn to make good use of the natural resources and variety resources, do well in the industrial planning, link well the industrial chain, do well in the internal circulation of the industry, reduce production costs and improve ecological and social benefits while increasing production and efficiency.

Section 4　The main models of ecological circular agriculture in foreign countries

I. Substance reuse model

1. Circular agriculture in Aidong, Japan

　　The core of circular agriculture in the Aidong area is the development of oilseed rape. The oil residue left after the utilization of rapeseed can be composted or treated as feed to get high-quality organic fertilizer or feed. At the same time, the waste cooking oil is recycled and reprocessed into biofuel. Circular agriculture in Aidong district effectively promotes the efficient regeneration of resources in the agricultural economic system, reduces the input of external resources and the discharge of agricultural wastes, realizes the rational recycling of resources and ecological environmental protection, and conforms to the basic principles of ecological circular agriculture.

2. "Green energy"agriculture in Germany

　　Through continuous research, German scientists found that mineral energy and chemical raw material substitutes can be extracted from some agricultural products to realize the recycling of agricultural products. These bioenergy and raw materials are green and pollution-free, and the German government began to attach importance to the development of such cash crops. German scientists have successfully developed green energy by producing ethanol and methane from beet, potato, oilseed rape and maize through targeted breeding. They got alcohol from Jerusalem artichoke plant, alkaloids from quill bean. Rapeseed is the most important energy crop in Germany at present. It can not only be used as chemical raw materials, but also can be used to extract plant diesel fuel instead of mineral diesel fuel as power fuel.

II. Reduction model

1. Precision agriculture in the U. S.

　　Precision agriculture pursues high yield and benefits with minimum input. The guiding ideology is to precisely manage the soil and crops, according to the specific conditions of each operation in the field, to maximize the optimal use of agricultural inputs such as fertilizers, pesticides, water, seed, etc. to obtain the maximum yield and economic benefits, to reduce the use of chemicals, to protect agricultural ecological environment. Precision agriculture is the "reduction" of circular agriculture. The United States is one of the earliest countries in the world to implement precision agriculture. As early as around 1990, the United States applied GPS technology to agricultural production. A farm in Minnesota conducted a precision agricultural technology experiment, and the yield of crops guided by GPS fertilization was about 30% higher than that of traditional balanced fertilization crops. After the success of the experiment, the production management of wheat, corn, soybean and other crops began to apply precision agriculture technology.

2. Water-saving agriculture in Israel

In order to maintain the continuous stability of regional water environment and ecology, circular agriculture in Israel is highlighted as a perfect water-saving agricultural system. Sprinkler irrigation, drip irrigation, micro-sprinkler irrigation and micro-drip irrigation are widely used in Israel. More than 80% of the croplands are irrigated with drip irrigation, 10% with micro-sprinkler and 5% with mobile sprinkler, completely replacing the traditional irrigation method of ditch flooding. The most effective is agricultural drip irrigation technology: first, water can be directly transported to the root of crops, compared with sprinkler irrigation, it can save 20% of the water; Second, the application of drip irrigation in the arable land with large slope will not aggravate soil erosion; Third, the purified water after sewage treatment (higher salt concentration than fresh water) used in drip irrigation will not cause soil salinization. Drip irrigation saves more than 30% of water and fertilizer than traditional irrigation, and helps recycle waste water. To open up water resources, Israel has invested heavily in sewage treatment and recycling. Israel plans to use all the recycled water after sewage retreatment for agricultural irrigation. At present, 80% of the municipal sewage treatment has been recycled, mainly for agricultural production, accounting for 20% of the agricultural water. In addition to being used for agricultural irrigation, the treated sewage is returned to the aquifer.

III. Resource model

1. Circular agriculture in Lingzhen, Japan

Lingzhen is an early and successful area in the development of circular agriculture. It is a circular agriculture model that transforms wastes from agricultural production and everyday life into organic fertilizer and develops waste resources. In 1988, the town passed the "Development of Natural agriculture Ordinance", which prohibits the use of pesticides, fertilizers and other non-organic fertilizers in agricultural production, and the agricultural products should be organic agricultural products without chemical fertilizers, pesticides added residues and pollution-free. Since then, Lingzhen dealt with the small-scale sewage sludge, poultry manure and the organic waste as raw materials for the fermentation, thus the methane gas was used to generate electricity. The remaining half solid waste residue was for solid-liquid separation, solid components used for composting and drying and liquid compoments reused or discharged after processing (emissions has been largely harmless to the environment). Thus they realized the high and harmless use of the waste resources. In addition, kitchen waste in Lingzhen is collected and processed to make organic fertilizer.

2. Permanent agriculture in the UK

"Permanent agriculture" is an important form of recycling waste resources in circular economy, characterized by the efficient allocation of elements on the basis of saving resources and not destroying the environment to achieve the maximization of favorable relations. Growers are saving energy by recycling everything such as cigarette butts collecting rainwater, turning manure into organic fertilizer and returning straw to the field. Permanent agriculture seeks to conserve land resources as much as possible, emphasizing the use of perennials and encouraging the use of self-regulation systems. When cultivating the land, they use techniques such as multi-variety planting and green mulching to maintain the land, monitor the local environment and construct green development plans. Permanent agriculture does not use man-made fertilizers and pesticides. It keeps pests at bay by growing a diverse range of plants and encouraging predators to enter the ecosystem, such as the first bud of legume, which releases nitrogen that disorients pests.

IV. Ecological recycling industrial park model

The scale of circular agriculture has 3 levels: sectoral level mainly refers to an enterprise or a farmer as a circular unit; social level means "circular rural area"; regional scale is based on the principle of ecology, through the integration of material, energy and information among enterprises, forming an ecological industrial park driven by leading enterprises and containing several small and medium-sized enterprises and farmers within the park.

Maya farm in the Philippines is a successful example of ecological recycling industrial park. Maya farm was originally just a flour mill. After 10 years of construction from the 1970s, it has formed an ecosystem with a virtuous cycle of agriculture, forestry, animal husbandry, by-products and fishing. A large number of wheat bran were produced in flour mills. In order not to waste the wheat bran, breeding farms and fish ponds were established. In order to increase income, meat processing and canning factories were established, and livestock and aquatic products were intensively processed. The farm has 25 000 pigs, 70 cattle and 10 000 ducks. In order to control the pollution of livestock and poultry manure and recycle the waste from processing plants, the farm has set up more than a dozen biogas workshops. The biogas produced every day can meet the energy needs of farm production and family life. Some livestock feed can be recovered from the biogas residue after gas production, and the rest can be used as organic fertilizer. The biogas liquid after gas production is treated and sent to the pond to raise fish and ducks. Finally, pond water and pond mud are taken to fertilize the field. The grain produced in the field is sent to the flour mill for processing and enters the next cycle. Maya farms do not have to buy raw materials, fuel, fertilizer from outside, but can maintain a high profit, and there is no waste gas, waste water and waste residue pollution, fully realize the recycling of materials.

Section 5 Several main models of ecological circular agriculture in our country

I. Circular model in farmland

1. Interplanting between different crops

For example, in the interplanting of soybean and rice, the two crops require different nutrients, so they can be exchanged between species. The nutrients that are not used by rice can be used by soybean, and the nutrients that are not absorbed by soybean can be transferred to rice, forming an interspecific cycle.

2. Returning and leaving the residues to the field

Generally, rice and wheat are given priority to use the straw. The harvested straw is usually left in the field as nutrients for the next crop or applied to the field by composting.

3. Farming and breeding in the field

For example, rice is grown in the paddy field, and the water is used to raise fish, ducks and other small livestock or special economic animals (such as bullfrogs, etc.) . The animals can eat weeds and pests in the field, and the excrement can be used as fertilizer for rice to provide nutrients.

II. Circular model between planting and breeding

1. Simple direct circulation

The planting of crops are treated as the feed of breeding animals , for example by growing feed crops such as corn, amaranthus, feed vegtables (forage), bitter cabbage, chrysanthemum (Juba), rumex, silphium perfoliatum, pearl millet (Bajra), Medicago Sativa and astragalus adsurgens. After the harvest they are used to feed cattle, sheep, pigs, chickens, fish, geese, ducks, etc., and animal waste are processed back to the field as crop fertilizer.

2. Biogas combined with planting and breeding

This is the more common model of ecological circular agriculture, such as raising pigs to produce waste, which can produce "three bigas-states" (biogas, biogas liquid, biogas residue) by the processing of biogas project. The biogas as an energy source material is used for production and living, the biogas liquid and residue are reused as fertilizers to crops in the field, and the crops can be used as pig feed.

3. Fungus base combined with planting and breeding

That is, the manure produced by the livestock and poultry mixed with the straw from the planting industry is used to produce the base material for the edible fungi. After the cultivation of edible fungi, the base material can be returned to the field as fertilizer for crops.

4. Biological treatment

The most famous case of this model is the use of earthworms in the Sydney and Beijing Olympics to deal with waste. The food waste produced in the two Olympic Games, and the excrement of horse racing in equestrian events, are swallowed and excreted by earthworms and converted into highly fertile biological organic fertilizer. In agricultural production, earthworms can also be introduced into the reprocessing of compost and the redigestion of sludge in the settling sludge tank of sewage treatment plants, and the organic manure produced can be used for returning to the field.

5. Aquaculture combined with planting and breeding

As early as the Zhou Dynasty, the technology of silk raising was recorded, and the development of eco-circular agriculture further developed it into three-dimensional culture, that is, mulberry trees were planted to produce mulberry leaves for silkworms, silkworm shit could be applied to fish ponds as bait for aquatic products, and pond mud could be applied to mulberry fields as base fertilizer.

III. Circular model of industrialization

This model is not limited to the circulation of various breeding varieties, but extends to the industrial entities. The industrial circulation model is established between relevant crop planting companies and breeding companies. For example, the entity company of tea planting buys organic fertilizer from the entity company of beef cattle and dairy cows.

IV. Integrated model of industry and agriculture

In the days when industry was still underdeveloped, municipal waste was mainly organic matter, which was generally collected and returned to the countryside as fertilizer. With the rapid development of industry and urbanization, the main means to deal with these waste is to landfill, and to incinerate, but many organisms cannot be collected to return to the agriculture. Through extensive publicity and relevant rules, organic matter would be effectively separated and directly applied to the field, or the industrial and agricultural recycling parks are set up. Waste in industrial production is processed into organic matter and returned to agricultural production, and agricultural production is further developed with the help of industrial production.

V. Circular models of agriculture and geochemistry

The material input and output of agriculture itself is part of the global material cycle, and afforestation and the protection of forests play an important role in the carbon cycle, nitrogen cycle, phosphorus cycle and sulfur cycle. Through afforestation and the protection of the existing forest, we can help to adjust the concentration of carbon dioxide in the atmosphere, and stablize the original material balance in nature.

Chapter 3 Recycling of Agricultural Residues Resources

Section 1 The concept, source and classification of agricultural residues

I. The concept of agricultural residues

The definition of agricultural residues in Chinese academic circle is not unified, its connotation and denotation being differently explained. Sun Zhendiao was the first scholar to define agricultural residues. On his basis, subsequent scholars elaborated on agricultural residues from different disciplines. According to academia, agricultural residue refers to organic materials discarded in the whole process of agricultural production, mainly including the plant residues produced in the process of agricultural and forestry production, animal residuse generated in the process of animal husbandry and fishery production, and the processing residues generated in the process of agricultural processing and rural household wastes, etc. Usually, the agricultural residue refers to crop straw and livestock and poultry dung, including four categories such as plant residue (residue in the process of production of agriculture and forestry), animal residue (residue produced in the process of animal husbandry and fishery production), processing residue (resideue produced in the manufacturing process) and the rural living garbage. According to the theory of resource abandonment, agricultural residue is the difference in substance and energy between resource input and output in the chain of agricultural production and reproduction, and is the part of substance and energy loss generated in the process of resource utilization. Some scholars thinks that agricultural residues include industry waste in the process of production and rural living wastes from the residents. From the perspective of circular economics, in the present technical conditions such as capital and labour, the residue which can be reused as raw materials in the agricultural production or processing, is the carrier of matter and energy, being a special energy, and it is an inevitable by-product in the process of agricultural production and product processing.

China Rural Technology Development Center of the Ministry of Science and Technology defines agricultural residues as: "In the process of agricultural production, it's the things which are thrown away and not used except the targeted products, and is inevitably a kind of non-product output in agricultural production. According to its different sources, it can be divided into the various crop straw in the planting industry, livestock and poultry dung and the residues when slaughtering livestock and poultry in the breeding industry, the residues produced in the agricultural and sideline products processing, and the agricultural film, which remains in the soil during agricultural production, is also one of the main agricultural residues."

At present, relevant laws and regulations in China do not define agricultural residue in a unified way. In different occasions and different cities, Chinese laws and official documents of the connotation and denotation of agricultural residue are used differently. But China has made some laws, departmental regulations and local regulations on the concept of agricultural residue. As shown in the provisions of the "livestock and poultry pollution prevention and control management method" passed by the National Environmental Protection Administration Bureau meeting in March 20,2001, the livestock and poultry residue refers to the dung waste, the house bedding material, waste feed and scattered hair and other solid residues. In October 2013, the 26th executive meeting of the State Council adopted "Regulations on the Prevention and Control of Livestock and Poultry Breeding", which extended the definition of "livestock and poultry waste" in 2001 (the Measures on the Prevention and Control of Livestock and Poultry Breeding), regulating that the breeding residue refers to the dung, dead bodies and wasted water, etc. The fourth session of the Standing Committee of the 11th National People's Congress of the People's Republic of China passed the Law of the People's Republic of China on Promoting Circular Economy on August 29, 2008. The fourth chapter, Article 34, describes the "comprehensive utilization of crop straw, livestock and poultry dung, by-products of agricultural processing industry, waste agricultural film, and the development and utilization of biogas and other bio-energy". Although there is no clear using the concept of "agricultural residue", actually its connotation was shown. The 56th executive meeting of Zhejiang province in September 2010 examined and passed the Methods of Treatment and Utilization of Agricultural Residue, which is the first local regulations on the definition of agricultural residue. It stipulates that agricultural residue refers to the waste produced in the production of planting and animal husbandry, including residue of livestock and poultry in breeding, crop straw, edible fungus cultivation residue, wasted agricultural films and other agricultural residue as determined by the people's government at or above the county level. At the 123rd executive meeting of

Ningxia autonomous region in September 2012, the 48th document was passed, named with the Ningxia Agricultural Residue Treatment and Utilization Method. Article 3 of the first chapter points out, agricultural residue refers to the waste produced in the process of farming, animal husbandry and other agricultural production, including livestock and poultry waste when breeding, crop straw, waste agricultural film, etc. In combination with the above-mentioned concept of agricultural residue and the actual situation in our country, agricultural residue can be defined as the substance and energy which did not have original value and whose owners have used or prepared or had to discard, in the production process of planting, forestry, animal husbandry, aquaculture and the activities related to the processing of its products, or in the daily life of rural residents or in the activities providing services for daily life. Agricultural residue is the general term of wastes discharged in the process of agricultural production.

II. The sources of agricultural residues

1. Agricultural residue from planting industry

Agricultural residue in planting industry mainly comes from farmland and orchards. It mainly refers to the material or energy that is discarded except fruits in the process of planting and forestry production or after harvesting activities, such as straw of grain crops, leftover leaves and vines of vegetables and fruits.

China is a big agricultural country, and there will be a large number of residues produced in the production of planting. There are many kinds and huge amounts of them, and the main part is the crop straw. In 2009, the total output of Chinese crop straw was 687 million tons, including 250 million tons of rice straw, 150 million tons of wheat straw, 265 million tons of corn straw, 27.26 million tons of bean straw and 25.84 million tons of cotton straw. Obviously we can see the crop residue in China is huge.

Since ancient times, straw has been used as living fuel for farmers and rough feed for livestocks, or returned to the field as fertilizer, or used for water conservation, drought resistance, insect control and insect removal. They can also be used as building materials, and a small amount is used to make paper and other industrial raw materials. Now, straw can also be converted into electricity generation, be processed into fuel to make gas, be used as edible mushroom base material to produce oyster mushroom and shiitake mushroom etc. It can be seen that the renewable value of agricultural residue in planting industry is high. According to the investigation, there are 215 million tons of China's straw which has not been utilized in 2009, with a huge straw resource utilization potential.

2. Agricultural residues from breeding industry

Agricultural residue in breeding industry mainly comes from the residues produced in the process of animal husbandry and fishery production, which mainly refers to livestock and poultry manure, sewage, waste feed, feathers and other residues, as well as aquaculture pond mud containing feed residues and pesticide residues produced in the process of aquaculture. Among them, livestock and poultry dung accounts for a large proportion of livestock and poultry residue, and different livestock and poultry produce different waste. China is the world's largest livestock and poultry industry, which produces a large number of manure every year. According to statistics, the production of the livestock and poultry manure in 2011 was 2.121 billion tons, among which the production of cattle is the largest, followed by sheep, pigs, broiler chickens and egg-laying chickens. The added quantity of livestock and poultry manure from 2008 to 2011 is equivalent to about 50% of the industrial waste production in 2010. Compared with the annual production of various crop straw about 650 million tons, the production of domestic livestock and poultry is about 3.26 times of straw. In conclusion, livestock and poultry manure accounts for the largest proportion of agricultural residue in China, which would lead to the greatest harm to environmental if improperly disposed. Pollution from livestock and poultry farming has become one of the most important sources of pollution in China, according to the First National Pollution Source Survey Bulletin jointly released by the Ministry of Environmental Protection, the National Bureau of Statistics and the Ministry of Agriculture in 2010. In recent years, the state has taken measures to control pollution from large-scale livestock and poultry breeding, and published the Regulations on Prevention and Control of Pollution from Large-scale Livestock and Poultry Breeding in 2013, so that there are laws for supervision and control. However, most farms fail to effectively treat and utilize livestock and poultry manure, and pile up untreated manure at will, leading to a large amount of nitrogen and phosphorus loss, resulting in water, soil and air pollution. A large number of pathogens in livestock manure and heavy metal additives in feed not only pollute the environment, but also affect people's health. But there has been a tradition of using livestock manure since ancient times. Livestock manure can be used as organic fertilizer for crop cultivation, which is not only beneficial to soil improvement, but also to save soil resources. Minor groups in the North have a tradition of using cow dung as fuel for cooking and heating. At present, livestock and poultry manure can be processed by advanced scientific means as feed for fish and pig breeding. Livestock and poultry manure can also produce biogas after centralized treatment, which can be used as energy for farmers' lives.

3. Residues from the agricultural product processing

Agricultural processing residue is the waste produced in the processing of agricultural products, including the residue from meat processing industry, sugar cane and beet residue from the sugar industry, cannery processing residue, wood processing residue and the fuelwood amount obtained by pruning during the tending management of various economic forests.

Annual production of such waste now exceeds 100 million tons. With the growth of food production, it is expected that by 2020, the annual production of this kind of waste will increase to 200 million, most of which can be comprehensively used. For example, the residue from the meat processing industry can be used to manufacture leather products, soap, animal glue, biological agents, down, bone meal, etc. Crop straw can be used as raw materials for paper and boards, and its lignin or cellulose can be further processed into chemical products. However, if the processing residue of agricultural products is not properly treated, the residue and wastewater will enter the environment. When accumulated to a certain extent, it will bring pollution and threat to the environment.

4. Rural household waste

Rural household waste is the waste produced in the daily life of rural residents, such as human feces, urine, leftovers and so on. After the reform and open policy, with the further development of rural economy, the living standard of farmers has been further improved. At the same time, the quantity and types of rural household garbage have gradually increased. The threat to environmental pollution and the impact on human health are gradually increasing. Rural residents lack of environmental awareness, inadequate rural management system, the lack of laws and regulations and backward infrastructure, etc. , have gradually become an important factor affecting the rural ecological environment.

III. The classification of agricultural residues

(I) Classification of agricultural residues according to their sources

① Residue from farming industry. It mainly refers to the residues generated in the production and harvesting process of grain, vegetables, melons, fruits, sugar and other agricultural products, such as straw, stubbles, weeds, fallen leaves, fruit husks, vine branches and other residues.

② Residue from breeding inustry. It mainly refers to the residues produced in the process of animal husbandry and fishery production, including livestock and poultry manure, feathers, feed residue, bedding material for the animal house, aquaculture pond sludge, etc. , but also including dead bodies of the livestock and poultry, sewage produced in the process of breeding, etc.

③ Residue from the agricultural product processing. It refers to the residues generated in the processing of agricultural products, in the processing of agriculture, forestry, animal husbandry and fishery. The residue produced in the primary processing of products mainly includes rice husk, corn cob, peanut husk, bagasse and so on.

④ Rural household waste. It mainly refers to the mixture of human feces, urine and household waste, including plastic bags, construction waste and household waste.

(II) Classification of agricultural residues according to their utilization value

① Resource agricultural residue. It refers to the part of the by-products of agriculture or agricultural processing industry that can be reused as raw materials when the current technology, capital and labor conditions permit, including resource agricultural production residue (waste straw, residue from the edible mushroom cultivation and animal residue such as livestock manure, etc.) and resource agricultural product processing waste (bagasse, slaughter blood and sewage, etc.)

② Non-resource agricultural residue. It refers to the residue generated in agricultural production and primary processing of agricultural products and the part of rural household waste that cannot be recycled under the existing technologies and funds, such as chemical fertilizers and pesticides, discarded agricultural film, pesticides, agricultural greenhouse gases, and discarded plastic bags.

(III) Classification of agricultural residues according to their chemical properties

① Organic agricultural residue. A general term for organic substances produced in agricultural production and farmers' life, characterized by resource utilization. It mainly includes plant residue (such as crop straw, fruit shell, weeds and leaves, fruit residue, etc.), animal residue (such as livestock and poultry manure), and household waste of rural residents (such as human feces, kitchen waste, etc.) generated in agricultural production or primary processing of products.

② Inorganic agricultural residue. It refers to the residues or garbage made of the inorganic composition that can't be reused, can't naturally degrade and need special measures to deal with in the environment, produced in the agricultural production process or from the rural residents, including packaging, chemicals, agricultural pesticide residues, waste household appliances and battery etc.

(IV) Classification of agricultural residues according to their forms

① Agricultural solid residue. It refers to the solid waste produced in the whole process of agricultural production and farmers' life, such as residues from livestock and poultry breeding, crop straw, agricultural plastic films, etc.

② Agricultural liquid residue. It refers to the liquid waste produced in the whole process of agricultural production or farmers' life, mainly including sewage, domestic wastewater produced in farmers' life, etc.

③ Agricultural gas waste. It refers to greenhouse gases such as CO_2, CH_4 and N_2O emitted during agricultural production and agricultural product processing, mainly from rural electricity, power of agricultural machinery, use of diesel fuel and fertilizer application.

(V) Classification of agricultural residues according to their nature

① Hazardous agricultural residue. It refers to those that cause harmful effects on the environment or human health, that have hazardous characteristics and require corresponding treatment to reduce or eliminate their hazards, such as fertilizer, pesticide packaging, plastic products with daily necessities, and agricultural films, etc.

② General agricultural residue. It refers to agricultural residue that does not have obvious harm to the environment or human health, can be naturally degraded, and does not need special treatment, such as crop straw, livestock and poultry manure, etc.

Section 2 Crop straw recycling

With the transformation from traditional agriculture to modern agriculture in our country and the rapid development of our society and economy since the reform and opening-up, great changes have taken place in the traditional way of straw utilization, especially with the agricultural scientific research and comprehensive utilization of agricultural residues, the gradual deepening of scientific research, technological progress and innovation opened up a new method, the new way for crop straw's utilization. People have realized the huge and potential economic value of crop straw, which can be turned into treasure and can be used from many channels. Crop straw can be degraded, so farmers can obtain additional economic benefits, and the government can achieve better social benefits, killing multiple birds with one stone.

I. The principles of crop straw recycling

1. Principle of recycling resources

Crop straw is an important by-product of crops, and it can also be an important resource for industrial and agricultural production. Therefore, the treatment of crop straw should start from the two aspects of resources and energy.

2. Principle of tailoring measures to local conditions

Crop straw recycling should be specifically analyzed as a specific problem, according to different output of different areas and local economic level, sci-tech level, to choose the appropriate crop straw resources utilization technology.

3. Principle of sustainability

The recycling of crop straw should be aimed at improving the comprehensive utilization rate, focusing on highlighting the utilization value of crop straw, reducing agricultural production costs, and promoting the sustainable and green development of agriculture.

II. The analysis of the recycling ways of crop straw

At present, the way of crop straw recycling is mainly reflected in fertilizer utilization, feed utilization, energy utilization and industrial utilization.

In our country, according to Liu Jiansheng's investigation in the eight provinces, in the comprehensive utilization of all kinds of crop straw, the proportion of straw used as fuel is dominant; The second is forage, but cotton straw is basically not used as feed, because its lignin content is high, which is difficult to be digested. Generally it is used as fuel, with the utilization ratio being as high as 83%; In the third place, straw returns to the field, which is used as fertilizer.

1. Straw as fertilizer

There are two ways of recycling straw as fertilizer: directly returning straw to the field and indirectly returning straw to the field. Direct straw return includes straw mulching return, straw turning and pressing return, and straw leaving high stubble return. The indirect straw return includes dry stalk overbelly return to the field, straw overbelly marsh fertilizer return to the field, dry stalk fungus bran, straw fungus bran marsh fertilizer return to the field, straw heap retting, dry stalk marsh fertilizer return to the field, external biological reactor, plant ash return to the field, etc. It is a rich fertilizer resource, because the organic matters, nitrogen, phosphorus, potassium, magnesium and other elements contained in crop straw are essential nutrients for crop growth. After harvest the crop straw return to field in suitable way, which will greatly increase the agricultural organic fertilizer, make the elements such as nitrogen, phosphorus and potassium in the soil increased. Especially the increase of potassium is most obvious, and the activity of soil and organic matter also increase to a certain extent, which plays an important role in improving soil structure. In the practice, most of the different utilization ways are combined with each other and recycle each other, and finally realize the efficient cascade utilization of energy.

2. Straw as feed

Crop straw as feed are mainly through biological method, physical method, chemical method, dry storage method and other ways, to change the length, thickness, hardness, etc., to change the straw into high-quality feed, to improve its palatability and digestibility, in order to feed cattle, horses, sheep and other large livestock, and return their feces to the field, namely transabdominal return to the field. This process plays a significant role in improving the nutrient composition of the straw feed, which is simple, labor-saving, time-saving, convenient for long-term preservation, and can balance the feed supply all year round. It not only solves the problem of lack of livestock feed in winter, but also saves the feed, which has broad prospects for popularization and application.

3. Straw as energy

Straw can be returned directly to the field. There are three ways of energy utilization of crop straw: direct combustion, conversion to gas fuel and conversion to liquid fuel. To be specific, the straw is used to make biogas, solid molding fuel, straw gasification, direct power generation, ethanol and other ways, among which straw gasification (gasification into biogas, water gas, etc.) has begun to get a larger scale of promotion and application in China.

4. Straw used in industrialization

The industrial utilization of crop straw is very extensive. Straw as a raw material can be used in building materials, paper, board production, light industry, textile and chemical industry and other fields. In addition, crop straw can also be converted into basic materials, mainly used for base material to cultivate edible fungus , base material for seedling, base material for flowers and trees, and base material for lawns, etc.

III. The prevention and control technology of crop straw pollution

(I) Straw returning technology

Straw returning to the field is applied to the farmland after the straw is fermented, or the straw is smashed and buried in the farmland for natural fermentation. Straw returning is one of the effective measures to improve soil and increase the content of organic matter in soil. Crop straw is the agricultural residue with the largest output in planting industry. In order to prevent environmental pollution caused by straw incineration, the agricultural cycle of "grain production-crop straw-returning to the field-organic fertilizer-crop production" can be realized through various forms of straw returning to the field, so as to achieve sustainable agricultural development.

In China, for several decades, fertilizer has been widely used in the countryside because of its rapid fertilizing effect, convenient application, and the greatly increased production of food products. At present, our country has become one of the countries using the most fertilizers in the world. However, the utilization rate of chemical fertilizer is low, which is about 30%. A large number of losses of chemical fertilizer not only cause a waste of agricultural funds, but also cause serious adverse effects on rural environment and agroecological environment, especially on farmland soil environment because of the improper use of chemical fertilizer. In addition, most of the areas in China didn't adopt effective measures to recycle the straw, and the continous use of the land year after year can not timely supply the available nutrients for the soil. So in the soil organic matter contents decreased, and the fertility decreased year by year. The focus on more growing than maintaining, more output than input caused such serious consequences as the soil hardening acidification, less crop nutrition, diseases and insect pests, which is not conducive to agricultural production, nor to the harmonious development of the ecological environment.

The straw returning to the field has a long history in our country. It can raise flowers by grass, press grass by grass to realize the combination of use and maitainance of the land. and to achieve the purpose of land fertility. At the same time, straw returning can reduce the application amount of chemical fertilizer, increase the contents of organic matters and available nutrients in the soil, and alleviate the contradiction of the imbalance of nitrogen, phosphorus and potassium ratio. It can also regulate soil physical properties, improve soil structure, form the ground cover, and regulate soil water evaporation and storage. It can reduce the incidence of diseases and insect pests, so as to meet the basic requirements of organic agricultural production, improve the quality of agricultural ecological environment, and avoid environmental pollution caused by straw burning. After straw mulching in farmland, the underground temperature can be increased by 0.5–0.7 °C in the low temperature season in winter, and the underground temperature can be decreased by 2.5–3.5 °C in the high temperature season in summer, and the soil moisture can be increased by 3.2%–4.5%. Weeds can be reduced by more than 40.6%. Therefore, crop straw returning to the field is an important technology to prevent and control straw pollution and promote agricultural circular economy.

There are many ways of crop straw returning to the field, which are the simple, easy and effective model based on the experience from agricultural production in different regions in our country.

1. Direct straw returning technology

Straw directly returning to the field is mainly the use of mechanical operations with high degree of mechanization, and the straw treatment time is short, rot required time is long, which is a simple treatment of straw with machinery.

(1) Mechanical direct returning technology

Mechanical straw returning technology is based on mechanical smashing, stubble breaking, deep ploughing and harrowing and other mechanized operations, and returns the smashed straw directly to the field, which can increase soil organic matters, improve crop yield and reduce environmental pollution. It is a comprehensive technology to compete for agricultural time. The mechanical technology of straw returning to the field is an effective way to realize "raising the field with the field", to protect the environment and to establish high and stable yield agriculture.

Straw smashing and returning technology is a mechanical operation to directly crush and return the upright or laid straw in the field, so that multiple manual returning processes can be completed at one time. This method can increase the production efficiency by 40 times. The use of crushing root stubble machine can make straw crushing and rotary tillage integrated, which can accelerate the decomposition of straw in the soil, so that nutrients can be absorbed by the soil. It can improve the soil structure and physical and chemical properties, increase soil fertility, and promote the continuous increase in crop yield and income. This form is mainly used in North China, where the water and heat conditions are good, the land is flat and the mechanization degree is high. The length of the crushed straw should be less than 10 cm, evenly spread on the farmland and combined with nitrogen fertilizer $300-600$ kg/hm^2. At the same time, in order to consolidate the soil and accelerate the rot of the crop straw, the first sufficient water must be irrigated after the whole land is finished.

The whole straw returning technology mainly refers to the mechanization of whole straw returning of wheat, rice and corn stalks, which can turn over and bury upright crop stalks in the field or lay them flat on the ground.

Mechanical direct returning technology is efficient, low consuming, time-saving, labor-saving, easy to be accepted by the majority of farmers. Moreover, this technology can protect the soil by covering the the surface with straw, which can improve the ability of soil water storage and water preservation, and achieve the sustainable development of agriculture with high efficiency, high yield, high quality and low consumption. But it also has shortcomings, such as in the hills, mountains, due to the small area of arable land, the use of machinery is not convenient, which affect the effect of returning to the field; and the early investment is relatively large, the cost is higher, and it is difficult to popularize.

(2) Straw mulch returning technology

Straw mulch returning technology is to directly cover the field with smashed crop straw or the whole straw, which can improve soil saturated water conductivity, increase soil water storage capacity, regulate soil water supply, improve water utilization, and promote the growth of aboveground crops. After the crop straw rots in the farmland, it can also increase the organic matters of the soil, supplement the nitrogen, phosphorus, potassium and other nutrients in the soil, and improve the physical and chemical properties of the soil. Straw mulch can also regulate soil temperature and effectively alleviate the damage of temperature change to crops.

When applying straw mulching technology, the cover thickness should be 3–5 cm, with uniform coverage greater than 30%, in order to ensure the successful completion of planting tasks such as sowing. The whole straw mulching method is suitable for arid areas and small area in the north, while the high stubble mulching technique is mainly used in wheat and rice planting areas in China.

At present, the straw mulching returning technology commonly used in our country countryside has the following several forms.

① Direct mulching and returning to the field. This method is very simple. The straw is directly used to cover the surface of farmland soil. Combined with no-tillage and sowing, the effect of water storage, water retention and yield increase in farmland is very obvious.

② Straw with high stubble mulching and returning to the field. This method is mainly applied to wheat, rice and other straw, which can be divided into two categories: high stubble smashing mulch and high stubble leisure mulch. It is suitable for rotary tillage sowing, hard tillage sowing, sowing seeds before rotary tillage sowing and so on. High stubble recreational mulching mainly includes rotary tillage mulching, deep loose mulching and whole stalk mulching. The specific operation is to leave the stubble height of 20–30 cm at the time of harvesting, and the crop straw returning quantity is about 2250 kg/hm^2. At the same time, 150–225 kg/hm^2 of nitrogen fertilizer is used, and then it is turned into the soil by tractor plow, and the irrigation in autumn and winter and the water protection in early spring are implemented. This method can turn and press the soil on the spot, save time and labor, and evenly return the crops to the field. But this form is not enough to make up for the consumption of land fertility, because of the small amount of crop straw return.

③ Returning straw with super high stubble in wheat field. In this way, rice seed is sown in the wheat field before wheat harvest, the rice seed germinates, and the high stubble wheat straw is returned to the field when the wheat is harvested, and the wheat field is directly transformed into rice field after irrigation. The returning straw with super high stubble wheat field is a combination of labor-saving cultivation, water-saving and drought nursery, no-tillage and straw returning and other cultivation techniques, which can fatten the soil, protect the environment, save labor and cost (without seedling raising and transplanting).

④ Zonal no-tillage mulching. This is a new type of conservative tillage technology, using zonal no-tillage drill machine to directly sow when the crop straw in the upright state, thus realize the ridge covering and the interridge seedings in the zonal no-tillage crop straw fields, which has the advantages of strong adaptability, less production procedures, low production cost, good application effect and so on.

(3) Mechanical rotary tilling and burying technology

This form is mainly suitable for corn stalks due to their low lignification and brittle wall, which is easy to break. When the corn is harvested, the corn stalks can be cut into long straw about 20 cm after twice rotary operation and rotarily ploughed into the soil by a mechanical rotary tiller. As the aerenchyma of corn stalk is developed, it is easy to soften in water and decompose quickly, so its nutrients can be used in the current season. If the amount of straw returned is calculated as 30,000 kg/hm^2, it is equivalent to 345 kg of Ammonium bicarbonate, 150 kg of potassium chloride and 975 kg of superphosphate per hectare, which can increase the yield of rice by 1.2 – 1.65 tons per hectare.

2. Indirect straw returning technology

(1) Straw stack retting technology

Straw stack retting technology is to make crop straw decomposed in sufficient high temperature, undertake artificial adjustment and control to it, add manure of livestock and poultry and a variety of trace elements, biological bacteria, and process into biological organic fertilizer. This method uses local materials and is simple to operate. It can solve the problem that crop straw is not perishable in arid and low-rainfall areas, and is especially suitable for small-scale applications where farmers are scattered. Straw retting is the main way to solve the shortage of current organic fertilizer source, it is also an important measure of improving soil and fertility in middle and low yielding fields.

According to the difference of composting conditions, straw retting can be divided into aerobic composting and anaerobic composting.

The aerobic composting technology can also be called high temperature composting technology, mainly in aerobic state, the use of aerobic microorganisms under high temperature conditions to degrade crop straw, accelerate the degradation of lignin, cellulose and hemicellulose in crop straw to form compost.

The anaerobic composting technology can also be called natural fermentation composting technology, mainly collecting the smashed straw and making it fully mixed with livestock manure and urine, closed without ventilation, making its natural fermentation. The fermented straw can accelerate the decomposition of humus, and be made into good quality organic fertilizer, as the base fertilizer to the field. In order to shorten the composting time, fermentation bacteria, nutrient solution and degradation bacteria can be added. This crop straw utilization technology is a traditional and widely used method in China, deeply welcomed by farmers.

According to the data, the fertilizer effect of 500 kg decomposed compost is equivalent to 15.2 kg urea and 24 kg phosphate fertilizer. With the long-term use of decomposed compost, it can not only reduce environmental pollution, but also improve the content of organic matter in the soil, and reduce the use of chemical fertilizer, improve the quality of agricultural products.

(2) Straw transabdominal returning technology

Crop straw transabdominal returning to the field has a long history in our country, and it is a straw utilization of high efficiency. Straw transabdominal returning technology is to ensilage, microsilage, ammonify crop straw, then feed the livestock and poultry as food. And after the livestock and poultry stomach digestion, the manure as fertilizer returns to the field. This method can not only achieve the purpose of increasing the value of animal husbandry and income, but also realize the return of organic fertilizer to the field, forming a virtuous cycle of "grain-straw-feed-livestock and poultry-organic fertilizer-grain", and really forming a grain saving structure of animal husbandry. Practice has proved that crop straw transabdominal returning to the field can use one thing with multiple function just like killing two birds with one stone. It can not only increase the output of animal husbandry, promote agricultural production, alleviate the contradiction between food supply and demand, but also improve the recycling efficiency of crop straw and reduce the pollution of crop straw to the environment.

(3) Plant ash returning technology

This technique is not desirable. In this technology, the user burns crop straw and uses the potassium contained in them to make potash fertilizer. Although the potassium content of crop straw is the highest, about 0.9%, and the burning can get a certain amount of natural potassium fertilizer, other organic matters and nitrogen fertilizer in the straw will be wasted. However, the main role of potash fertilizer is single, which only plays a certain role in crop growth, such as stalk toughness and preventing crop lodging during the growth period. However, there are many nutrients needed for crop growth, only potash fertilizer can not be sufficient.

The most important point is that the burning straw in the open field will produce a lot of smoke, cause air pollution, and influence on traffic safety, etc., which is the main straw pollution to the environment, so we don't advocate this form, and even the national and local government published some policies across the country to deal with the prevention and control of environmental pollution.

(4) Biogas residue returning to the field technology

Biogas residue returning technology, namely "straw gasification, residues returning to the field", is a biomass thermal energy gasification technology. After straw' gasification, the combustible gas (biogas) generated can be used as the rural domestic energy supply, and the waste residue formed after gasification can be returned as fertilizer after treatment. After the straw is not completely burned, it can be turned into plant ash that retains nutrients and returned to the field as fertilizer. The production practice proved that the biogas residue returning technology was a good way of crop recycling. The biogas residue and biogas slurry produced after straw fermentation are high quality organic fertilizers, non-toxic, harmless, high quality and effective, rich in nutrients. It is a good choice for producing pollution-free agricultural products and organic food because of its high humic acid content and slow fertilizer effectiveness.

(5) Straw fungus bran technology

In this way, crop straw is used to cultivate edible fungi, and then the fungus brans are returned to the field. This technology can achieve economic benefits, social benefits and ecological benefits, which can not only save costs, but also reduce fertilizer pollution and protect the ecological environment of farmland.

(II) Straw as feed technology

The countryside in China has used crop straw as livestock feed since ancient times, but the direct use of straw as feed is not the best choice. Because the crude fiber content of straw is high, the content of crude protein, soluble sugars, carotene and various mineral elements is relatively low, and the content of silicate is also high, which makes the palatability of the feed straw is poor, animal intake is low, and the digestibility is low. If straw is fed directly as feed to livestock and poultry without processing, it will not only increase the consumption of the feed, but also cannot meet the physiological requirements of livestock and poultry. Therefore, in order to improve the nutritional value of straw, they must be processed reasonably, and the lignin in straw can be degraded by physical and chemical methods, and are converted into biological protein feed containing rich thallus protein, vitamin and other components, so as to improve its digestibility, so that the feed utilization rate of straw can be improved.

1. Improvement method of processing the crop straw as feed

(1) Physical methods

Physical methods to improve the use value of straw are mainly carried out through mechanical processing, radiation treatment, steam treatment and other ways. In the implementation process, different methods are mainly adopted according to the proportion of straw in the diet, the types of livestock and poultry, and economic conditions etc.

① Mechanical processing. This method is mainly through mechanical processing to shorten the length of crop straw, so that livestock and poultrys' straw intake, digestibility and utilization of metabolizable energy have changed. The advantage of this method is that it is easy to feed, prevent picky feeding of livestock and poultry, and reduce waste.

② Radiation treatment. This method uses X-ray to irradiate wheat straw, barley straw and other crop straw to improve the digestibility of straw feed in and out of the body.

③ Steam treatment. The digestibility of straw feed can be improved by hydrolyzing chemical bonds of straw with high temperature water vapor. However, steam treatment consumes too much energy and is difficult to popularize.

(2) Chemical methods

Chemical methods mainly use chemical agents to act on straw, so as to change the internal structure of straw, which is conducive to the decomposition of rumen microorganisms, and achieve the purpose of improving the nutritional value and digestibility of straw.

There are many chemical preparations used for crop straw treatment, including acid preparations such as formic acid and acetic acid, salt preparations such as NH_4HCO_3 alkaline preparations such as NaOH and NH_3, and other varieties of chemical methods most commonly used are NaOH treatment and ammonification treatment.

① The NaOH treatment. This method first originated in the early 20th century, there being "wet" and "dry" two kinds. "Wet method" is a solution with NaOH ratio 10 times the volume of straw, soaked the straw for a certain time, washed the residual alkali with water, and the remaining straw was used to feed livestock and poultry. The "dry method" was proposed in the late 1960s, which mainly involves spraying a high concentration of biological NaOH solution on straw, thoroughly mixing the solution to penetrate into the straw, and feeding directly to livestock and poultry without washing. These two methods will pollute the environment, and for the livestock and poultryto eat a large number of straw feed treated by this method for a long time, it will cause mineral imbalance in the body, affectting the health of livestock and poultry. Therefore, since the 1970s, this method has been gradually phased out.

② Ammonification treatment. This method mainly uses the low content of nitrogen and ammonia in the straw to meet and produce ammonolysis reaction, which can destroy the ester bond between lignin and polysaccharide chain to form ammonium salt, which is used as nitrogen source for ruminant microorganisms, which is the key to strengthening feed digestion.

(3) Biological treatment

Biological treatment of straw for feed processing has the advantages of low energy consumption, low cost and good effect. There are two main biological methods to treat straw as follows. One is to use straw as a substrate for single-cell culture, which can directly cultivate single-cell organisms that can decompose fibers on straw; the other is to hydrolyze the polysaccharide of straw into monosaccharide by chemical or enzymatic action, and then cultivate yeast to produce high-quality feed. The second is to decompose lignin, destroy the compound structure of cellulose-lignin-hemicellulose in order to improve straw digestibility. The most widely used biological treatment is silage.

(4) Compound treatment

In the actual production, a single treatment method can not achieve good results, and we often need to use a combination of various methods. For example, adding concentrated food in the process of silage is the combination of physical treatment and biological treatment; The treatment of straw with alkalization and microbial fermentation is the combination of chemical treatment and biological treatment.

At present, heat injection technology and alkalization-fermentation treatment are ideal techniques for straw processing, but they are suitable for popularization in economically developed rural areas. Each of the above methods has its own advantages, so it is necessary to choose the appropriate method according to the specific local conditions.

2. Crop straw silage technology

Silage technology is a biological treatment technology. It uses the activities of lactic acid bacteria and other microorganisms to change the carbohydrates such as saccharides in silage into lactic acid and other organic acids through fermentation, so as to increase the acidity of silage, inhibit the activity of mold in anaerobic silage environment, and ensure the long-term preservation of silage.

Silage technology works mainly on the crops with water content of about 60%, with a certain amount of sugar, such as corn stalk, sorghum stalk, peanut vine, etc.

There are many kinds of silage equipment, mainly silage cellars, silage bags, silage towers, etc. The most common is silage cellars. The requirements of silage equipment are airtight, compression resistant, load-bearing and convenient for loading and unloading. Concrete and plastic products can be used as raw materials for the equipment. Take the silage cellar as an example, its architectural structure is mainly brick and concrete, generally rectangular semi-underground type, with a slope at one end for easy transportation: the width of the cellar is generally 2.5–3 m, the depth is not more than 3 m, and the length depends on the scale of breeding. The silage cellar should be located in dry, well drained, high terrain with hard soil, sheltered from the wind and sun, with no manure and far away from the livestock barn.

Silage methods mainly include above-ground storage method, cement pool storage method, cellar storage method and soil cellar storage method. Silage technology requires about 70% water content of raw materials, so in the process of operation, the raw materials with high water content should be properly dried, or mixed with raw materials with less water content for adjustment. When the water content of raw materials is low, the materials should be evenly sprayed with water or mixed with more water-containing feed which should be timely harvested, so as not to affect the yield of raw materials or the quality of silage, or lead to the failure of silage. Since the growth and reproduction of lactic acid bacteria from straw requires a moist, anaerobic environment with a certain amount of sugars, the raw material of crop straw should be cut into segments 2–3 cm long during silage, which should be pressed and compacted layer by layer during loading, and air should be discharged to create an anaerobic environment and prevent fermentation failure. Silage feed can be fed to livestock and poultry in 30 to 50 days. when taking the feed, we should generally start from the end of the dark side, take material step by step, after taking material each time, the remaining material should be covered with plastic cloth immediately, and tightly pressed, preventting air from entering to make them mouldy and bad.

Silage additives include ammonia, urea, formic acid, propionic acid, dilute sulfuric acid, hydrochloric acid, formaldehyde, salt, molasses and live dry bacteria.

The factors affecting the quality of silage include crop variety, growing period, soil and fertility, climate and topography, etc. The main factors that determine the degree and mode of silage fermentation are dry matter content (the key to the success of silage), water-soluble carbohydrates, buffering capacity, nitrate content, oxygen and silage preservation capacity. In short, to improve the quality of silage, it is necessary to carry out good management of crop straw before and after harvest, harvest at the appropriate time, and keep the appropriate moisture of raw materials. The straw that have been harvested should be cut short as soon as possible, packed into the cellar, compassed and sealed, and the management should be strengthened after the cellar is opened. The faster the harvesting process, the higher the quality of silage.

The identification of the silage quality is mainly based on sensory identification, and the silage quality is evaluated by color, odor, taste and so on. Good quality silage is yellowish green or green, with a weak acidity of fruit or lees, soft texture, distinct veins and unaltered stems.

3. Crop straw microsilage technology

Crop straw microsilage treatment technology is to make straw fermented under anaerobic condition with the help of lactobacilli, which can not only inhibit or kill various microorganisms, but also degrade soluble carbohydrates in straw to produce mellow taste and improve the palatability of feed.

Microsilage technology needs to add microbial additives, such as live dry bacterias for straw fermentation, yeast, etc. , which can convert cellulose, hemicellulose, lignin and other organic carbohydrates in straw into sugars, lactic acid and some other volatile fatty acids to improve the utilization rate of straw.

In the process of implementing the microsilage treatment technology, we should pay attention to the following factors: the sugar of raw material of straw should be high; The moisture content of crop straw is high, 55%–65%. there should be sealed anaerobic environmental conditions.

4. Crop straw ammonification treatment technology

Ammonification of crop straw is a method to treat straw with liquid ammonia or urea under closed conditions. Beginning from the 1980s, our country began the research and test of straw ammonification, to the 1990s it has been popular throughout the country, with the total amount of ammonification straw reaching 10 million tons, ranking first in the world. Straw ammonification treatment technology is widely used because of its low cost, simple method and harmless to the environment.

The ammonification treatment technology can increase the digestibility of straw by 15%–30%, increase the nitrogen content by 1.5–2 times. The nutritional value of treated straw is greatly improved, the palatability is good, and the food intake of livestock and poultry is increased.

(1) The main ammonia source of crop straw

The main ammonia sources of straw ammonification in China include urea, liquid ammonia, ammonium bicarbonate and ammonia water.

① Urea contains 46.67% nitrogen, which absorbs water in the process of straw ammonification, and is decomposed into ammonia and carbon dioxide under the appropriate temperature and the action of urease. Urea does not need special equipment in the process of transportation and use, which is more suitable for farmers to use, but because of the low content of urease in straw, urea decomposition is not complete, it is necessary to add about 1% bean cake powder as urease source in the ammonification process, and increase the ammonification temperature.

② The liquid ammonia. Liquid ammonia decomposition effect is better than urea, with low cost. However, liquid ammonia needs to be stored in high-pressure containers, and the local government needs to establish special equipment stations and provide ammonia bottles for storing liquid ammonia.

③ Ammonium bicarbonate NH_4HCO_3. Ammonium bicarbonate contains about 16% nitrogen and can be decomposed into ammonia, water and carbon dioxide at a certain temperature.

④ Ammonia water. Ammonia water's concentration is 18%–20%, which is also commonly used ammonia source, but more suitable for the places with convenient transportation conditions and close to the ammonia plant.

(2) Main methods of ammonification of crop straw

① Ammonification pool method. The ammonification pool is rectangular or cylindrical, and the site is located in the sunny, sheltered, high terrain with hard soil, easy to manage and transport. Ammonification material should be smashed crops or segments 1.5–2 cm long. Use warm water to disolve the urea into a solution 3%–5%, evenly spray on the straw, and stir evenly, to ensure that each layer of straw is sprayed to the urea water solution, and step on it, finally cover pool mouth with plastic film, adn seal it surrounded by soil.

② Ammonification plastic bags method. The size of the plastic bag is determined according to personal convenience. Select a strong multi-layer plastic bag, spray the cut straw evenly with the prepared liquid ammonia solution, seal the mouth of the bag after filling, and put it in a sunny, sheltered and dry place. It should be noted that in order to ensure the ammonification process, the plastic bag should be sealed well.

③ Ammonification cellar method. The equipment selected in this way has the form of cellar, or pool, etc. The size of the cellar depends on the specific conditions, and the cellar can be built as underground or semi-underground. Site selection should also be in the sun, wind sheltered, high terrain with hard soil, convenient management and transportation, close to livestock and poultry houses, and convenient to take. This method requires that the cellar can not leak gas, can not leak water, and the cellar walls are smooth. After the ammonification cellar is built, the required ammonification raw material is crop straw 1.2–2 cm long, which are sprayed evenly with urea solution. After filling the raw material, straw with a thickness of 5–20 cm should be covered on the raw material, and then covered with soil with a thickness of 20–30 cm and compassed. When sealing the cellar, the raw material should be 50–60 cm above the ground to prevent rainwater infiltration. The sealing of the cellar should be observed frequently during the ammonification process.

The above three straw ammonification forms are relatively common methods used in our country, which are suitable for small-scale production of individual farmers.

Both the cellar and the pool can be multi-purpose, can be used for both ammonification and silage, with high utilization rate, low construction cost, long use time, and because the pool and cellar are fixed size, it is easy to measure the amount of crop straw.

④ Stacking ammonification method. This straw ammonification method needs to choose the site of dry terrain with no rodent infestation. The preparation work is to attach a layer of plastic film about 0.2 mm thick on the ground. The length and width need to be determined according to the size of the straw stacking. Generally speaking, the stacking height should not be higher than 2.5 m. On the plastic film 15–20 kg of straw should be stacked and bundled, plus ammonia and adjust the moisture conten. 70 cm of plastic film should be allowed for free use, and then cover the plastic film on the stack, the upper and lower film edge wrapped up, buried or sealed with heavy pressure.

⑤ Ammonification furnace method. In this method, the ammoniated straw is heated to 70–90 ℃ in the ammoniating furnace and kept warm for 10–15 hours. Then, the heating is stopped and the closed state is kept for 7–12 hours. After opening the furnace, the residual ammonia is distributed for 1 day before it can be used for feeding. The structure of the ammonia furnace can be brick and concrete structure or steel structure, and it is necessary to install automatic temperature control device, track. To make special grass car along the track transportation ammonia furnace will have one-time investment, high cost, but the advantage is durable, one furnace a day. Ammonia treatment time is short, and it is not affected by seasonal weather and other factors, with high production efficiency.

⑥ Vacuum ammonification method. This method is mainly used in Australia and other developed countries. In this method, the straw was put into the container and a part of the air was extracted by vacuum pump, and then liquid ammonia was injected by ammonia pump.

(3) Matters needing attention for ammonification of crop straw

Firstly, the quality of raw materials has a great influence on the improvement of straw ammonification. Next, the requirement of the straw moisture content is 30% or so, if the moisture content is too high, it is not convenient to transport, and easy to be mouldy. Moreover, the economic consumption of ammonia is related to the digestibility of straw. Therefore, when treating straw with liquid nitrogen and urea, the nitrogen content should be calculated according to the respective nitrogen content, which is required to be within the range of 2.5%–3.5% dry matter of straw. In addition, environmental changes and ammonification time are also factors that need to be paid attention to. In general, the ambient temperature is inversely proportional to ammonification time. Finally, when used to feed, take the material freely, but after taking the material you must pay attention to the sealing. Straw ammoniated feed can not be directly fed to livestock and poultry, but need to be dried for 1 to 2 days before feeding.

5. Crop straw granulation treatment technology

Crop straw granulation treatment technology needs the help of machinery to smash the crop straw and knead them into a certain length, and then according to the formula of various raw materials, mix them for a certain time. Finally a specific type of granular mechine was used to make granular feed.

The advantage of this method is that it is easy to add vitamins, additives and other ingredients into the feed, improve the nutritional value of feed, achieve nutritional balance, and improve palatability. This technology is simple, practical, with obvious feeding effect, and the investment is not much, more suitable for farmers to use. It can effectively solve the restriction of the straw on the scale development of animal husbandry, avoid the disadvantages of most of the straw utilization technology in local processing. The production modes have automated efficient production mode, semi-automatic production mode and household production mode.

6. Crop straw thermal spray treatment technology

Crop straw thermal spray process is to feed raw material (straw, chicken manure, cake, etc.) into the feed thermal spray machines, fill the machine with hot saturated steam, after a certain time, with the high pressure heat treatment of material. And then suddenly decrease the pressure for the material, force the material from the machine to explode in the atmosphere, so as to change its structure and some chemical composition. And after disinfection and deodorization process, the material was made into a more nutritious feed. This is a pressure and thermal processing process, which requires a special thermal injection device and a unique process flow to complete.

The straw after thermal spray processing, digestion rate can be increased by 50%, the whole feeding rate can be raised to more than 90%. It can also detoxificate rapeseed meal, cotton seed meal, etc to make its utilization improved 2–3 times. This method can also disinfect the chicken duck and livestock manure for deodorization and sterilization to make it a normal protein feed.

7. Crop straw bulking treatment technology

The bulking treatment technology of straw has been widely used in the production and processing of straw feed. Under the action of high temperature and high pressure inside the bulking machine, the straw is bulked, and finally achieved the effect of ripening and bulking. Its working mechanism is that the straw is sent to the extruder through the screw extrusion way, in this process, the screw rolls to promote the material flow in the direction of the axis.

In the friction of the screw, barrel and the straw material, the material is strongly squeezed, stirred and sheared, finally achieve the purpose of homogenization. The pressure and temperature inside the machine will rise accordingly, so that the straw material which has been crushed inside is matured and become paste from powder, and it will be ejected from the hole of the extrude. At the moment when the material is ejected from the machine, the temperature and pressure around the material decreases rapidly, and the material is bulked and loses water under the action of this pressure difference, forming loose and porous bulked feed.

(III) Straw as energy technology

Biomass energy is the energy stored in biomass by solar energy in the form of chemical energy, which can directly or indirectly come from photosynthesis of green plants, and is the only renewable energy that can be stored and transported. As the main source of biomass energy resources, straw is the world's fourth largest energy after coal, oil and natural gas. Compared with fossil energy, biomass energy is clean, environmentally friendly, renewable, dispersed and rich in resources. Therefore, through the physical, chemical and other methods, dealing with the straw cellulose, hemicellulose and other main components to achieve the purposeful transformation is an important topic of straw pollution prevention and recycling.

1. Straw direct burning technology

(1) Crop straw burning for heating technology

Straw direct burning is the main way of straw energy utilization, is also the traditional energy conversion method, with low cost and easy to promote. In our country, the straw is mainly used for farmer household cooking energy, the direct combustion in this year is accounted for more than 99% of the total amount of straw energy utilization. According to the data, the calorific value of straw is about 15,000 kJ/kg, which is 50% of that of standard coal. The calorific value of different straw is different.

Straw direct burning system for heating is a heating system with straw as fuel and special furnace as the core. The system is composed of water boiler by direct straw burning, supporting straw collection and pretreatment system and heating pipeline, etc. It can provide hot water and winter heating energy for township departments, primary and secondary schools, relatively concentrated township residents and economically developed natural villages. But this method has low efficiency, environmental pollution and other problems, so it is difficult to popularize.

(2) Crop straw burning for power generation technology

Crop straw direct burning for power generation technology refers to putting the straw raw material into the boiler for direct burning to produce high pressure water vapor, through the turbine to do work, drive generators to generate electricity. At present, this kind of straw burning for power generation technology mainly includes water-cooled vibrating grate boiler power generation technology and fluid bed combustion power generation technology.

The problem that straw direct burning power generation technology needs to pay attention to is to solve the problem of bed material sintering, alkali corrosion from heating surface high temperature and ash accumulation. This is because the ash amount after burning is very small, it is difficult to form bed material, so the river sands are mostly used as bed material, but in fact, burning agglomaration phenomenon will occur for temperature is the main factor. This phenomenon occurs because biomass ash is rich in alkali metal (Na, K), oxides and salts. Compounds of these elements react chemically with SO_2 in the sand to form low melting point co-crystals, which flow along gaps in the sand, agglomerating sands and disrupting fluidification.

2. Straw gasification technology

(1) Straw gasification for centralized gas supply technology

This technology is a new technology of rural energy construction that our country attaches great importance to, starting from 1996. It is based on the abundant straw in rural areas as raw material, after pyrolysis and reduction reaction to generate combustible gas, and then through the pipe network to farmers' homes for cooking and heating use, improve the original fuelwood-based energy consumption structure of farmers.

Because biomass biogas cannot be liquified at normal temperature, so it needs gas transmission pipe network to be sent to the users, so the straw gasification central heating system should set up the gasification station on the basis of the villiages (gas tanks set in the gasification station), lay the pipes, by which the biomass fuel gas is sent to the user's home. And the scale can be dozens of households to hundreds of families. Central heating system includes feedstock processor (mill, etc.), feeding device, gasifier, purification device, fan, gas storage tanks, safety device, pipe network and customer gas system and other equipments. Because of the different gas characteristics, the biomass gas needs a special stove.

Now our countryside has the uniform planning, which is advantageous to the popularization and implementation of this technology. Because of the small scale of the project, the laying of the main pipe network is eliminated and the distance of the pipeline network is short, which reduces the transportation cost and the requirements for pipes, thus greatly reduces the investment cost of the project. Straw gasification centralized gas supply technology has improved farmers' life style, improved their living comfort, saved the amount of labor and time used for cooking, made straw used as a resource, saved the consumption of wood, and protected the environment. However, the crude gas produced after the gasification of straw contains tar and other harmful impurities, if with improper treatment, it will cause pollution to the surrounding air, water and soil.

(2) Crop straw gasification for power generation technology

Crop straw gasification for power generation technology firstly refers to the biomass thermal chemical reaction in anoxic condition into fuel gas (CO, H_2, CH_4), and then the combustible gas after conversion are taken out by fan, after cooling and removal of dusts, tar and impurities. The gas would supply the internal combustion engine or small gas turbine to drive electric power generator. At present, straw gasification power generation is mainly used in small scale projects.

3. Crop straw fermentation for biogas technology

Crop straw fermentation for biogas technology is mainly based on straw as raw materials, in the conditions with tight air and a certain temperature, humidity, pH, through methane bacteria fermentation to produce biogas. Biogas is a kind of renewable clean energy, mainly being methane, its content being 55%–70%, the calorific value is 20–25 MJ/m^3 with high combustion thermal efficiency. One user pool can be equivalent as 0.8 ton of standard coal each year, which can save the farmers' more than 50% of the life energy. At the same time the use of agricultural residues such as straw, livestock and poultry dung can not only reduce the farmers' use of coal, save the economic costs, but also the use of renewal, biogas liquid or residues to plant vegetables and other crops can increase crop yield, improve crop quality, reduce the production cost, and improve the fertilizer effect, which can realize the comprehensive utilization of crop straw and embodies the benefits of circular economy. In terms of scale, the technology can be used by individual farmers, or laid pipelines can used for centralized gas supply.

There are two kinds of crop straw fermentation technology. The first is to use the biogas pool directly. This method is mainly to put straw and weeds, shrubs, branches and other agricultural and forestry residues mixed livestock manure and urine directly into the pool, under the condition of air isolation, adjust the appropriate temperature and humidity, through the fermentation of microorganisms to produce biogas. Straw for biogas needs to be pretreated. Generally, the physical method is taken, that is, the straw is smashed or cut, and then the biogas is produced under the fermentation of anaerobic microorganisms. In order to ensure the normal production of biogas, strict anaerobic is needed, that is the biogas pool must be closed, emptied withou air. Since straw is not easy to be directly used by microorganisms or enzymes, it is necessary to add nitrogen-rich raw materials during fermentation to reduce fermentation start-up time and improve biogas production. This requires a normal C/N ratio, which generally controls the ratio of livestock manure and crop straw to 2 : 1. Control the temperature well, ensure that in the temperature range of 25 – 40 °C degrees, the higher the temperature, the better the fermentation, the more biogas produced. The pH value is 6.5 – 8, which can be adjusted by ammonium bicarbonate and lime water.

In addition, there is another way to pretreat the smashed crop straw, and put them into the pigsty according to the requirement, and the pigs will pile up the straw. By dry and wet fermentation, bamboo cage are used for air, then return to the field.

4. Crop straw for solid biomass fuel technology

Crop straw solid biomass fuel technology is to crush agricultural residues such as crop straw, and then under certain pressure and temperature, through solid molding equipment to compress and shape straw, mainly granular, rod and block three kinds. The technology can soften the cellulose, hemicellulose, and lignin of the straw biomass in 200 – 300 °C, use the compression molding machine after drying and crushing the loose biomass fertilizer under the condition of ultrahigh pressure, by the heat generated by the friction between the machine and biomass residues and the friciton between biomasses, as well as the external heat, softening the cellulose and lignin. After extrusion molding, a new type of fuel with certain shape and specification is obtained. At present, this technology can improve the mass and calorific value of biomass per unit volume, its combustion efficiency is more than 80%, there are little production of SO_2, ammonia nitrogen compounds and dust, and it reduced air pollution and CO_2 emissions. It is convenient for transportation and storage, and can realize commercialization, which has good social and environmental benefits.

In a broad sense, biomass fuel technology can be divided into three forms: wetting compressing molding technology, heating compression molding technology and carbonization molding technology.

(1) Wetting compressing molding technology

This technology is commonly used in raw materials with high water content. After soaking the raw materials in water for several days, the water is squeezed out, or the raw materials are sprayed with water, and the binder is added to mix evenly. Simple levers and wood molds are used to squeeze out the water in the decomposed agricultural residues and compress them into forming fuel. In this way, the fuel density of the molding is low, and the wet molding equipment is relatively simple and easy to operate, but the equipment's wear rate is high, drying cost is high, the combustion of the product is poor.

(2) Heating compression molding technology

The process of heating compression molding technology includes raw material crushing, drying, mixing, molding, cooling packaging etc. If the production process is different, the working principle of the compression molding equipment is also different, mainly using screw extrusion molding equipment, piston stamping molding equipment and die grinding particle molding equipment.

The screw extrusion molding equipment adopts the continuous extrusion method, and the forming fuel is usually hollow fuel rod.

Piston stamping molding equipment doesn't need heating, and press the raw material by stamping. This molding has large forming density, the water content requirements are wider, but the productivity is low, product quality is not stable, and the products of the fuel are solid fuel rod or fuel block.

The die grinding particle moulding equipment is mainly used for producing pellet fuel, it does not need external heating, and the friction heat produced by extrusion molding material can make the material soften and bond. In this way, the moisture content of the raw material is required to be 10% ~ 40%.

(3) carbonization molding technology

Carbonization molding technology has two main forms: one is the carbonization before molding. The biomass raw materials are carbonized into powder granular charcoal, add a certain binder, and then use compression molding machine to squeeze it into a certain specification and shape. Second, the molding before carbonization, that is using compression molding machine to compress the loosely crushed raw materials into a certain density and shape, and then using the carbonization mechine to make the fuel rod into charcoals.

5. Crop straw liquification technology

The straw liquification technology is to convert straw lignin and cellulose into alcohol, combustible oil or other chemical raw materials through physical, chemical or biological methods. At present, there are three forms, namely direct liquification, high temperature and pressure liquification and microwave liquification.

① Direct liquification. This refers to the thermochemical reaction process of converting biomass into liquid state under the condition of medium low temperature, high pressure and the participation of catalyst, with the participation of reducing gases such as H and CO, which can be divided into two categories: retaining the macromolecular structure of the raw materials and destroying the macromolecular structure of raw materials by reaction products. Ethanol production from straw requires pretreatment in advance.

② Liquification at high temperature and pressure. This is a thermochemical process that takes place under high pressure, 300 to 500, catalyzed by a catalyst. This technology consumes a lot of energy and requires high pressure of equipment. It is mainly used in making diesel from crop straw.

③ Microwave liquification. This is the use of microwave radiation to make small molecules of polar substances have physical effects, so as to accelerate the reaction, change the reaction mechanism or start a new reaction channel technology.

(IV) Straw used in industrialization technology

The industrialization of crop straw is widely used in construting materials industry, light industry and textile industry. Currently, our crop straw is mainly used for paper making, accounting for 2.3% of total straw amount, also used in the production of constructing materials. After technical processing, straw can also produce furfural, wine and xylitol, etc.

1. Paper making technology

Straw is an important raw material in China's papermaking industry. In the 1990s, the amount of straw used in papermaking reached more than 20 million tons. Papermaking technology in our country has 2000 years of history, one of the country's "Four Great Inventions". At present, China basically uses wood pulp to make paper. According to the needs, it can have two methods of mechanical pulping and chemical pulping. Due to the shortage of wood resources, non-wood plant fiber pulping technology has been actively developed in the world in recent years. Most of the Jiangnan areas in our country use the cellulose part of straw as the raw material to make paper pulp. However, in the process of production, straw pulping and papermaking will produce "black liquid". When the wastewater discharged into the water cycle without treatment, it will not only cause a huge waste of resources, but also cause pollution of water sources because the "black liquid" contains substances difficult to degrade. Small paper mills in rural areas are unable to build and use sewage treatment equipment due to lack of funds and backward technology and equipment, so the state prohibits small paper mills from using straw pulping. At the same time, pulp and paper making with crop straw as raw material and waste liquid treatment technology are the difficult problems to be overcomed in the straw recycling technology.

(1) Pulping and papermaking technology of waste liquid reduction treatment

The process of this technology is as follows: ① extract the fibers of crop straw; ② wash and cut the fiber; ③ feed and add 8%–15% sodium hydroxide; ④ slowly press and knead; ⑤ make paper pulp. The principle is to increase the amount of straw fiber by 30%, reduce the amount of alkali by 8%–12%, and use rubbing machinery to make fiber friction heated to make pulp under high temperature conditions.

This technology can reduce the amount of black liquid wastewater, and the process of gentle pressure and kneading can also make fiber dispersion, avoid evaporation and dispersion process, also reduce the discharge of waste liquid.

(2) Pulping and papermaking technology with biological enzymes

This technology mainly uses microorganism and enzyme to treat crop straw. Then pulps and papers are made by simple physical or chemical methods, which are characterized by low cost and low risk of environmental pollution.

(3) Pulping and papermaking technology by high-speed spinning fiber

This paper making technology is to cut the fibers of raw straw material into pieces, mix water and fiber into the machine, disperse the fiber through high-speed rotation of the machine, and then rub the dispersed fiber, wash them to make pulps.

2. Crop straw for construction, packaging materials technology

Crop straw has been used as building material since ancient times. In ancient times people used thatch as a substitute for tiles to cover the roofs of houses. In rural areas, people mix crushed straw with soil to build houses. The fiber of straw can be used as both a covering material for the house and a building material for reinforcement.

(1) Crop straw as building materials technology

Crop straw are widely used in the field of building materials. It can be made into composite board, fiberboard and other products, and can also be made into gypsum base and cement base. At present, China applies straw in the new building materials of saving energy, recyling the wastes, and protecting the environment with high strength and high construction efficiency. Now the straw-made glass fiber can reinforce the new building materials such as composite materials, gypsum board etc, which have become the leading building materials.

Nowadays, the production of building materials using crop straw mainly includes straw-molded wall material, straw-extruded wall material, lightweight insulation lining material and structural board combined wall material, etc. The main technological processes can be summarized as integrated process and particle board process.

The integrated process mainly uses straw to make artificial fiber board. This technology is suitable for wheat and rice straw, no binder is needed in the processing process, and the thickness is 20–80 mm, which can be used as wall material.

The technology of particle board mainly uses straw to make artificial fiber board, including hard board, light material and composite material. This process is mainly used for building materials, packaging materials, furniture, and interior decoration.

(2) Degradable packaging materials technology

Plastic packaging material is one of the environmental pollution sources, which is consumed a lot every year. Now our country is developing the "green package" with the idea of sustainable development. Smashed straw and adhesives as raw materials, after mixing, cross-linking, foaming and other processes, they can be made into a cushioning packaging materials, with degradability, and they can reduce environmental pollution. For example the use of straw as the main raw material to made new type of pollution-free plant fiber foaming packaging can not only achieve degradation in a short period of time, but also degradation can be used as feed ingredients after degradation, to realize the recycling use of crop straw. The straw and corn starch as raw materials can produce the biodegradable packaging materials of polylactic acid, which can be used in the fields such as packaging materials, fiber and nonwoven fabric, Its strength, resistance and buffering properties are comparable to those of polystyrene.

(3) Disposable degradable tableware production technology

Every year in China there is a great demand for disposable tableware, so the use of straw to produce disposable tableware has the advantages of being cheap and fine, good performance, no pollution, and being degradable and so on. It has a broad market, has good economic value, social benefits and environmental benefits.

Disposable straw tableware is mainly the use of crop straw as natural plant fiber, smashed into materials, added safe non-toxic molding agents with edible health standards. After processing they can be made into completely degraded green tableware. The product is not only non-toxic and harmless, waterproof and high temperature resistant, with high strength and no deformation, but also can be decomposed by itself, and then recycled into feed or fertilizer.

3. Straw as industrial raw material technology

(1) Technology of high purity silicon tetrachloride for electronic industry

The straw of rice, wheat and maize is rich in silicon, and the technology of producing silicon tetrachloride using rice straw has been developed to meet the demand of silicon in the electronic industry. First the straw should be burned or carbonized, then the carbon obtained after straw treatment would be mixed with chlorinated carbon compounds (or hydrochloric acid) and carbon compounds, finally it will turn to super high purity $SiCl_4$ at the boiling point of 56.8 °C.

(2) Lignin adhesive production technology

Using crop straw as the main raw material, the adhesive can be produced by chain reaction between lignin and formaldehyde. The advantages of the product are high adhesive strength, water resistance, low cost, biodegradable and so on.

(3) Staws for ceramic glaze technology

In order to prevent glaze flow in the process of ceramic glazing, straw ash, silica and clay would be added, especially rice straw and tree ashes are rich in silicon, which can be added to the glaze to obtain high-quality white opaque glaze.

(V) Other technologies for the utilization of crop straw

1. Fungus base technology

Crop straw is rich in carbon, nitrogen, minerals, hormones and other nutrients, which is very suitable for the growth of edible fungi. Crop straw has abundant resources with large yield and low cost, so they are suitable for the production of edible fungus base material.

At present, the use of crop straw to cultivate edible fungi is very mature, and the investment is less, but the effect is fast, with low technical content, without being limited by the objective factors, so it is easy to promote. Now our country can make use of straw to cultivate 20 many kinds of edible fungi, such ordinary products as mushrooms, needle mushroom, and other valuable fungi such as hericium erinaceus (monkey head mushroom) and lucid ganoderma. In addition, the mushroom residue after straw cultivation, due to the biodegradation of the bacteria, the content of nitrogen, phosphorus and other elements also increased significantly, so it can also be used as a high-quality fertilizer for agricultural cycle production.

However, when choosing straw for edible fungus base material we should strictly carry on the sanitary disinfection on the environment in accordance with the "Food Safety Law" requirements, to ensure a good production environment. In addition, we should choose good crop straw as raw materials, such as the choice of fresh, mould-free straw, to ensure the normal growth of edible fungi. In addition, formulations should be carefully selected according to different fungi to ensure the yield and quality of edible fungi.

2. Crop straw weaving technology

The straw recycling is also on use of straw to make hand-woven handicrafts, which has aesthetic and practical value, can be used in furniture, interior decoration and other fields.

Baskets, wall hangings, futons and curtains woven with crop straw such as wheat straw and corn husk have the green trend of returning to nature. It is loved by urban consumers and satisfies people's ideal of returning to nature. In particular, some crop straw crafts with local characteristics and minority characteristics have higher cultural value, so it has higher economic value.

Section 3 Recycling of residues from livestock and poultry breeding industry

I. Recycling and comprehensive utilization technology of livestock and poultry residues

Livestock and poultry manure is rich in organic matter, nitrogen, phosphorus, potassium and other nutrients, which can be regarded as precious resources. If it is properly treated and reasonably recycled, it can play an important role not only in ecological environmental protection, but also for agricultural production and economic efficiency. At present, the comprehensive utilization of livestock manure is mainly from three aspects: fertilizer, energy and feed.

(I) Livestock and poultry residues as fertilizer

Livestock and poultry manure contains abundant organic matter and lots of nitrogen, phosphorus, potassium and other elements. When used in crop cultivation, it can improve soil structure, improve soil fertility, increase the output value of crops, and improve their quality. It is a "farm manure" used by Chinese farmers for a long time.

1. Composting technology

The residues of livestock and poultry has been composted since ancient times to make fertilizer for farmers or directly used for crop cultivation. Composting is a process that degrades organic matter in solid manure and makes it mineralized, humified and harmless under the condition of artificial control of moisture, carbon/nitrogen (C/N) ratio and ventilation. Composting is one of the most effective ways to dispose of livestock&poultry residues. In the process of composting, the high temperature can release the nutrients in the manure and generate the important active substance–humus, which is conducive to improving the soil fertility. It can play a role in regulating and improving the soil. At the same time, high temperature can also kill all kinds of pathogenic microorganisms and weed seeds in the manure to make it harmless. The compost is widely used because it reduces the odor of the final product, and it is dry and easy to package and store. However, there are also many problems in composting. For example, because the degree of microbial activity in the process of composting directly determines the cycle and quality of the composting, it is necessary to strictly control the parameters such as moisture, carbon/nitrogen (C/N) ratio, temperature and pH value, so it requires high manual operation. In addition, composting requires a large site and a long time (4–6 months), and ammonia is released during the process, so odor evaporation cannot be completely controlled.

At present, there are two kinds of composting technology: aerobic composting and anaerobic composting.

(1) Aerobic composting technology

Aerobic composting is to absorb, oxidize and decompose wastes with aerobic bacteria under aerobic conditions. Through their own life activities, microorganisms oxidize part of the absorbed organic matter into simple inorganic matter, and release the energy needed for microbial growth and activity, while the other part of organic matter is synthesized into new cytoplasm to make the microorganism grow continously and continue to produce more organisms. The temperature of aerobic compost is generally 50–65℃, the highest can reach 80–90 ℃, so it is also called high temperature compost.

The aerobic composting process requires the following three stages.

① Medium temperature stage (30–40℃). This stage is the initial stage of the composting process and also the stage of heat production. Thermophilic microorganisms are more active and make use of the soluble organic matter in the compost to carry out vigorous life activities. This process takes 1–3 days.

② High temperature stage (45–65). At this stage, the residual and newly formed soluble organic matter in the compost continues to be oxidized and decomposed, and the complex organic matter also begins to decompose strongly. This process takes 3–8 days.

③ The cooling stage. At this stage, the remaining organic matter that is difficult to decompose decomposes further, and the humus increases continuously, entering the stable and mature stage. The oxygen demand decreases, the moisture content decreases, the porosity of compost increases, and the oxygen diffusion capacity increases. Natural ventilation is required, and this process takes 20–30 days, with the characteristics of low operating cost. After fermentation of compost, high quality organic fertilizer without odor, insects (eggs) and pathogens can be obtained.

The factors influencing the composting process are as follows: oxygen supply should be appropriate, and the actual required air quantity should be 2–10 times of the theoretical air quantity; The moisture content of the material is 50%–60%, and 55% is the best, when the decomposition rate of microorganism is the fastest at this time. The C/N ratio should be appropriate. Water has two functions: one is to dissolve organic matter and participate in the metabolism of microorganisms; The second is to adjust the compost temperature, when the temperature is too high water evaporation can take away some of the heat.

(2) Anaerobic composting technology

Anaerobic composting technology uses the characteristics of anaerobic or facultative microorganisms growing and propagating with raw sugars and amino acids in manure materials to carry out lactic acid fermentation, ethanol fermentation or biogas fermentation. Biogas fermentation was the main method for manure with water content higher than 80%, while lactic acid fermentation was the one for manure with water content lower than 80%. At present, it is also useful to carry on the fermentation compost treatment of the chicken manure with EM bacteria.

Anaerobic fermentation technology does not need to turn over the stack, does not need ventilation, which is energy saving, relative low cost, and easy to operate. At the same time through anaerobic treatment, a large number of soluble organic matter can be killed, as well as a large number of infectious bacteria.

2. Bio-organic fertilizer technology

Biological organic fertilizer is to combine the beneficial microorganisms and organic fertilizer to form a new and efficient microbial organic fertilizer. The raw materials of biological organic fertilizer mainly include poultry manure such as chickens and ducks, livestock manure such as pigs, cattle and sheep, other animal feces, straw, agricultural products residues after processing and so on. Its production process generally includes raw material pretreatment, inoculating microorganisms, fermenting, drying, crushing, screening, packaging, weighing, etc. The method of ingredients varies according to the source of raw materials, fermentation methods, microbial species and equipment.

Biological organic fertilizer is rich in beneficial microbial flora, strong nutritional function, good adaptability, and rich in various nutrients. And small size is easy to apply, suitable for large-scale production. The production of organic fertilizer from the livestock and poultry residues can overcome the disadvantages of inconvenient transportation, storage and use of high water content manure, and is safe and harmless compared with other methods. And because biological organic fertilizer can improve the utilization rate of fertilizer, improve soil fertility, increase crop output, improve its quality, it has good economic benefits suitable for pollution-free agricultural production needs. The market prospect of producing organic fertilizer from livestock and poultry manure is bright, the economic benefit is good, and it also has good social and ecological value.

3. Bioconversion technology

Bioconversion technology is the use of earthworms to digest livestock and poultry residues, usually dealing with the residues with a water content of more than 85% and high organic content, such as pig, cow, sheep and other poultry residues. The relevant studies showed that the available nitrogen, phosphorus and potassium in soil increased by 15.68, 10.71 and 24.30 mg/kg respectively, when earthworms are applied in soil compared with no use of them. In pig manure treated with earthworms, the organic nitrogen was more converted to inorganic nitrogen, and nitrogen volatilization was reduced. The contents of mineral nitrogen and available phosphorus in the undecomposed cattle manure inoculated with earthworms were significantly increased, and the activities of alkaline phosphatase (alp) were more increased, the carbon and nitrogen contents as well as the urease activity are more reduced, compared with the natural composting mature residues and the undecomposed cattle manure without the inoculation of earthworms. It can be seen that livestock manures which are processed by earthworms and maggots and then applied, can improve the fertility of manure, improve soil structure, increase soil permeability, prevent soil surface compaction, and improve soil fertility. Considering the whole ecosystem, this biological method can not only solve the pollution problem, but also improve the economic benefits of the livestock industry.

In addition to the livestock manure treatment technology mentioned above, the rapid drying method, bulking method, microwave method and other technologies can also be used to produce high efficiency and quality fertilizer.

(II) The energy conversion of livestock and poultry residues

It is mainly realized by direct combustion, biogas technology and power generation.

1. Direct combustion

Direct burning is mainly used in grassland areas, where herders collect dried cattle, horses and other animal manure and burn them directly as fuel for cooking and heating. Jilin Province uses animal manure, straw, coal ash and other materials to produce cow manure coal as energy. This is the most simple way to turn livestock residue into energy, but with the improvement of people's living standards, and this way is easy to produce a lot of smoke, air pollution, and health problems, this fuel is gradually replaced by other fuels.

2. Biogas technology

At present, China's large-scale livestock and poultry farms increase year by year, and most of the farms use water flushing method for cleansing, causing the manure with much water. For this kind of livestock manure, biogas technology is more applied.

Biogas method mainly uses anaerobic fermentation technology to produce combustible gas such as methane through anaerobic fermentation of biogas bacteria under certain conditions of temperature, moisture, pH and so on, in accordance with a certain proportion of manure, garbage, weeds and sewage. This method can kill the microbial pathogens in livestock and poultry manure, reduce the amount of biological sludge, achieve harmless production, achieve the purpose of environmental purification; It can also achieve the multilevel utilization of resources, that is, the comprehensive utilization of "three types of biogas products", that is, biogas can directly provide energy for farmers as gas fertilizer, biogas slurry can be directly used to fertilize fields, fish, etc., biogas slag can make efficient and high-quality organic fertilizer. "Three types of biogas products" can be regarded as a means of agricultural production, as fertilizer, feed, bait, which can be used for crop seed infiltration, pest control, improving the yield and quality of crops, fruits, and keeping fresh the storage of agricultural products. More importantly, through the link of "biogas", the planting and breeding are linked together to form a virtuous recycle system of ecological agriculture with multi-level and efficient utilization. In addition, as an emerging energy source, biogas has a wide range of uses. In addition to being used as a fuel for life, biogas can also be used for energy production.

China's current use of biogas technology to deal with livestock and poultry residues mainly has two kinds, one is agricultural users biogas ecological model and the other is the large-scale biogas project model.

(1) Agricultural users biogas ecological model

Agricultural users model has successful experience across the country, such as the "pool, pigsty, toilet, solar greenhouse" four-in-one ecological model, with biogas as link to adapt to the special environment of the northern winter, the "pigs-biogas-crops" energy ecological engineering, "devices-gas-pool" ecological engineering, etc.

The establishment of energy ecological engineering model based on biogas pools in rural areas can make livestock and poultry residues get good treatment, not only can increase farmers' income, but also can greatly improve farmers' production and living conditions, which is conducive to the improvement of rural ecological environment, and has profound significance to promote the sustainable development of rural economy.

(2) Large-scale biogas project model

Biogas engineering model mainly includes small biogas engineering model and large&medium-sized biogas engineering model, which is determined and optimized by the breeding model and the scale of breeding farm in different regions. This method can make rational use of residues and turn them into treasure.

Although at present, the use of biogas technology processing livestock and poultry residues has been mature, in practice, because of the construction of the pool and its corollary equipment investment is huge, the operation of the pool is strongly influenced by temperature, season, small biogas project is successful, but the operation condition for large and medium-sized project is not successful, which can not play the best economic benefits and environmental benefits.

3. The power generation

Livestock and poultry residues can be incinerated pollution-free and then used for power generation. The ash produced in the incineration process can also be used as high-quality fertilizer. The British Fibrowatt company uses chicken manure as fuel to generate electricity. Fujian Shengnong Group burns the mixture of rice husk and chicken manure to generate electricity. The annual consumption of chicken manure and rice husk mixture is about 250 000 tons, equivalent to saving about 88 000 tons of coal, which not only creates economic benefits, reduces environmental pollution, but also saves non-renewable resources such as coal and natural gas.

4. Thermochemical transformation of livestock and poultry residues

As a kind of biomass energy, livestock and poultry residues belong to renewable energy. The development and utilization of renewable energy has been paid more and more attention by the international community. China is a great country of energy consumption, and conventional energy reserves are relatively insufficient, so diversified energy configuration is the only way to solve our energy problem. Renewable energy is abundant on the reserves in our country, the development and utilization of new energy is significant to our country's energy strategy safety and sustainable development of the environment and economy. Thermochemical conversion is the main technology for developing biomass energy and is also the focus of research in many countries. Its basic principle is to heat the biomass, make its cracking at high temperatures (pyrolysis), the gas after pyrolysis mixed with the gasification medium (air, oxygen, water vapor, etc.) would have oxidation reaction and combustion, the resulting is a mixture of gases containing a certain amount of solid fuel such as charcoal, liquefied oil, biological oil, or biomass gas such as CO, H_2, CH_4, etc.

(III) Livestock and poultry residues as feed

Livestock poultry residues is not only high quality organic fertilizer, but also better feed resources itself. The crude protein content in livestock poultry residues is almost 50% higher than that in the feed of livestock and poultry. The livestock and poultry residues contain 17 kinds of amino acids, accounting for 8%–10%. In addition, manures also contain crude fat, crude fiber, phosphorus, calcium, magnesium, sodium, iron, copper, manganese, zinc and other nutrients (Table as follows).

Among them, chicken manure is the most rich in nutrients, with crude protein content accounting for 25% of the dry matter of chicken manure, equivalently 57%–66% of the bean cake. Moreover, it has a complete variety of amino acids, and is rich in minerals and microelements. Therefore, chicken manure can become a high-quality and efficient feed resource. In the United States, farmers feed cows directly with feed mixed with chicken manure and bedding grass, and the results are the same as those of soy cakes.

Harmful substances such as heavy metal elements, pathogens, parasites and others exist in livestock and poultry residue, so it needs to be properly treated to kill pathogenic bacteria, improve the digestibility and metabolism of protein, and improve palatability before it can be used as feed. Feed of livestock and poultry residue has been commercialized for many years in foreign countries, and the research work in China has been carried out for many years, and its commercialization has been realized in some areas. At present, the methods of livestock and poultry residue as feed are as follows.

1. Direct use as feed

This method is the simplest, requiring only chemical treatment of the residues and then directly used in animal feed. The raw material of this method is mainly chicken manure. Because the chicken's intestine is short, the digestion and absorption of feed is poor, about 70% of the nutrients in the feed are not digested and absorbed before being discharged from the body, so the chicken manure is rich in nutrients, and can be used as feeds for pigs, cattle. However, chicken manure contains complex components such as parasites, urea, and pathogens, so it needs to be treated with chemicals in advance to prevent cross-infection between livestock and poultry and spread of infectious diseases. The steps of processing drying chicken manure are simple, easy to operate, a wide range of raw materials, low processing cost, and the dried chicken manure contains rich nutrients, which can completely replace some refined, roughage and calcium, phosphorus and other additives, can greatly reduce the cost of feed, improve economic benefits, and can promote the development of aquaculture.

2. The ensiling method

The carbohydrate content in poultry manure is low, in order to adjust the proportion of feed and manure, control the moisture content, it can not be silaged alone, but to silage together with bluegrass feed, to prevent the crude protein loss in feces too much. This kind of feed has sour flavor and high palatability. Moreover, the storage method can kill microorganisms and pathogens in feces and improve the safety of feed.

3. The drying method

Drying method is the most common method for recycling livestock manure. This method can make livestock and poultry manure dry and dehydrated, can deodorize and kill eggs, can meet the requirements of health and epidemic prevention and production of commercial fertilizer. This method mainly uses thermal effect and spraying machinery. Drying method is mainly used to deal with chicken manure, and the advantages of the treatment is high efficiency, simple equipment, small investment, easy to promote. However, chicken manure is difficult to keep fresh in summer and has odor, so it is necessary to add additives with good odor effect such as lactic acid bacteria during processing for odor treatment, so that the cost increases. At present, the drying technology mainly includes natural drying by sunlight, rapid drying by high temperature, microwave drying, drying puffing and so on.

(1) Naturally drying by sunlight

This method is mainly in natural conditions or in the greenhouse to smash the manure, screen, remove the impurities, place them in a dry place, by the sunlight drying treatment. After drying they can be used as feed or fertilizer. This method is small in investment, low in cost and easy to operate, but it is small in scale, long in time, occupies a large area, and is greatly affected by weather factors. In addition, ammonia is easy to volatilize in the process of treatment, and the odor is large, which not only affects the fertilizer effectiveness, but also poses a threat to the environment, so it is not suitable for intensive livestock and poultry farms.

(2) Rapid drying by high temperature

This method uses machinery to separate solid and liquid feces and dry them, through high temperature, high pressure, heat, sterilization, deodorization and other processes to produce organic fertilizer. It is mainly to treat chicken manure, at the same time, it is also one of the more widely used methods to deal with livestock manure in China. Drying machine is mainly rotary drum dryer. The fresh chicken manure contains water content of 70% – 75%, through high-speed drying, it can achieve drying, deodorization, sterilization, storage effect. The advantages of high temperature rapid drying method are not affected by weather, can be produced in great mass, and can be dried rapidly, etc. , suitable for large livestock and poultry farms. However, it has the disadvantages of large input, large energy consumption, a large amount of odor in the drying process, and large water consumption.

(3) Microwave drying

Microwave drying is a treatment method that uses microwave to produce high temperature and quickly reduce the moisture content of wet livestock and poultry manure to below 13%. In the drying process, it can achieve the effect of disinfection, killing bacteria and eliminating odor, but this method has a large loss of nutrients and high cost.

(4) Drying puffing treatment

Drying puffing treatment is to make use of thermal effect and spraying mechanical effect, to make livestock and poultry manure puffing, loose. It can not only deodorize them but also can thoroughly sterilize, exterminate eggs, to meet the requirements of health and epidemic prevention, commercial fertilizer and feed. The disadvantages of this method are large investment, high energy consumption during drying and expansion, especially in summer, it is difficult to keep chicken manure fresh, and there is still odor in mass processing, and the cost is high, which leads to the application of this technology is limited. It is reported that if a farmer who raises 100 000 laying hens buys a puffing dryer that can handle 10 tons of chicken manure a day, the cost will be recovered in 7 – 8 months and the net profit will be 500 000 to 800 000 yuan per year.

4. Decomposing method

The decomposing method uses lower animals such as earthworms and flies to decompose manures in order to provide actin and dispose of livestock and poultry manure. Earthworms and maggots are very good protein feeds. With a protein content of 10% to 14%, earthworms can be used as live feed for aquaculture, as well as feed for pigs, cattle and sheep, while maggot dung can be used as fertilizer. This method is relatively economical and has significant ecological benefits, but it is difficult to operate technically and has harsh requirements on temperature, so it is difficult to produce and popularize throughout the year.

II. Treatment and comprehensive utilization technology of wastewater from livestock farms

The waste water of livestock and poultry farm mainly includes manure and urine of livestock and poultry and the washing water. The solid content is the highest, containing a large number of organic matter, such as nitrogen and phosphorus, suspended matter and microbial pathogens. The concentration of organic matter is high, and it is easy for biochemical treatment. However, the sewage discharge of livestock and poultry farms varies greatly due to the different feeding methods, management level, livestock house structure and dung cleaning methods. The large-scale pig farm waste water treatment is very difficult, the reasons are: ① most pig farms are using leaky seam type houses and water flushing, with large amount; ② the washing time is relatively concentrated, and the impact load is very large; ③ the amount of manure and sewage is large and concentrated, and the agricultural production is seasonal, and the surrounding farmland can not absorb all; ④ The concentration of organic matter in wastewater mixed with solid and liquid is high, and the viscosity is high. The relevant data shows that: the annual production of 10 000 pigs farm, the daily sewage volume is 73 tons, and the urine volume is about 1.05 tons.

The main treatment methods of livestock and poultry waste water are as follows.

(I) Ecological return to farmland

It is a traditional treatment method to return livestock and poultry wastewater as fertilizer directly to the field for agricultural cultivation, which is simple, cost-effective and widely used. This method can not only prevent the livestock wastewater from being discharged to the environment, but also recycle the useful nutrients in the wastewater to the soil-plant ecosystem, reduce the amount of chemical fertilizer, and realize the recycling of breeding wastewater. However, this method also carries certain risks, such as the risk of spreading microbial pathogens, and the possibility of soil contamination when applied improperly or in excessive amounts.

(II) Physical processing technology

1. Solid-liquid separation technology

Solid-liquid separation is a pretreatment step of livestock and poultry wastewater, which is mainly to separate the solid and liquid by settling, filtering, compressing, centrifuging and other methods. First, the solid waste is separated from the waste water by sedimentation, and then the solid waste is further treated by filtration to reduce the subsequent treatment load and cost.

Generally, 40%–65% of the solids can be removed and 25%–35% of the biochemical oxygen demand (BOD) can be reduced by using solid-liquid separation facilities such as screen.

At present, the commonly used separation equipment has rotary screen, inclined plate screen, belt filter and extruded separator. Solid-liquid separation in large-scale livestock and poultry farms is mainly achieved by mechanical separation through sieving and extrusion. The solid-liquid separation technology has low cost, simple process and equipment structure and convenient maintenance. However, both solid and liquid after separation need further treatment to meet relevant requirements.

2. Medium adsorption method

This method mainly uses adsorption medium materials with large adsorption capacity to adsorb and pretreat the nitrogen and phosphorus in livestock and poultry wastewater. Different adsorbents can be selected according to different kinds of pollutants in wastewater, which can achieve the purpose of treating some pollutants. Wang Yaping et al. applied attapulgite clay in the treatment of wastewater, and the ammonia and nitrogen removal rate reached 75.1%. Hang Xiaoshuai et al. (2012) studied the adsorption and removal performance of three kinds of red clay on phosphorus in wastewater. The phosphorus removal rate of the three kinds of red clay reached 90% in wastewater with a phosphorus content of 35 mg/L, and 85% in wastewater with a phosphorus content of 50 mg/L, which was significantly better than that of activated carbon. Yu Hupeng et al. (2009) made a new attapulgite adsorbent according to the ratio of attapulgite to rice husk 9 : 1 for adsorption and treatment of $NH_4^+ - N$ in the wastewater, and the highest removal rate of $NH_4^+ - N$ could reach 87%. Gao Meng et al. studied the trapping effect of modified chitosan on Cu^{2+} and Zn^{2+} in wastewater, and the results showed that modified chitosan had a good treatment effect on actual wastewater in complex system, and the residual concentration of Cu^{2+} and Zn^{2+} could meet the national discharge standard.

3. Chemical oxidation method

Chemical oxidation is the oxidation decomposition of organic matter and inorganic matter in water by using oxidants with high oxidation potential energy to produce strong oxidizing free radicals. This is a new water treatment technology. Hyunhee Lee et al. use Fenton oxidation method to treat livestock and poultry wastewater with COD concentration as high as 5000 – 5700 mg/L. When the Fe^{2+} concentration is 4700 mg/L and H_2O concentration is 1.05 times of the initial COD concentration of wastewater, after 30 minutes of reaction, the removal rate of COD can reach more than 80% or even 95%. Kaan Yetilmezsoy et al. used Fenton oxidation method to treat wastewater digested by UASB (upflow anaerobic sludge bed). The removal rates of COD and chroma in anaerobic effluent reached 95% and 96% respectively.

Electrooxidation combined with electroreduction can effectively remove the organics and heavy metals. Ouyang Chao et al. (2010) used electrochemical oxidation method to treat pig wastewater. After 3 hours of reaction, the $NH_4^+ - N$ removal rate in pig wastewater could reach 98.22%, but the COD removal rate was only 14.04%. In addition, the electrochemical method can simultaneously remove pollutants such as antibiotics, hormones and heavy metals from wastewater. Li Wenjun et al. (2011) used UV/H_2O_2 combined oxidation method to treat wastewater containing antibiotics (sulfamethoxazole, sulfadimethaxine, sulfamethazine etc.). Under the conditions of UV wavelength of 254nm, antibiotic concentration of 2.0 mg/L, H_2O_2 dosage of 7.0 mmol/L, pH value 5.0, after One-hour reaction, the removal rate of 5 antibiotics in wastewater can reach more than 95%.

4. Biological treatment technology

Biological treatment technology mainly includes natural treatment technology, anaerobic treatment technology, aerobic treatment technology and anaerobic-aerobic combined treatment technology.

(1) Natural treatment technology

Natural treatment technology is the traditional treatment method of livestock and poultry wastewater, mainly using the physical, chemical and biological comprehensive action of the natural water, soil and creatures to purify sewage. The purification mechanism mainly includes filtration, retention, precipitation, physical and chemical adsorption, chemical decomposition, biological oxidation and biological absorption. Its principles involve the symbiosis of species, the principle of material recycling and regeneration, the principle of coordination between structure and function, and the mechanism of multi-level retention, storage, utilization and transformation of nutrients in the ecosystem. This method has the characteristics of low investment, simple process and less power consumptionmode, but the purification function is restricted by natural conditions.

The main natural treatments include oxidation pond, soil treatment and constructed wetland treatment. Oxidation pond, also known as biological stabilization pond, is a structure that uses natural or artificial pond for biological sewage treatment. The purification process of sewage is similar to the self-purification process of natural water body. The sewage stays in the pond for a long time, and organic pollutants are degraded by the metabolic activities of microorganisms in the water. Dissolved oxygen is provided by algae through photosynthesis and reoxygenation on the pond surface, and also by artificial aeration. Oxidation pond is mainly used to reduce organic pollutants in water, increase the content of dissolved oxygen, properly remove nitrogen and phosphorus in water, and reduce the degree of eutrophication of water.

Soil treatment, different from seasonal sewage irrigation, is a perennial sewage treatment method. It is to put sewage in the land soil, use the ecological system of soil, microbes, plants, carry on a series of physical, chemical and biological purification process on the pollutants in wastewater, make the water quality purified, and through the system of nutrients and water recycling make green plants grow, so as to realize the circulation, innocuity and stabilization of waste water.

Constructed wetland can remove suspended matter, organic matter, nitrogen, phosphorus and heavy metals in wastewater through precipitation, adsorption, barrier, microbial assimilation and decomposition, nitrification, denitrification and plant absorption. In recent years, more and more attention has been paid to the research of constructed wetland. Ye Yong et al. used mangrove plants such as Bruguiera gymnorhiza and Kandelia candel to treat nitrogen and phosphorus in livestock and poultry wastewater, and the results showed that the two plants had better removal effect on nitrogen and phosphorus. Liao Xinti and Luo Shiming respectively used vetiver and windmill grass as vegetation to build constructed wetlands. The removal rates of pollutants in the wetlands were different with different seasons. The removal rates of COD_{Cr} and BOD_5 could reach more than 90% and 80% respectively.

Due to the low investment and low operating cost of natural treatment, it is a relatively economical treatment method when there is enough land available. The natural ecological treatment of livestock and poultry wastewater is suitable for domestic conditions, especially suitable for the treatment of small livestock and poultry farms, which has a broad application prospect.

(2) Anaerobic biological treatment technology

Anaerobic biological treatment of livestock and poultry sewage is a treatment method that uses anaerobic microorganisms to purify organic matter in sewage under anaerobic conditions. In the absence of oxygen, anaerobic bacteria in sewage decompose carbohydrates, proteins, fats and other organic acids into organic acids, and then under the action of methanogens, further ferment to form methane, carbon dioxide and hydrogen, so that the sewage can be purified.

The anaerobic biological treatment system is mainly composed of anaerobic reactor, biogas collection system, purification system, storage system, usage system and supporting pipelines, biogas slurry and biogas slag collection and treatment system. The selection and design of anaerobic reactor type can be determined according to the type and process route of livestock and poultry pollutants. The BOD load of anaerobic biological treatment of livestock and poultry wastewater is high, generally 3.5 $kg/(m^3 \cdot d)$, and the removal rate can reach more than 90%. Anaerobic biological treatment produces biogas through anaerobic fermentation, which can reduce the content of COD and BOD in sewage and realize recycling. This technology has less investment, less energy consumption, no need for special management, and low operating cost, so it has been widely used in livestock and poultry wastewater treatment, especially in the field of high concentration organic wastewater treatment. However, because the anaerobic effluent by this technology is difficult to meet the discharge standards, it must be used in combination with other technologies.

The construction model of biogas project in livestock and poultry farm using anaerobic biological treatment technology is mainly "ecological model". This model can make all the livestock manure sewage into the treatment system, the feed TS (total solid) concentration can reach more than 10%, can be used according to the specific situation of anaerobic tank such as full mixed anaerobic tank, the biogas slurry and biogas slag produced can be used for comprehensive utilization, as organic fertilizer for planting industry. The biogas produced can be used as fuel for farmers or to generate electricity. This method is a good ecosystem linked by biogas, which is worth popularizing. However, the construction of this "ecological model" project requires that there should be enough farmland around the livestock and poultry farms to absorb a large amount of biogas slag and slurry.

(3) Aerobic biological treatment technology

The basic principle of aerobic biological treatment technology is to use microorganisms to decompose organic matter under aerobic conditions and synthesize self-cells at the same time. Aerobic microorganisms take organic pollutants in sewage as substrates for aerobic metabolism. After a series of biochemical reactions, they release energy step by step, and finally stabilize as low-energy inorganic substances to achieve the requirements of harmlessness. At present, the commonly used aerobic biological treatment methods for livestock and poultry wastewater mainly include activated sludge process, biological filter, biological turntable, SBR and A/O etc.

Aerobic technology is used to treat livestock wastewater, and the combination of hydrolysis and SBR is the most popular technology in this field. Sequencing Batch Reactor (SBR) process, namely sequencing batch activated sludge process, is an intermitten activated sludge process improved and developed based on the traditional FillDraw system. It converts the sewage treatment structures from space series to time series. In the same structure, it completes the recycle of water filling, reaction, precipitation, drainage, and static disposition. When SBR combined with hydrolysis method was used to treat livestock and poultry wastewater, the removal rate of COD_{Cr} was high. The removal rate of total phosphorus was 74.1%, and the removal rate of high concentration ammonia nitrogen was more than 97%. In addition, other aerobic treatment technologies have been gradually applied to livestock and poultry wastewater treatment, such as Intermittent Drainage Extended Aeration (IDEA), Circulating Activated Sludge System (CASS), Intermittent Cyclic Extended Aeration System (ICEAS).

The aerobic biological treatment can remove most of the nitrogen, phosphorus, organic matter and so on, generally used in the latter steps in the anaerobic digestion treatment.

(4) Anaerobic-aerobic combined treatment technology

Due to the complexity of livestock and poultry wastewater properties, changeful ingredients, higher organic load and nitrogen and phosphorus content, the single process is not ideal in the economic costs and the treatment effect, so a combination method of system processing is often used in order to make up for the inadequacy of a single method. One of the commonly used method is the combinations of anaerobic-aerobic treatment technology. Neither anaerobic technology nor aerobic technology can achieve the discharge standard of livestock and poultry wastewater when treated alone. But in combination with their respective advantages, the combined anaerobic-aerobic processing can both overcome the disadvantage of energy consumption and a big area of aerobic treatment, and overcome the defect of not meeting the discharge demand after the anaerobic treatment, which has the advantages of low investment, low operating cost, good effect of purification, high energy environment comprehensive benefits, particularly suitable for large-scale livestock and poultry farm wastewater treatment. Therefore, most economically developed and intensive scale livestock and poultry farms use anaerobic (anoxic)-aerobic combined treatment process.

For example, Hangzhou Xizi farm adopts anaerobic and aerobic combined treatment process. After treatment, the COD_{Cr} content in the wastewater is about 400mg/L, and the BOD_5 content is 140mg/L, basically meeting the wastewater discharge standard. Li Jinxiu et al. used ASBR-SBR combined reactor system. ASBR, as a preprocessor (anaerobic), was mainly used to remove organic matter, and SBR (aerobic) was used for biological nitrogen removal. Membrane bioreactor (MBR) is a new biochemical reaction system which combines membrane separation technology with bioreactor. It uses membrane to replace the traditional secondary sedimentation tank, has the advantages of stable effluent, high concentration of activated sludge, strong ability to resist impact load, less residual sludge, compact device structure, less land occupation and so on.

The representative of the combined treatment method is the "environmental protection model", which has high requirements. The farm must implement strict clean production, make dry and wet separation, and flush sewage and urine. Sewage should be strictly pretreated to strengthen solid-liquid separation and precipitation and control SS concentration. Upflow anaerobic sludge bed reactor (UASB) is used in the anaerobic digester, and the COD of the anaerobic effluent is controlled at 1000 mg/L. Aerobic treatment can use SBR, and the sludge produced in the treatment process can be made into organic fertilizer or as bacteria. Natural treatment methods such as oxidized pond and constructed wetland can be used for post-treatment. In this way, the investment is less, the operation management cost is low, the energy consumption is less, the sludge quantity is less, the impact on the surrounding environment is smaller, but the land occupation requirement is larger, the temperature requirement is higher.

Livestock and poultry wastewater is relatively an organic wastewater difficult to treat mainly because of its large displacement, low temperature, mixed solid and liquid in wastewater, high organic content, small volume of solid, difficult to separate, and relatively concentrated washing time, so that the treatment process can not be continuous. Because COD and BOD in wastewater seriously exceed the standard, the amount of the suspended matter is large, nitrogen and phosphorus content is rich, ammonia and nitrogen content is high and difficult to remove, it is difficult to meet the discharge requirements by simply using physical, chemical or biological treatment methods. Therefore, the wastewater treatment of general livestock farms requires the use of a combination of treatment methods. According to the characteristics and utilization of livestock and poultry wastewater, the above different treatment technologies can be adopted.

Chapter 4 Utilization Model of Ecological Circular Agricultural Resources

Section 1 Pig-biogas-grain circular model

I. The principle and model connotation of systemic ecology

The pig-biogas-grain circular model refers to the production model of planning, designing, organizing, adjusting and managing livestock and poultry production, according to the principles of ecology and ecological economics, so as to maintain and improve the quality of ecological environment, maintain ecological balance, and maintain the coordinated and sustainable development of livestock and poultry industry. Under the rational arrangement of food production, grow grass and breed livestocks, fertilize the land using the manure of livestock and poultry, combine the farming and breeding to realize the sustainable development of our aquaculture industry.

II. The characteristics and key technologies of the model

The destruction of soil structure and the decline of soil fertility are closely related to the shortage and imbalance of water resources, fertilizer resources and energy resources, which have become the restricting factors for the development of "high yield, high efficiency and high quality" agriculture. Ecological model of pig-biogas-grain circulation will transform the single traditional structure of grain production to the binary structure of food crops and livestock and poultry breeding. The key to four circles of "planting, breeding, biogasing and fertilizing" is to change monocyclic technology for combination chain technology, integrate many advantages in one, improve the quality of the environment of agricultural production, and reasonably adjust the relationship in the process of production. The pig-biogas-grain cycle transforms excreta (waste) from one production process into inputs (raw material resources) of another, thus achieving a waste-free process (zero emissions) in agricultural production, that is, the wastes are recycling. Grain straw is returned to the field and coarse products as feed enter the large-scale breeding farm, and livestock and poultry manures enter the fermentation tank to generate biogas for energy circulation and solve the fuel problem. Biogas slurry returns to the field to reduce the amount of fertilizer, and biogas slag is used to produce special organic fertilizer, entering the agricultural production circulation system. We make full use of the resources in all aspects of the circulation system, reduce the amount of pesticides and fertilizers, and produce green and high-quality rice. To establish an ecological agricultural industry system based on large-scale intensive livestock farms, we should base on food crop production and make livestock industry as the leader, biogas energy as the link, organic fertilizer production as the driver, to form a virtuous cycle of feed, fertilizer, energy and ecological environment.

1. Key breeding technologies in pig-biogas-grain model

① Breeding and biological environment construction. In the process of livestock and poultry breeding, we should use advanced breeding technology and biological environment construction, to achieve the quality of livestocks, no pollution. Through dry cleaning technology on the livestock house and disease control technology, we should create good growth environment for the livestock and poultry, so as to achieve no disease or less disease.

② Solid-liquid separation technology. For large-scale livestock and poultry farms, the residues are rinsed with water, that is the manure and urine are discharged by water flushing, which not only pollutes the environment, but also wastes water resources, and is not conducive to the utilization of nutrient resources. The solid-liquid separation equipment was used for solid-liquid separation, the solid part was composted at high temperature, and the liquid part was fermented by biogas. At the same time, to reduce water consumption, dry cleaning technology is adopted as far as possible.

③ Sewage treatment and comprehensive utilization technology. We should use advanced solid-liquid separation technology, and utilize sewage treatment technology for the liquid part, such as oxidation pond, biogas fermentation and other aerobic and anaerobic treatment technology in the non-planting season to treat. Discharge must meet the standards. In crop growing season, water and fertilizer resources in sewage can be fully used to irrigate farmland.

④ Harmless high temperature composting technology of animal husbandry manure. Advanced solid-liquid separation technology and high temperature composting technology and equipment are adopted to produce high quality organic fertilizer and commercial organic and inorganic compound fertilizer.

2. Key technologies for efficient utilization of biogas slurry in pig-biogas-grain model

① Biogas slurry immersion sterilization technology. Different technical measures are adopted for different crop varieties in the planting field, mainly from the aspects of soaking time, biogas slurry dilution, biogas slurry dosage and so on.

② Application technology of biogas slurry in rice field. According to the law of fertilizer and water demand for rice growth, biogas slurry irrigation was carried out by controlling the concentration of biogas slurry at different times.

III. The technical operation specifications

After the large-scale breeding farm is built and the biogas process is completed, the advices of the biogas slurry utilization under the condition of rice and wheat rotation is mainly as follows.

1. Pipeline erection measures

Due to the high concentration of biogas slurry, it should be used after dilution. According to the existing water channels in the rice field, biogas slurry pipes should be arranged, control valves should be installed, and the application should be performed with irrigation water.

2. Measures for soaking rice in biogas slurry

Fresh seeds produced in the previous year or the same year shall be used for soaking and sterilization. Before soaking, seeds shall be screened, debris and chaff shall be removed, and the drying time shall not be less than 24 hours; Biogas slurry that has produced gas by normal fermentation for more than 2 months is selected. The seeds are placed in a bag that can filter water, and the bag is suspended in the liquid supernatant of the biogas pool. The temperature of biogas slurry is required to be above 10 ℃, and the pH is 7.2–7.6.

Conventional rice varieties were soaked once. The immersion time in biogas slurry: 48 hours for early rice, 36 hours for middle rice, 36 hours for late rice, and 6 hours more for glutinous rice, and then washed with water for bud acceleration.

The conventional rice varieties with poor stress resistance should be soaked by diluting biogas slurry twice with water, and the soaking time is 36–48 hours, and then clean water for bud acceleration.

The hybrid rice varieties should be soaked with intermittent biogas slurry, three soaking and three drying, and washed with clean water to accelerate the germination. The hybrid early rice was 42 hours, and each time soaking 14 hours, drying for 6 hours; For hybrid medium rice, it was 36 hours, each time soaking 12 hours, drying 6 hours; Hybrid late rice was for 24 hours, each time soaking 8 hours, drying 6 hours.

3. Irrigation management of combined application of biogas slurry fertilizer in paddy field

Rice should be transplanted in time and densely planted reasonably. Generally it is sown in mid-May, transplanted in early June, the seedling age is controlled at about 20 days, and the leaf age is 3–4 leaves. By machine transplanting, the distance between plants and rows was controlled at 12 cm × 30 cm, and about 18 500 holes were transplanted per 667 square meters, with 70 000 to 80 000 basic seedlings.

If the biogas slurry was applied instead of chemical fertilizer in the whole process, and diluted with irrigation water, the quantities of biogas slurry applied in the whole rice growing season ranged from 21 to 27 tons (equivalent to pure nitrogen 16 to 20 kg /667 square meters). The application ratio of biogas slurry in different growth stages was: base fertilizer: seedling fertilizer: stalk fertilizer: spike fertilizer = (3 : 3 : 2 : 2)–(4 : 2 : 3 : 1) is more suitable.

4. Pest control and management in rice fields

① Rice blast (spike plague). At bud break and full head stage, 900 grams of 75% tricyclazole and 900 kg of water were applied to prevent rice blast per hectare.

② Rice smut disease. At round bract stage and full heading stage, the rice smut disease was controlled twice per hectare with 1500 g of 40% dimethachlon or 1500 g of 20% Jinggang amycin, mixed with 900 kg water to spray evenly.

③ Bacterial blight. In areas with frequent occurrence of bacteria blight, 100 grams of bismerthiazol and 60 kg of water were used to spray for prevention and control at booting stage. For flooded fields and storm-hit fields, spray 100g of bismerthiazol with 60 kg of water to prevent bacterial blight after the rice leaves are dry and it is clear.

④ Rice planthopper. The control index of the pest density of 800 in the field can base on the following agents for control: 75% Emmerol 8000 times liquid for spray control; 25% Daoshijing (planthopper pestcide) 800–1000 times liquid; 15% Jinhaonian (gold good year, imidacloprid + buthiogram Budweiser) 30–40 mL, mixed with 60 kg water for spray prevention; 50%–70% imidacloprid 3–5 grams, mixed with 60 kg water for spray prevention.

5. Application management of biogas slurry in wheat fields

The wheat seeds should be sun-dried for 2–3 days in sunny days to improve the water absorption performance of the seeds. The biogas slurry with long fermentation time and good decomposition for normal use was used to soak seeds. The seeds should be soaked one day before the sowing, and the soaking time should be based on the liquid temperature, generally soaking 6–8 hours in 17–20°C biogas slurry. After soaking the wheat seeds for 6–8 hours, take out the seed bag, rinse with water, drain the water in the bag, and then spread the seeds flat, and wait for sowing until the surface moisture of the seeds is dry.

Biogas slurry as base fertilizer is evenly sprinkled on the field surface before ploughing, and then ploughed into the bottom layer of the soil, and topdress with urea at jointing stage. Wheat is sown in early November, and the sowing amount is 15 kg /667 square meters and usually about 4 tons of biogas slurry and 10 kg of urea are applied, which is mainly applied by mechanical ditch or spraying.

6. Wheat pest control management

① Control of wheat aphids with biogas slurry. Biogas slurry sprayed on wheat can prevent and control wheat aphids. The method is as follows: 14 kg of biogas slurry and 0.5 kg of washing powder solution (the solution is prepared according to the ratio of washing powder and water 0.1 : 1) are used to make biogas slurry compound agent and sprayed. 450 kg per hectare for the first day, and another spray for the next day. Spraying time is best in the morning on a sunny day.

② Biogas slurry control wheat scab. The best effect is to spray 750 kg of biogas slurry solution per hectare for one time in full bloom stage and for another spray every 3 – 5 days.

Section 2　Grass-rabbit-manure-rice circular model

I. The principle and model connotation of systemetic ecology

Grass-rabbit-manure-rice circular model is a cyclic agroecosystem. Grass, rabbit, rabbit manure and rice are the ecological factors of the system, and constitute a food chain and cycle chain. According to ideas of "breeding centralization, waste recycling, pollution reduction and ecological administration", the model focuses on the grass varieties, rabbit varieties, rice varieties, comprehensive collocation and implementation of technologies of crop layout, rice and grass cultivation technique, rabbit manure recycle to realize the energy cycle and substance flow throughout the cycle, and establish a dynamic, balanced ecological cycle model that can fully achieve high efficiency, high quality and high yield, and eventually produce high quality agricultural and animal husbandry products (rice, rabbits, etc.).

II. The characteristics of the model

The biggest characteristic of the model is that it fully realizes the grass-rabbit-manure-rice cycle, with no waste. Breeding rabbits, which is the core of the circulation model, through the study of the dry wet seperation of the manure, rabbit urine into the fermentation tank and the oxidation pond, with returning as liquid fertilizer directly, and the rabbit manure combined with straw compost to breed earthworms, will directly produce wormcast and organic fertilizer to the field. The rice and pasture could be grown rotation, pasture breeding rabbits, which formed a small ecological cycle chain.

III. The operating technical specifications

1. Key technologies of rabbit breeding

① Variety selection. Select suitable good rabbit variety for breeding, such as New Zealand big ear rabbit, etc. Caged breeding of meat rabbits shoud be adopted, for its characteristics of less activity, uniform feeding, easy management and fast growth. Rabbit cage is generally made of wire horizontal type or column type, and is provided with a trough for food and water, and grass rack. The cage floor is made with bamboo or wood board, the width requirements of each board being 2.0 – 2.5 cm, and the distance between the board being 1 cm. On the cage size, breeding rabbit cage should be larger than that of commercial rabbit, generally male and female rabbits and reserved breeding rabbit each required area of 0.25 – 0.40 square meters, and the ordinary rabbit is 0.12 – 0.15 square meters, requiring neat and compact layout in the house, and in line with the health and epidemic prevention regulations.

② Feeding method. When green feed is fed, proper supplement of fine feed can make meat rabbits grow fast and not easy to get sick, to improve high feed utilization rate, shorten feeding cycle, improve the slaughtering rate and increase economic benefits. The weight of meat rabbits can reach 2.3 – 2.7 kg after 70 – 80 days of feeding with fine feed and sufficient water.

③ Regular deworming and drug administration for prevention. Fasciola hepatica, ascaris, nematodes, coccidiodes, enteritis and other common diseases will affect the growth of meat rabbits. Regular deworming and prevention of meat rabbits at different growth stages can ensure the healthy growth of meat rabbits.

④ Regular immunization and disinfection. Rabbit fever, pasteurellosis and other epidemic diseases have a great impact on the production of meat rabbits. In weaning period, 4 months of age and other different growth periods, we should pay attention to the vaccine immunization of rabbit fever, pasteurellosis and other epidemic diseases, and often wash the rabbit house, keep clean and hygienic. At the same time, the rabbit farm should be regularly disinfected every 15 or 30 days, and a strict disinfection system should be established. In case of an epidemic, isolation and blockade measures should be taken in time. Regular immunization and disinfection work is very important to reduce the occurrence of diseases and stabilize production.

2. Harmless treatment technology of rabbit manure and urine

The solid-liquid dry and wet separation technology was adopted for rabbit manure and urine. Rabbit urine was directly fed into the sewage tank through the pipeline for fermentation treatment. After the treatment reached the standard, it was mixed with rainwater and directly discharged into the farmland as liquid fertilizer. The rabbit manure is transported to the greenhouse and mixed with rice straw for composting, breeding earthworms and producing organic fertilizer.

3. Green management technology of rice planting

① Variety selection. High-quality rice varieties with good agronomic characters, strong resistance to disease and insects and moderate growth period were selected.

② Sufficient base fertilizer. According to the lifetime nutrient requirements of different rice varieties, about 1000 kg of rabbit manure and earthworm organic fertilizer was applied every 667 square meters, and 15–25 kg of rice BB fertilizer as base fertilizer every 667 square meters. 15–20 kg of ammonium bicarbonate every 667 square meters was applied as topdressing fertilizer.

③ Timely transplanting, reasonable dense planting. According to the production period of young frogs, the sowing date of rice is generally sown in the middle of May and transplanted in the first ten days of June. The seedling age is controlled at about 20 days and the leaf age is 3–4 leaves. The plant and row distance are controlled at 12 cm × 30 cm, and there are about 18 500 holes per 667 square meters, with 70 000 to 80 000 basic seedlings.

④ Reasonable irrigation. Rice was irrigated frequently in shallow water from planting to tillering stage with a depth of 2–3 cm. When the number of ears was reached, rice was moderately stopped irrigating. The shallow water layer was maintained in booting stage, and the dry and wet conditions alternated in earing and flowering and seed filling stage.

4. Management technology of grasses

Grass species suitable for rabbits are: ryegrass, alfalfa, etc.

(1) Ryegrass

① Selection of sowing date: ryegrass likes warm and humid climate. The temperature of seed germination is above 13 ℃, and the seedlings can grow better at more than 10 ℃. Therefore, the sowing period of ryegrass is longer, both autumn sowing and spring sowing being appropriate. Autumn sowing is generally available from mid-to-late September to early November. With the delay of sowing date, due to the decrease of temperature after sowing and the late emergence of seedlings, tillering occurred later and less, the number of fresh grass harvesting decreased, and the yield decreased.

② Selection of sowing quantity: in a certain area, the sowing quantity is small and the growth will be good, but the density is too small, it will affect the total production of fresh grass per unit area, especially in the early stage. On the contrary, if the sowing quantity is too large, the yield of fresh grass is not necessarily high, and the individual growth and development are also affected. Therefore, only reasonable dense planting can give full play to the production potential of ryegrass and increase the yield per unit area. Generally, 15 – 22.5 kg per hectare is the best sowing quantity.

③ Sowing method: ryegrass seeds are small, requiring shallow sowing. Rice stubble field soil water content is high, the soil is heavy, and when autumn sowing it often continuously rains, or due to the autumn harvest labor tension it delays sowing and affects the emergence of seedlings. In order to make ryegrass germinate quickly and neatly, use 150 kg of calcium magnesium boron zinc fertilizer per hactare, 300 kg of fine soil mixed with seeds per hactare to sow.

④ Ryegrass harvesting: ryegrass regeneration ability is strong, which can be repeatedly harvested, therefore, when using ryegrass as feed, it should be timely harvested. The number of ryegrass harvesting was mainly affected by sowing date, temperature during growth and fertilization level. Autumn-sown ryegrass grows well and can be harvested multiple times. In addition, the higher level of fertilization will make ryegrass grow faster, so they can be harvested in advance, at the same time increase the number of harvesting. On the contrary, poor fertility, slow growth of ryegrass can not reach a certain amount in a short period of time, so it can not be harvested and utilized. Harvest at the right time, that is, when the ryegrass grows to 25 centimeters or more, if the plant is too short, the fresh grass yield is not high, and the harvesting operation is difficult. The stubble height at each harvest is about 5 cm to facilitate the regeneration of ryegrass.

⑤ Seeds of ryegrass: mature ryegrass seeds have strong pelleting ability. Therefore, when ryegrass ears change from green to yellow, the spikelets in the middle and upper part turn yellow, and the glumes below the spikelets are still yellowish green, harvest should be timely. In order to prevent of grain falling when harvesting, it is best to harvest in the morning when there is dew or rain. And we should cut and put them lightly, dry them timely, thresh them, sun dry and clean them. If the farm work is tight, labor is insufficient, or the weather is not good, they can not be timely threshed and dried, the ryegrass should be hung in a dry and ventilated place to prevent mould and ensure seed quality.

(2) Purple alfalfa

① Preparation before sowing: Sun dry the seeds 2 – 3 days before sowing can break dormancy, improve germination rate and seedling uniformity. For the fields that have never been planted with alfalfa, the seeds should be treated with rhizobium. It can improve the seedling formation rate, seedling nodulation rate and grass yield of alfalfa. It can also increase soil fertility and promote the growth of later crop. Common inoculation methods are old soil mixing seeds, nodule bacteria agent mixing seeds, etc.

② Sowing amount and sowing date: If mechanical strip sowing is used, the sowing amount per hectare is 15 – 22.5 kg, and when broadcast sowing, the amount per hectare is about 30 kg. Purple alfalfa can be divided into spring sowing, summer sowing and autumn sowing according to need. Spring sowing: in the spring after the thawing of the land, and sowing at the same time with the other crops, and the spring sowing alfalfa have a good development in the year. Summer sowing: if the spring is under drought, soil moisture is poor, they can be sown in summer after rain. Autumn sowing: Autumn sowing should be no later than mid‑August, otherwise it will reduce the overwintering rate of seedlings.

③ Sowing method: there are strip sowing and broadcast sowing, according to the specific situation. For the seed field, we should use spot sowing, using the method of hole sowing or strip sowing, generally the row spacing 50 cm, hole spacing 50 – 70 cm. For the grass field, we can use strip sowing or broadcast sowing, and the general row spacing is about 30 cm. When sowing, the utility model should be plowed shallow first, then sow, harrow, and finally press for moisture retention. After sowing, soil moisture content should be pressed in time to ensure full touch between seeds and soil, which is conducive to water absorption and germination of seeds. Pression also has the role of raising soil moisture, which is conducive to keeping the superficial soil moist where the seeds are located.

④ Reasonable fertilization: Combined with ploughing before sowing, apply more than 7.5 tons of farmyard fertilizer per hectare, 750 kg of high‑quality superphosphate, 150 kg of potassium sulfate, 225 kg of urea as base fertilizer. 300 kg of superphosphate or 75 kg of diammonium phosphate was applied per hectare after each harvest. Topdressing fertilizer is generally carried out combined with irrigation in spring after the seedlings turn green, in the branching stage, and in the budding stage. In the first year, before the formation of rhizobia at seedling stage, 75 – 150 kg urea was applied per hectare combined with irrigating.

⑤ Disease and insect control: summer is a high incidence period of alfalfa diseases and insect pests, the main diseases are rust and downy mildew, 25% Fenxiuning powder 1000 – 1500 times liquid for spray control; Insect pests mainly are aphid, slime, leaf rope, beet armyworm, thrips, stink bug and so on. The specific alternative is to choose 4.5% cypermethrin emulsion, 2.5% deltamethrin emulsion, 10% imidacloprid emulsion and other insecticide for spray control. It is strictly prohibited to use highly toxic and high residue pesticides in chemical control. According to the harvest time, determine the reasonable safety interval, to prevent environmental pollution and harmful residues in the plant to cause livestock poisoning.

⑥ Timely harvest: the yield of the first two crops of alfalfa accounts for about 70% of the annual yield, with good quality, good commodity. Grass fields should be harvested in time from the first flowering period to the full flowering period, and stubble 4–5 cm should be retained when harvesting to ensure adequate nutrient accumulation. Minimize drying time in the field after harvesting. After binding, choose a ventilated shelter to air dry them.

Section 3　Grass-sheep-manure-grain circular model

I. The principle and model connotation of systemetic ecology

Grass-sheep-manure-grain circular model is a cyclic agricultural ecosystem, in which pature, sheep, sheep manure and urine, and rice are the ecological factors of the system, forming a food chain and recycling chain, which can realize the whole energy cycle and material flow in a cycle, to establish a dynamic and balanced ecological circular agriculture model that can fully realize high efficiency, high quality and high yield, produce high quality products (rice, Huyang sheep, etc.) , achieve ecological, high quality, safe and efficient, and have good ecological and social benefits.

II. The characteristics of the model

The most important feature of this model is to form an environmentally friendly agricultural ecological model through the cycle chain of grass-sheep-manure and grain. The model focus on the standardized ecological breeding sheep, which is the core of the circular model, adopts advanced high leak bed plate and faeces and urine collection processes. By separation of the manure and urine, the sheep urine collected by sewage pipeline into the wastewater collection pool, then pumped into three-graded ponds for anaerobic fermentation disposal, then pumped into the water storage tank returning directly to field according to the need of crops. After harvesting and ploughing ryegrass (sowing in autumn and harvesting in summer) , before sowing rice, the treated sewage as liquid fertilizer is pumped into the underground pipe with manure pump, mixed with water, and then poured into the farmland as base fertilizer. At the tillering stage, the rice was irrigated once or twice as topdressing fertilizer.

The sheep manure combined with straw after high temperature fermentation, are made into high quality organic fertilizer for the grass or rice, providing nutrients for them, and the grass or staws can feed sheep, the sheep manures return to field, which is benificial for the grass and rice planting. Thus they produced high quality sheep and rice, forming ecological circulation chain with good economic benefit and social benefit.

III. The operating technical specifications

1. Key technologies of Huyang sheep breeding

The following work should be done on the basis of selecting suitable Huyang sheep varieties.

① Building houses. Sheep house should be selected in the high dry sunny leeward places, with good ventilation and heat preservation. The sheep house adopts the production technology of high bed leaky seam board and manure & urine collection, and they should be bred in different pens. There should be enough troughs and activity space in the pens. Each sheep should have 1.5 to 2.5 square meters of activity space, with 2 to 3 meters of feeding area between pens. The layout of the house is neat and compact, in line with the health and epidemic prevention regulations.

② Separating rooms. In accordance with the factory production mode, the sheep of different ages, different breeds and different body conditions should be raised in separate rooms, and special delivery rooms and lamb houses, meat sheep rooms, female sheep rooms, ram sheep rooms and sick sheep isolation rooms should be set up, and corresponding feeding and management measures should be taken.

③ Mating. Selecting the ideal hybrid combination and suitable mating period for mating, the mating rate in estrus period reaches more than 90%, the total conception rate reaches more than 95%, and the initial mating weight of ewes should reach more than 70% of the adult body weight for mating. Pregnant ewes should be well protected, and make the fetus develop well. Do not feed moldy, spoiled, frozen or other abnormal feed. Do not let them drink water and icy residue on an empty stomach. In daily management, there should be no drastic action such as scare and drive, especially when sheep go in or out of the door or supplement feeding, to prevent mutual extrusion and avoid abortion. Late gestation ewes should be given supplementary feeding and should not be vaccinated.

④ Delivering the lambs. According to the local climate, delivering lambs is carried out in the warm shed, with soapy water or 2%–3% lysol to clean the ewe udder and hind body, after delivering timely broke the umbilical cord, and do such jobs as birth weight, feeding colostrum, observation the elimination of meconium.

⑤ Feeding. Use coarse material as the meal, refined feed as a supplement. The best scale feeding is planting alfalfa, ryegrass for "grass planting and sheep raising", so that all the year round we can have fresh grass. After autumn harvest, grass can also be made into hay or silage for sheep. In addition to a certain amount of corn, the refined feed should be proportioned with soybean meal, bran, fish meal, bone meal and other protein feeds. Lambs should be fed colostrum within 30 minutes after birth and trained to consume refined feed and hay from 5 to 7 days of age. Within 1 month of age, 0.05 – 0.1 kg of refined feed and 0.1 kg of hay were supplemented daily. At the age of 1 to 2 months, 0.15 – 0.2 kg of refined feed, 0.3 – 0.5 kg of hay, and 0.2 kg of silage were supplemented daily. At the age of 3 months, they were fed 0.2 – 0.25 kg of refined feed, 0.5 – 0.8 kg of hay, and 0.2 – 0.3 kg of silage. Generally they should be weaned at 1 – 2 months of age. For the growing sheep, coarse and refined feed are used, and commercial refined feed and coarse feed are selected. The daily feed amount of meat sheep should be adjusted according to the different growth stages of sheep, generally 2.3 – 2.7 kg per day. In order to adjust the distribution of feeding amount, it is necessary to adjust the feeding amount according to the standard of basically no leftover feed in the trough, and to ensure the timing and quantity of feeding, with a special person in charge. The composition of refined feed was 55% – 75% corn, 5% – 15% wheat bran, 10% – 25% soybean meal (cake), 0.5% – 1.5% limestone powder, 1.0% – 1.5% salt, 0.5% – 1% premix additives, selenium and vitamin A. The coarse feed (roughage) consists of grass, hay, or silage corn. In the first month of fattening period, the ratio of refined and roughage accounted for 60% and 40% respectively, and the ratio in the later period was 1 ∶ 1. The body weight will meet the standard requirements in six months, generally ram should reach 45 – 50 kg, ewes 35 – 40 kg.

⑥ Disease prevention. Raising Huyang sheep in groups is easy for them to get sick, mainly lamb dysentery, streptococcal disease, infectious pus scar, nematode, sheep lice and so on. To carry out the policy of "prevention is more important than treatment", we should also do a good job of vaccination and comprehensive prevention and control of key diseases. In March, June and December de-worming should be carried out respectively. Abamectin injection (0.2 mg/kg body weight, subcutaneously) is used for ectoparasites, and thiophenimazole (15 to 20 mg/kg body weight, orally) is used for internal parasites. In spring and autumn each year, the vaccine (rapid epidemic disease, cataplexy, lamb dysentery, enterotoxemia) was injected once, and 5 ml was injected into muscle or subcutaneously. The immune period for enterotoxemia was 6 months, and for other diseases it was 1 year.

⑦ Comprehensive management. 3 days after the birth of the lamb, ear number or ear tag should be made; Within 10 days after birth, the tail was severed by ligation at the third and fourth caudal vertebrae. The non-breeding male lamb is castrated by ligation or by operation 1 – 2 weeks after birth, which is beneficial to improve the quality of the meat (reduce the smell), and can make the temperament docile, easy for management and fast growing. In addition, the breeding ram, ewe also should regularly be bathed on the hoof and hoof trimmed; After shearing hair in spring, the medicine bath is done for one time, and the numbers can be increased appropriately in areas with more serious parasitic diseases. To regularly (1 week or so) rotate the selection of different types of disinfectants, such as 20% lime milk, 2%–5% alkali, 10% bleach solution, 3% formalin, 10% poison (Baidusha), etc., to disinfect the sheep house, sports field, feeding trough, drinking vessels, feeding tools and enclosure cake, as far as possible to make clean the sheep rooms, sheep body, trough, and utensils. The dead sheep should be buried or burned to prevent the spread of infectious diseases. Large-scale disinfection should be carried out once a year in the spring and autumn. The liquid medicine in the disinfection pool at the gate and entrance of the farmland should be changed frequently, and the effective concentration should be maintained. Unrelavent staff are declined.

2. Sheep manure and urine harmless treatment technology

For Huyang sheep breeding, the high-bed leak board technology is adopted to separate dry and wet manure and urine. The dry manure is collected and shipped to the company's organic fertilizer plant to mix rice straw for aerobic fermentation compost, producing high-quality organic fertilizer. Sheep urine (urethral fluid) and cleaning sewage are leaked into the sewage collection tank, pumped into the tertiary treatment tank for anaerobic fermentation and harmless treatment, put into the sewage storage tank for temporary storage, and directly returned to the field according to the needs of crops. After harvesting or plowing of ryegrass (autumn sowing and summer harvesting) and before sowing rice, the harmless treated urines are pumped into the field as base fertilizer. In the tillering stage of rice, it is necessary to irrigate topdressing for 1–2 times according to the actual situation.

3. Rice green production management technology

① Variety selection. High-quality rice varieties with good agronomic characters, strong resistance to disease and insects and moderate growth period were selected.

② Sufficient application of base fertilizer. According to the nutrient requirements of different rice varieties, 1000–1500 kg of sheep manure organic fertilizer was applied every 667 square meters, supplemented by 20–25 kg of rice BB fertilizer (ammonia : phosphorus : potassium = 24 : 8 : 10) as base fertilizer. In the tillering stage, topdressing urea 10–15 kg or ammonium bicarbonate by 15–20 kg, earing fertilizer should based on the leaf color situation, apply 5 kg of urea as topdressing, if the leaf color is too green, earing fertilizer should not be applied, so as to avoid late mature or lodging. When topdressing, we also can reduce the dosage of chemical fertilizers, appropriately utilize the harmless manure and urine, 120–150 tons per hectare for irrigation.

③ Timely transplanting, reasonable dense planting. According to the stubble arrangement, determine the sowing date of rice. Generally sowing in mid-May, transplanting in early June, the seedling age controlled at about 20 days, the leaf age of 3-4 leaves. They can be transplanted by machine, and plant and row spacing is controlled at 12 cm × 30 cm, about 18 500 holes per 667 square meters, and the basic seedlings are up to 70 000-80 000.

④ Reasonable irrigation. Rice was irrigated frequently in shallow water from planting to tillering stage with a depth of 2-3 cm. When the number of ears was reached, irrigation was moderately stopped, the shallow water layer was maintained in booting stage, and the dry and wet conditions alternated in heading and flowering and filling stage.

4. Production and management technology of grass

The ryegrass is suitable for Huyang sheep.

① Selection of sowing date. Ryegrass prefers a warm and humid climate. The suitable temperature for seed germination is above 13 °C, and the seedlings can grow well at more than 10 °C. Therefore, the sowing period of ryegrass is longer, both autumn sowing and spring sowing. Autumn sowing is generally available from mid-to-late September to early November.

② Selection of seeding quantity. The suitable sowing quantity per hectare is 15-22.5 kg, and the reasonable dense planting can ensure the good development, give full play to the production potential of ryegrass, and increase the yield per unit area.

③ Seeding method. Ryegrass seeds are small, requiring shallow sowing. In order to make ryegrass seedlings come out quickly and neatly, we should mix 150 kg/hm^2 of calcium magnesium phosphate fertilizer with 300 kg/hm^2 fine soil before sowing.

④ The harvesting of ryegrass. Ryegrass' regeneration ability is strong, so they can be repeatedly harvested. When the ryegrass grows to 25 centimeters or more, it should be harvested; If the plant is too short, not only the fresh grass yield is not high, but also the harvesting operation is difficult. The stubble height at each harvest should be about 5 cm to facilitate the regeneration of ryegrass.

⑤ Fertilization of ryegrass. Generally, ryegrass does not need fertilizer alone. Combined with the harmless treatment of manure and urine from Huyang breeding, after each fresh grass harvest, the manure and urine are pressurized with a pump and then irrigated or sprayed, with the consumption of 150-225 tons per hectare.

⑥ Seeds of ryegrass. The mature ryegrass seeds are easily to fall, so when the ryegrass ear change from green to yellow, the middle and upper spikelets turn yellow, and the glumes below the spikelets are still yellowish green, it should be harvested in time. In order to prevent grain falling during harvest, it is best to harvest in the morning when there is dew or rain. When they can not be timely threshed, they should be hung in dry and ventilated place, in case of mouldy to ensure seed quality.

Section 4 Rice-vegetable-earthworm circular model

I. The principle and model connotation of systematic ecology

Rice-vegetable-earthworm circular models change the rice and wheat rotation to rice and vegetable rotation, through the rice and frost resisting vegetables to complete a rotation of production cycle. Straw and a certain amount of organic fertilizer are applied, with the earthworms, the most valuable soil animal on the earth. Rice straw, vegetables leaf, organic fertilizer and soil humus are integrated as the earthworms' bait, using earthworms' life activity to regulate the soil microorganisms. This not only can make the earthworms' swallow of straw and vegetables leaf residue accelerate the straw recycle, but also can change the soil physical property by earthworms' activity, adjust soil microbial activity, speed up the straw decomposing, and realize the straw returning. This model can change the short transition period of rice and wheat stubble and the straw can not be fully decomposed to the field, so as to affect many problems in the process of rice and wheat cultivation, especially the effect of soaked field on wheat straw before rice planting. Moreover, it can not only achieve the three harvests of rice, vegetables and earthworms, but also can treat straw and residual leaves on the spot, reduce the amount of chemical fertilizer, improve the output efficiency of unit land, and achieve the sustainable and green development purpose of increasing income and saving expenditure.

II. The characteristics of the model

① Enrich the production mode, increase economic benefits. Establishment of the rice-vegetable-worm cycle model such as "utilization in the idle winter, filling in the spring shortage, usage in both winter and spring, comprehensive development, perennial cultivation" according to local conditions, increased unit land output and improve the income of farmers.

② Improve the farmland environment and enhance the demonstration and leading effect. The straw and other agricultural residues can be recycled and utilized, which improves the utilization rate of resources and improves the ecological environment. This model cultivates a number of scale subjects, plays a positive role of demonstration and leading, and has

Chapter 4 Utilization Model of Ecological Circular Agricultural Resources

remarkable social and ecological benefits.

III. The operating technical specifications

1. Rice technical operation code

1) Variety selection. Rice varieties or combinations with high yield, high quality, low stalk and strong stress resistance were selected, such as "Hua You 14" and "Han You Xiang Qing".

2) Seed treatment. Taking the first day when the average daily temperature was stable through 15 °C as the initial sowing date; the seeds were sown on March 20 every year. Before sowing, the seeds were selected and sun-dried, and then the seeds were intermittently soaked to accelerate germination for 24 – 48 hours. Germination was divided into three stages: the first stage of high temperature for germination, at 32 – 35 °C for 10 hours; the second stage of suitable temperature for root shoots, at 24 – 30 °C for 24 hours. The third stage is to dry them to promote germination.

3) Rice field preparation. The field should be land with loose soil, flat terrain, low groundwater level, convenient drainage, clean water, and they are applied with 15 tons/hm^2 of decomposed livestock and poutry residuce, 300 kg/hm^2 of compound fertilizer for base fertilizer. Seedling with wet method, the bedding width 100 – 130 cm, trench width 30 – 40 cm, trench depth about 15 cm, flat, straight, thin.

4) Fertilizer and water management in rice field. Apply of 30 kg/hm^2 of urea in the heading stage of one leaf, 37.5 kg/hm^2 of urea in the heading stage of two leaves, and 60 – 75 kg/hm^2 of urea for "marrying" fertilization 4 – 5 days before transplanting. Before the leaves emerge, no irrigation on surface, being wet is ok; After the leaves emerged, the shallow water often drains to regulate the temperature, air and fertilizer to promote tiller. If there is rice seedling throwing, water should be cut off 2 – 3 days before seedling throwing.

5) Rice transplanting. After the first crop harvesting, machine was applied to till and bury, filled with enough water, so that the vegetable leaves, stems and so on can be fully submerged to decompose. 1 – 2 days before transplanting seedlings, till the land and rake it flat. The seedling age should be 20 – 25 days, and the requirements are shallow seedling (2 – 3 cm), stable (not floating seedlings), uniform (uniform number of seedlings), straight (not oblique insertion). The planting density is about 300 000 holes/hm^2 for "Han You Xiang Qing", and about 225 000 holes/hm^2 for "Hua You 14".

6) Field management.

① Topdressing. In accordance with the principle of "stabilizing at the beginning, taking care in the middle and protecting in the end", the vegetable field can be applied less nitrogen fertilizer and more potassium fertilizer. 6 – 10 days after transplanting, apply rice fertilizer 225 kg/hm^2 and potassium chloride 180 – 225 kg/hm^2; at heading stage 0.2% potassium dihydrogen phosphate can be applied in combination with spraying .

② Field water management. In tillering stage the field should be irrigated with shallow water, drainage and irrigation combined, timely drying the field under the sun, and light drying for several time. When the total number of seedlings in the field reached 80% of the expected number of ears, the field should be dried. In the booting stage, the method of frequent irrigation with shallow water and alternating wet and dry conditions is adopted, and it is advisable that the latter water does not meet the former water. The water should be kept shallow from heading to full heading stage at later stage. Water should be cut off 7 days before harvest and prevent too early dehydration and affect grain filling.

③ Pest and disease control. Agricultural control is the main method, and rice population optimization control technology is adopted to enhance the resistance of rice plants. The main pests and diseases in this season are rice sheath blight, rice striped stem borers (suppressalis) , rice yellow stem borers (scirpophag incerfulas) and rice leaf folder, etc. Rice sheath blight can be controlled with 30% difenoconazole propiconazole 300 mL/hm^2 with water 600 – 750 kg/hm^2 to spray, or 6% Jinggang A osthole wettable powder 800 g/hm^2. The rice striped and yellow stem borers, and the rice leaf folders can be dealt with 10% aveflurphthalide (Daoteng) suspension agent, plus 5% fipronil (Regent, Ruijinte) suspension agent 450 – 600 mL/hm^2, or 5% fipronil (Ruijinte) suspension agent 300 mL/hm^2 + EnSpray 99 (green glam) 500 mL/hm^2, or 40% Chlorantraniliprole·thiamethoxam 120 – 150 g/hm^2 for spray control.

④ Harvest. Rice can be harvested when the grain maturity reaches 95% – 100%.

2. Cauliflower technical operation code

1) Variety selection. Choose varieties of disease-resistant, high quality, high yield, frost resisting such as Chang Sheng 90 Tian.

2) Seedling breeding.

① Seedling field preparation. Select fertile soil with convenient drainage, rich in organic matter, and isolated sandy soil seperated with the production field to be the seedling field. Apply high quality organic fertilizer 3 tons/hm^2, potassium sulfate compound fertilizer 150 kg/hm^2 (nitrogen ∶ phosphorus ∶ potassium = 50 ∶ 50 ∶ 50 kg respectively) , calcium magnesium phosphate fertilizer 375 kg/hm^2, and then deep plouing 20 – 25 cm. 3 days before sowing, prepare the bedding with the width 100 cm, height 30 cm, furrow width 30 cm.

② Sowing. A strip ditch deep 1 cm was dug every 3 – 4 cm, and then evenly sow along the ditch, the grain distance is about 3 cm. After sowing, cover it with 1 – 1.5 cm thick sand (soil) , and spray with water, and then use 2 layers of shading net to cover the bed surface,

Chapter 4 Utilization Model of Ecological Circular Agricultural Resources

later, every morning and evening directly spray the shading net with water for one time.

③ Seedling management. When about 70% of the seeds emerge, remove the shading mulch in time. Control the water after all the seedling out, and prevent the blight and vain growth. When the seedling grows to 2–3 true leaves, if the seedling is weak, small, and the leaf color is pale yellow, the urea can be applied 37.5 kg/hm^2. In the seedling stage, it is mainly to prevent the damage of cabbage worm, diamondback moth (plutella xylostella), and jumping beetle, which can be prevented by 10% Removing all (Chujin) 3000 times liquid, or 15% Dopont Avatar 4000 times liquid for spray control. At 4–5 true leaves, the seedlings were transplanted and fixed planted.

3) Fixed planting. After rice harvest, the field should be timely deep ploughed, combined with full base fertilizer. Generally apply commercial organic fertilizer 4.5 tons/hm^2, calcium magnesium phosphate fertilizer 300 kg/hm^2, potassium sulfate compound fertilizer 300 kg/hm^2 and Borax 15 kg/hm^2. It is better to make high bed, in case of flooding. After the field was finished, the healthy seedlings with the age of about 25 days without disease and insects were selected. In the afternoon when the sunlight was mild, double row planting was adopted with the density of 30 000–37 500 plants/hm^2 and plant spacing of 40–50 cm.

4) Field management.

① Irrigation and drainage. After fixed planting, water once every evening to keep the soil moist. After the rejuvenation period of the seedling, topdressing can be combined with water spray. Adequate water is required for flowering, but during intertillage and fertilization, the soil should be slightly dry for weeding and fertilizer absorption. In case of excessive rainfall, we should timely clear the ditch for drainage.

② Topdressing. At the beginning of growth and the vine growing period, broccoli need more nitrogen and appropriate phosphorus, potassium fertilizer. In the full growth period, in addition to sufficient nitrogen fertilizer, we must increase phosphorus, potassium fertilizer and boron, magnesium, manganese, molybdenum and other trace elements. The first topdressing was carried out in 7–10 days after fixed planting, with 75 kg/hm^2 of urea. The second is at the 30 days after planting, when ploughing and weeding, apply topdressing with 150 kg/hm^2 of potassium sulfate compound fertilizer; The third topdressing in 60 days after planting, about the beginning of the bud before applying urea 75 kg/hm^2, potassium sulfate compound fertilizer 150 kg/hm^2; In the later stage, 0.2% borax was sprayed outside the root for 1–2 times.

③ Leaves breaking. Cauliflower flowering period if not breaking the leaves, the color is easy to become yellow, with poor commodity is poor. Therefore, when the diameter of the flower bulb is 7–8 cm, the old leaves should be broken off to cover all the flower bulbs.

④ Pest control. The main diseases are downy mildew, sclerotinia, black rot, soft rot and so on. Downy mildew disease at the beginning could be sprayed with 64% antitoxin alum, or 58% Ledomir, or 72% kelu 500–600 times liquid Sclerotinia at the beginning and on rainy days sprayed with 50% puhein 1000 times liquid or 50% procymidone (sukeling) 1500 times. For the black rot, soft rot after transplanting should be treated with 80% Bibei 500 times liquid, or 77% Keshade 500 times liquid, or 47% Jiaruinong (Kasumin+Bordeaux) 800 times liquid to spray. The main insect pests are aphid, whitefly, micronoctuini, spodoptera litura, etc. To control aphids and whitefly 10% imidacloprid 2500 times liquid can be used or 3% Nongbulao (farm not old) 5000 times liquid . 10% Chujin 3000 times liquid, or 15% Dopont Avatar 4000 times liquid can be used to control micronoctuini and spodoptera litura.

⑤ Harvest. Too early harvest when the flower bulbs not fully grown will lead to low yield. If the flower bulb is harveted too late, the flower bulb will be loose and the bud will be coarse and loose, which will affect the quality and value. Therefore, the flower bulb should be harvested when it develops to a proper size and the small buds have not been released. Cut off the base of the flower bult with a knife, remove the excess leaves, cover it with a net bag, label it and put it into the plastic basket, timely send it to the nearby factory for pre-cooling, and make a good record for later processing or sales.

3. Technical operation specifications for earthworms

1) Variety selection. Such Eisenia foetida as Taiping 2, North Star 2, Beijing striped worm and so on.

2) Breeding conditions and preparation. Feed preparation. Combine "three" feed collocation. "Three" feed, namely animal feed: pig, cow manure, etc. Plant feed: all kinds of straw, leaves, weeds, etc. Fruity feed: watermelon rind, rotten fruit, orange, etc. The first and the third types of feed accounted for 70%.

Outdoor breeding method. Choose sunny humid place with easy irrigation and drainage. The center was first filled with 1m wide fermented bait, followed by bait containing young worms to make the total thickness 23 cm and finally covered with straw or straw curtain. When the temperature reaches 15 °C or above, the breeding begins. When the temperature drops to 10 °C, breeding should be transferred to indoor preservation.

Temperature and humidity conditions. ① The optimum temperature is 10–30 °C. In winter, just a little cover, do not let earthworms hibernate and it can not be exposed to the sun and rain. ② The humidity requirement of earthworms is not high, the relative humidity is 60%–70%, generally the fresh fermented cow manurecan can be directly put in. If the pile is too long and dry, spray some water, and the water is appropriate when grasping the material in hand, you can see water droplets between the fingers but not dripping. pH value is 6–8.

3) Management work. Earthworms have a wide range of feeding habits. They feed on almost all plant residues, humus, decaying animals and domestic garbage. Feed earthworms as diverse as possible, avoid a single food. The bait must be treated. After removing impurities, straw, weeds, bagasse, etc. , are chopped and piled into 1m high cones with pig, cow and chicken manure, which are covered with straw curtains or plastic films for fermentation. If the raw material is too dry, spray some water to be wet. It is appropriate when grasping the material in hand, you can see water droplets between the fingers but not dripping. After fermented for 5 times, the material temperature generally rise to about 70 °C , turn the pile once, for it is conducive to even fermentation. After a few days the temperature goes down, then spray water again, and turn a few times. After about 3 weeks, the bait was fermented. The ideal fermentation bait should be dark brown, with no foul smelling, loose, and not sticky to the hands.

Generally, observe once every 20 days, and feed once, the method is the same as above. The feeding area can be doubled every 40 days. Generally every 40 days is a cycle, in a year you can raise 9 batches.

Section 5 Vegetable - fish co-cultivation model

I. The principle and model connotation of systematic ecology

The Vegetable - fish co-cultivation model is to achieve scientific co symbiosis through ingenious Ecological design based on the original two completely different farming technologies of aquaculture and dry farming vegetable cultivation, so as to achieve the ecological co growth effect of fish farming without changing water without water quality problems, and vegetable planting without fertilization. The vegetable fish symbiotic system is an agricultural technology that involves the nutritional physiology, environment, physicochemical, mechanical and electrical aspects of fish and plants. In traditional aquaculture, as fish excreta accumulate, the ammonia nitrogen in the water increases, and the toxicity gradually increases. In the vegetable fish symbiosis system, the water for aquaculture is transported to the hydroponics system, and the nitrifying bacteria nitrify the ammonia nitrogen in the water to nitrite, and then continue to be nitrified by the nitrifying bacteria to nitrate, which can be directly absorbed by plants as nutrition. The vegetable fish symbiosis enables animals, plants and microorganisms to achieve a harmonious Balance of nature relationship, thus forming a virtuous circle in a small area, maximizing the production of aquatic products and vegetables, and also minimizing water pollution. It is a low-carbon production model of sustainable recycling and zero emissions, and the most effective way to effectively solve the agricultural ecological crisis. The symbiotic system is a product of the organic combination of high-density aquaculture technology and vegetable soilless cultivation.

II. The characteristics of the model

① The Vegetable - fish co-cultivation model can prevent and control diseases rarely seen fish diseases occur.

In addition to the isolation conditions of the facility itself, which are conducive to disease control, the plant roots secrete an exclusionary micro-toxin that inhibits certain pathogens in animals, which reduces the use of pesticides.

② Improve the yield and quality of vegetables and fish.

Vegetables absorb nitrogen, phosphorus and other nutrients in the water, purifying the water quality, facilitating the growth of fish and promoting yield improvement; at the same time, fish manure is used as a high-quality fertilizer for vegetables, reducing the use of fertilizers and improving the quality of vegetables.

③ It can be self-evident.

Vegetable-fish co-cultivation is detached from soil cultivation mode, which can avoid heavy metal pollution of soil, so the heavy metal residues of both vegetables and aquatic products in vegetable - fish co-cultivation are much lower than traditional soil cultivation. And because of the presence of fish in the vegetable - fish co-cultivation, no pesticides can be used, which would cause the death of fish and beneficial microbial populations and the collapse of the system. If the vegetables produced by the vegetable-fish symbiosis farm are delivered with roots, it is easy for consumers to identify the origin of the vegetables, avoiding the doubt of consumers whether this vegetable comes from the wholesale market.

III. The operating technical specifications

1. Floating frame production process

① Production method of PVC pipe floating frame. Using PVC pipe (diameter 110 mm) at a distance of 1 m × 2 m or 1 m × 4 m specification, with four ends connected and sealed with elbows, and the upper and lower layers fastened with two types of polyethylene mesh to form a floating bed. The main function is to make vegetables float and grow on the water surface, isolate herbivorous fish from eating vegetables, and control the direction of stem and leaf growth. The density specification of the lower layer mesh is moderate, which can fix vegetable roots, prevent fish from eating vegetables, and facilitate water exchange. Farmers can choose suitable floating frames based on the actual conditions of their ponds and the principles of convenient movement, cleaning, production, and harvesting.

② Method for making bamboo floating frames. Choose bamboo with a diameter of more than 6cm, connected end-to-end, and fixed with bamboo strips according to the length of the bamboo to form a quadrilateral or triangular shape to prevent deformation. The upper and lower layers are woven with bamboo strips into grids of different sizes, which is not only convenient for planting and fixing vegetable seedlings, but also to prevent fish from eating them. The specific shape can be determined based on the pond conditions, material size, and convenient and flexible operation.

2. Selection of Cultivated Vegetable Types

Crops grown on water not only refer to vegetables to eat, but also include ornamental flowers, edible forage, etc. Vegetables and fruits with well-developed roots, water resistance, and strong water purification ability should be selected for cultivation. Flower plants use their well-developed roots and large absorption surface area to purify water quality. The commonly selected varieties include water spinach, water celery, water cabbage, green cabbage, lettuce, water chestnuts, loofah, pumpkin, watermelon, etc. Through experiments, it has been found that planting water spinach, water celery, water chestnuts, and loofah has a good effect because of their vigorous growth, high yield, and good water purification effect.

3. Selection of planting area

The ratio of planting and breeding area to pond area is related to the Balance of nature between organisms and the recycling of material and energy. For example, how many fish excrete feces can provide nutrients for how many vegetables, and how many vegetables can produce the best ecological effect on water purification, which is also the foundation of vegetable fish symbiosis technology. A precision aquaculture pond with a depth of 1.5 to 2 meters produces over 1200 kg/hm^2 of fish per unit area, with a breeding cycle of more than 3 years. The area of planting vegetables such as water spinach accounts for about 5% of the pond, and good results can be achieved. The proportion of Vegetable farming should be reasonably determined according to the water quality of the pond, the size of the water body, and the number of cultured fish. Generally, the proportion of planting can be properly increased for ponds with relatively high aquaculture density, relatively small water body, and long aquaculture cycle, but should be controlled within 10%.

4. Vegetable cultivation techniques

The main method of planting is transplantation, such as direct cultivation, nutrient cup cultivation, and mud ball cultivation using PVC standard floating frames. The direct cultivation method refers to directly inserting the plant stem into the lower layer with a dense mesh at a spacing of 20-30 cm (water spinach can be planted on the shore after the temperature is basically stable and above 15 °C in May, and water celery can be planted in October). Afterwards, attention should be paid to replanting the floating seedlings. This method is fast, simple, and has a high survival rate. The nutrient cup cultivation method mainly adopts a flower and grass cultivation cup, in which nutrient solution or soil (pond mud) is inserted and placed on a floating frame at a spacing of 20-30 cm (melon plant spacing of more than 1 meter). This method has a high survival rate, but is relatively cumbersome and costly, making it suitable for melon planting. Melon vines are attached around the pond or secured to it with a rope net. The mud ball cultivation method mainly refers to directly inserting plant stems into small mud balls (pond mud is sufficient), and placing them on floating racks at a spacing of 20-30 cm. This method has a relatively high survival rate, simple and convenient operation, and low cost.

Each breeding unit can flexibly choose suitable planting methods based on their own labor force. The survival rate of each planting method is as follows: nutrient cup cultivation method > mud ball cultivation method > direct cultivation method; The order of labor costs is: nutrient cup cultivation method > mud ball cultivation method > direct cultivation method.

5. Prevention and control strategies for diseases, pests, and weeds

The prevention and control strategies for diseases, pests, and weeds in fish and vegetable fields are the same as those in eel and vegetable fields. After raising fish in vegetable fields, the fish can eat some pests in the fields, but chemical control still needs to be combined. The toxicity of pesticides to fish can be classified into three levels: high, medium, and low. Carbofuran, deltamethrin, fenvalerate, triazophos, meothrin, DDT, hexachlorocyclohexane, rotenone, sodium pentachlorophenol, propanil and butachlor are highly toxic pesticides; dichlorovinylphosphate, dipterex, 1605, monocrotophos, isoprocarb, Daowenjing, iprobenfos, isoprothiolane, nitrofen, thiobencarb, alachlor, prometryn etc. are toxic pesticides; bisultap, methamidophos, carbendazim, validamycin, buprofezin and glyphosate are low toxic pesticides.

Chapter 5 Development Model and Technology of Modern Eological Circular Agriculture

In the practice of green low-carbon circular agriculture, many different types of development models and supporting technologies have been created, and a large number of good examples and cases have emerged, playing a positive role in demonstration and driving. For example, the model and technology of three-dimensional structure ecosystem of agricultural planting and breeding space, the model and technology of system of agricultural resource saving and recycling, the model and technology of green low-carbon circular agriculture of courtyard, the model and technology of green leisure agriculture and agricultural science and technology (ecological) parks, etc. To summarize and promote these different types of development models and technologies, and to make the vast number of farmers and producers use for reference, application and promotion in the process of implementing green, low carbon and circular agriculture, have a very important significance and role in promoting the development of green, low carbon and circular agriculture. At the same time, these models and technologies need to be continuously innovated and improved in practice, and more new models and technologies should be developed to lead and promote the sustainable development of green, low-carbon and circular agriculture.

Section 1 Three-dimensional planting model and technology of farmlands

I. Three-dimensional planting model and technology of farmlands

On the basis of inheriting and developing China's fine traditional planting experience, the farmland three-dimensional planting model makes extensive use of the existing natural resources, production conditions and modern agricultural science and technology, and scientifically carries out intercropping, interplanting, mixed planting and multiple planting of different crops and varieties according to the farmland ecological law. So that it can effectively improve the utilization rate of light, heat, water, fertilizer, gas and other natural resources in the land, effectively make the best use of everything, establish a good agricultural ecological economic complex, so as to obtain better economic benefits, social benefits and ecological benefits.

1. Three-dimensional planting model and technology of grain and vegetable (grain)

Most of these models are developed and popularized in intensive farming areas and urban suburbs with better water and fertilizer conditions. It mainly takes wheat interplanting corn, wheat interplanting millet, wheat interplanting sweet potato, wheat interplanting rice and other grain crops with two harvest a year as the main crops. On the premise of ensuring the continuous and stable increase of grain crops, it involves cash crops with high economic value or melons and vegetable crops to improve the economic benefits per unit. Such patterns need careful planing from wheat seeding, carefully determine the planting width, and the size of the bed surface, and the size of the interplanting rows, and the variety and sowing time of the crops and time decide. One year ago, the planting planning of crops was set between the two main crops, such interplanting and intercropping higher economic benefit crops as melons, vegetables, beans. Generally in combination with wheat autumn sowing, Intercropp overwintering vegetables, such as spinach, rape (tai cai) , garlic, onion, Chinese onion, etc. In early spring, Chinese cabbage, rapeseed, small red carrot, early cabbage, kidney bean and other fast-growing vegetables are interplanted in wheat fields. Before wheat harvest, in addition to interplanting (or transplanting) the main crops of the second season, such as corn, millet, sweet potato, soybean and other food crops, the wheat field can also be interplanted (or transplanted) with watermelon, ginger, potato and other high-benefit crops. After wheat harvest, summer sowing crops intercrop with soybeans, mung beans, autumn vegetables, etc. For example, wheat interplanting corn intercropped with soybeans, wheat interplanting corn intercropped with summer millets, wheat intercropped with Ganlan (wild cabbage) and corns intercropped with soybean, wheat intercropped overwintering vegetables such as spinach, Mancai (similar to rape) , garlic, onions and interplanted with corn and Chinese cabbage, wheat intercropped with overwintering vegetables and interplanted with corn and cucumber, wheat intercropped with overwintering vegetables and interplanted with corn and tomato, wheat intercropped with overwintering vegetables and interplanted with ginger, wheat intercropped with garlic (or onion) and interplanted with corn intercropped with autumn vegetables (radish, mustard, Chinese cabbage, etc.) , wheat interplanted with corn which is intercropped with gourd, wheat interplanted with potato or interplanted with corn which is intercropped with Chinese cabbage, etc. This type of model can ensure that the grain does not reduce production or less production, and can harvest more than 1-2 seasons of vegetables (or other crops), and the economic benefits of more than 500 yuan can be increased per mu, thus achieving high yield and efficiency.

The main technologies of this model include: planting and field management technology of wheat, corn, radish, mustard, Chinese cabbage, spinach, Taicai, garlic, onion and other vegetables, the rational space allocation and interplanting technology among crops, fertilization technology, and design of row and plant spacing ratio, etc.

2. Three-dimensional planting model and technology of grain-cotton-vegetable

This kind of model is mainly developed and promoted in the area where the spring cotton can harvest only once in a year, by reforming the farming system and using the winter idle season to grow one-season wheat or intercrop with overwintering vegetables, or on the basis of the two-ripening system of wheat intercropped with cotton, interplant overwintering vegetables or melons by using the cotton intercropping rows. Under the premise of ensuring the increase of cotton yield or the double increase of wheat and cotton, the economic benefits can be increased by interplanting melons and vegetables. On farming system, after the cotton woods are taken out, we should finish soil preparation and fertilization, determine the cultivation specifications and sow the late wheat. Set aside the cotton interplanting rows, combine the wheat, and intercrop the overwintering vegetables such as onion, garlic, spinach, or in the early spring of the second year interplant pakchoi, rape, carrot etc. Atfter the harvest of vegetbables interplant cotton from mid to late April. For example, wheat which is intercropped with garlic is interplanted with cotton, wheat which is intercropped with onion is interplanted with cotton, wheat which is intercropped with overwintering vegetables (or autumn mustard) will be interplanted with watermelon and cotton, wheat which is intercropped with overwintering vegetable will interplant with cotton which is intercropped with mung bean, wheat which is intercropped with mustard will interplant with cotton which is intercropped with pepper.

The main technologies of this model include: planting and field management technology of wheat, cotton, vegetables and melons, rational allocation of space between crops and interplanting technology, fertilization technology, design of row and plant spacing ratio, etc.

3. Three-dimensional planting model and technology of grain-oil-vegetable

This kind of model is mostly developed and promoted in peanut producing areas with better water and fertilizer conditions. On the basis of wheat interplanted with peanut double cropping system in one year, interplant melons or vegetables between interplanting rows of peanuts, which can increase economic benefits by more than 300 yuan per mu. In terms of farming system, it mainly started from the autumn planting, raise ridges between beds, sow wheat in the beds interplant overwintering vegetables such as spinach, Taicai on the ridges, the next spring after the harvest of vegetables, interplant spring peanuts or midsummer peanuts, and interplant watermelon, pepper and other autumn vegetables. For example, wheat which intercropped with overwintering vegetalbes is interplanted with peanuts and watermelon, wheat which is intercropped with garlic seedling is interplanted with watermelon and peanuts intercropped with autumn vegetables, wheat is interplanted with watermelon and peanuts which is intercropped with corn, wheat intercropped with overwintering vegetables (or rape) is interplanted with peanuts which are interplanted with sesame.

The main technologies of this model include: planting and field management technology of wheat, peanut, rape, sesame, vegetables and melons, fertilization technology, rational allocation of space between crops and interplanting technology, row spacing ratio design, etc.

4. Three-dimensional planting and breeding model and technology of paddy fields

With rice as the main crop, planting various kinds of cash crops in winter and spring or breeding various high-grade aquatic products in paddy fields can not only maintain a high yield of rice, but also greatly improve economic benefits compared with the two-cropping system of rice and wheat. Generally chosen patterns are as follows, "onion (snow peas, garlic, potato)-single cropping rice" "melons (strawberry, cucumbers, peppers, eggplant) cultivated in simple greenhouse cultivation-single cropping rice" "winter and spring vegetables-green glutinous corn-late rice" "winter and spring vegetables-watermelon, sweet melon (green bean)-late rice" "winter and spring vegetables-single cropping rice + crabs, shrimp, carp)", etc.

The main technologies of this model include: planting and field management technology of vegetables such as rice, potato and snow peas, garlic, cucumber and pepper, and management technology of aquaculture fish such as crab, shrimp and crucian carp.

5. Three-dimensional planting model and technology of cotton and vegetable

With cotton as the main crop and multiple interplanting with various cash crops in winter, spring and autumn, this model can not only maintain the high yield of cotton, but also greatly improve the economic benefits of cotton field. The models selected are "potato (pickle) interplanted with cotton" "green soy beans (vegetable) interplanted with cotton snow peas (radish, garlic)" "cotton interplanted with garlic (zucchini, watermelon)" and so on.

The main technologies of this model include: planting and field management technology of cotton, vegetables and potatoes, fertilization technology, reasonable spatial allocation and interplanting technology among crops, and design of row and plant spacing ratio, etc.

6. Ecological model and technology of comprehensive agricultural development

This model is a comprehensive development model integrating planting, breeding, processing, management and ecological protection. The main technologies are three-dimensional planting technology, breeding technology, processing technology, biogas technology, comprehensive utilization of resources technology and so on.

The above three-dimensional planting model is a new and comprehensive applied agricultural science and technology based on the experience of increasing production of traditional intercropping, interplanting and multiple planting, and through a large number of experiments, demonstrations and production practices. The three-dimensional planting model of farmland is an important way to increase agricultural income and make farmers rich scientifically. In line with the characteristics of China's national conditions, such as large population and small land, intensive farming and intensive management, it has strong vitality and broad prospects for development. Through three-dimensional and high effevtiveness cultivation, the vast number of farmers can make use of local natural resources and production conditions, reform and apply a variety of three-dimensional cultivation models according to local conditions, give full play to their wisdom and talent, constantly tap the potential of increasing production of land, and accelerate the development of green, low-carbon and circular agriculture.

II. Three-dimensional aquaculture models and technology in water areas

The three-dimensional pond models mainly include lotus root planting, fish and duck breeding, shrimp and crabs mixed breeding.

(1) Three-dimensional model and technology of lotus root and fish planting

Lotus root planting in the pond is to choose a pond with sufficient water source, pollution-free water, convenient irrigation and drainage. The pond area is 10–20 mu, the pool depth is 1–1.5 m, the bottom sludge is 20–40 cm thick, and the pond base is more than 70 cm high. Pond depth is generally maintained at 15–50 cm. In spring, lotus root is planted in the pond, and eel, loach (mud fish) and silver carps are bred in the pond to form a co-culture of lotus root and fish. In high temperature season, we should timely change water or add new water, and do a good job of anti-freezing in winter. For the large scale breeding area, eels and loaches can be caught by eel cages and other tools in September. Eel and loaches can also be caught by turning mud piece by piece with both hands before winter, and can be sold around the Spring Festival. Each mu of lotus root pond can produce about 200 kg loach, production of lotus root 2000–3000 kg.

The main technologies of this model include eel, loach, carp and other aquaculture technology, lotus root planting and management technology, pond management technology and so on.

(2) Three-dimensional model and technology of fish and duck breeding

Three-dimensional breeding of fish and duck in pond is a combination model with high effectiveness. Fish and duck in the pond is a unique symbiotic phenomenon. The water surface can be used for ducks to move in the water to oxygenate fish, and the plankton at the bottom is used for ducks, and duck dung is used for fish. It is a kind of symbiotic co-existing three-dimensional breeding technology. It not only saves the bait input for raising the fish, improves the output and income of fish farming, but also provides the activity place for raising ducks. Ducks bathing in water can promote the metabolism of ducks, improve the health level, and improve the quality of duck eggs. It also solves the problem of the pollution of duck residue, which is conducive to improving the environment of raising ducks. The implementation of duck fish aquaculture, the two complement each other, promote each other.

The main technologies of this model include three-dimensional breeding technology of fish and duck, pond management technology and so on.

(3) Three-dimensional model and technology of shrimps and crabs breeding

Shrimp and crab breeding generally requires the pond area to be 5 – 10 mu, the depth to be 1.5 m, with sufficient water, good water quality, no pollution sources, pond inlet and drainage convenient and independent, diagonal arrangement. It should be equipped with water filtration device. In winter and spring the pond should be cleard, be dried for disinfection treatment. According to the weather conditions, timely plant water grass, mainly bitter grass, appropriately transplant black algae, water peanuts, water hyacinth, etc. , with the surface coverage being 60% – 70%. In the first month of lunar year, the young fish and crabs should be released. The size of the crab is not less than 5 g, the stocking density is 400 – 500 per mu, and the size of the shrimp is 400 – 800 per kg, the stocking density is 7 – 8 kg per mu. The seedling should be disinfected when released. Always use quality and safe fishery feed, and the harvest of crabs to the market would be in time according to the size of crab, breeding density, bait cost and crab market. At the end of the year finish the catch and clear the pond , so as to get ready for next year's breeding.

The main technologies of this model include the planting technology of aquatic plants such as black algae, peanut and water hyacinth, the mixed three-dimensional breeding technology of shrimp and crab, and the pond management technology.

(4) Three-dimensional circular model and technology of "growing vegetables on water and breeding fish in water"

"Fish and vegetables" three-dimensional circular model is to build a bamboo raft on the fish pond, choose water spinach and other aquatic leafy vegetables, with floating bed as a carrier to grow. Water spinachs in the breeding pond do not need fertilization and also do not need to apply pesticide, for the fish and shrimp residues are the best fertilizer. Water spinach grown on floating beds have better taste and freshness than those grown on dry land, and their yield is several times higher than those grown on dry land. The most important thing is water spinach on the surface can improve the water quality environment of the pond, not only save the cost of aquaculture, but also increase the benefit. Not only can we harvest green, safe aquatic productsis conducive to human health, but also we can get fresh and delicious vegetables, which can kill two birds with one stone.

The main technologies of this model include aquatic leafy vegetables planting technology, fish breeding technology, pond management technology and so on.

(5) Three-dimensional circular model and technology of freshwater pearl and native duck breeding

Build piggery near a lake or reservoir, and build duck room to raise ducks, breed fish and mussels in the three-dimensional pond, water mussel (clam) in the bottom, the upper water area for growing pearl clam, pearl processed into a necklace, hydrolyzed pearl powder, pearl skin cream, shells made into arts and crafts and mineral feed, which composed into a breeding and processing comprehensive model. Breeding project should base on the proportion: 5 pigs per mu of water area, 0.3 sow pig; $0.5m^2$ baby clam pond; 20 ducks for the 0.67 hm² fish pond. Pig manure, duck manure into the pond for breeding fish and pearl, 1kg pearls and 50 kg fresh fish can be produced per mu. Under the artificial control and regulation, a water three-dimensional production system has been formed with the food chain as the link of water and land material flow and energy recycling. This comprehensive three-dimensional breeding model has greatly improved the economic benefits.

The main technologies of this model include pig, duck, fish, clam and pearl breeding technology, pearl processing technology, and scientific management of water areas.

(6) Three-dimensional model and technology of duck-fish-turtle breeding

The three-dimensional model of duck-fish-turtle is a combination of fishery and animal husbandry according to the water area. Put the duck in the water and raise them duck on the surface, where the water grass, bamboo leaves and insects became a natural feed for ducks. Each duck costs 4–5 yuan less than other feeding, and due to free raising, the duck meat is delicious, with higher sales price than the ordinary ones. At the same time, each duck is a "fertilizer", using duck manure as bait for fish, the fish production can be increased by about 20%. Turtle can also take away the water surplus, for they eat the poor quality of small fish and shrimp, thus improve the quality of fish. To raise fish and turtles in water, many varieties of them should be adopted for mixed breeding, mixed different fishes with different diets and different living levels, so as to improve the utilization rate of water and the rational use of bait.

The main technologies of this model include duck, fish and turtle breeding technology, and scientific management of water areas.

(7) Three-dimensional breeding model and technology of seawater pond

The three-dimensional breeding model of seawater pond is to raise shrimps in the water, raise crabs in the bottom, raise shellfish with the bottom mud, which formed a three-dimensional breeding model of shrimp, crab, and shellfish. Specifically, in the water of a seawater pond, shrimp is raised in the upper water, swimming crab is raised in the bottom, and mixed color clams or leeches are raised in the bottom mud of the pond. This breeding model is not only beneficial to ecological complementarity, but also can make comprehensive use of water and improve economic benefits.

The main technologies of this model include three-dimensional culture technology of shrimp, crab and shellfish, disease control technology, scientific pond management and so on.

(8) Three-dimensional breeding model and technology of multi-species in fresh waters

Fresh water area model is to breed carp, grass carp, crucian carp and other species of fish in reservoirs or ponds. Different fish species can be bred in different water layers according to their physiological characteristics, which can make full use of the space, time and bait resources in the water for comprehensive breeding, and achieve the purpose of high yield, high effectiveness, ecological and environmental protection, which is an ideal model for freshwater aquaculture.

The main technology of this model is the breeding technology of different fish species and the scientific management technology of water area.

III. Three-dimensional model and technology of forest and land industry

1. Three-dimensional ecological model and technology of forest, fruit and grain economy

This model generally refers to the agroforestry systems in the world, which is to use the difference and complementary relationship between forest&fruit and crops on the use of resources in time and space, to grow grains, economic crops, vegetables, medicinal herbs and melons in the middle of the fruit trees, to form the different types of agroforestry planting model, and the main production of three-dimensional planting, generally can obtain higher comprehensive benefits than single-planting.

The main technologies of this model include three-dimensional planting, interplanting and intercropping technology, and the supporting technologies include reasonable dense planting and cultivation technology, water-saving technology, balanced fertilization technology, and integrated pest control technology.

2. Ecological model and technology of forest & fruit-livestock & poultry

The basic structure of this model is "forest and fruit industry + livestock and poultry industry". In the woodland or orchard, a variety of economic animals are bred, growing within free range, feeding on wild natural food, supplemented by necessary artificial feeding. Products will have higher quality and safety than intensive farming, close to organic food. Generally in hilly and mountainous areas, the development of forest & fruit industry or forest & grass industry on the slopes, the establishment of poultry houses in woodland or orchards, and the direct placement of livestock & poultry manures into woodland or orchards, formed a three-dimensional ecological agricultural system of "forest (fruit), grass and poultry". Grass planted in the forest can use the forest space, can conserve water and soil, in addition, it can also reduce the growth of miscellaneous trees, weeds. In order to avoid being harmed by weeds in the process of grass planting, autumn herbage sowing was chosen in the season of tree deciduous fall. Livestock and poultry are placed in free area or in pens in the forest, fed with mown forage. In the process of free growing in the forest, livestock and livestock eat grasses and apply manure to the forest at the same time, which is conducive to the growth of forage and seedling in the forest. The main model are "forest-fish-duck" "chicken breeding in rubber forest" "chicken breeding in mountain forest" "chicken (rabbit) breeding in orchestra" and other models. In this kind of model livestock and poultry provide organic fertilizer for trees, pecking at pests, and the trees create a suitable growing environment for livestock and poultry, which has contributed to a good effect of planting and breeding coordinated development.

The main technologies of this model include forest and fruit planting technology, animal breeding technology and the proportion of planting and breeding, and the relevant technologies include feed formulation technology, disease control technology, grass cultivation technology and soil fertilizer technology, etc.

3. Ecological model and technology of agricultural forest network

This is an agroforestry system model. It is mainly applied on plains, in order to ensure the stability of crop production, reduce agricultural meteorological disasters, improve the farmland ecological environment conditions, through the standard planning and designing, use the grid construction of roads, ditches rivers and part of the forest or piece of forest construction. Fast-growing poplars are given priority to, supplemented by the willow trees, ginkgo trees, etc. And through thinning to ensure reasonable density and forest coverage rate, the forest network system is gradually formed matching the farmland ecosystem. For example, the farmland forest network in Huang-Huai area.

The main technologies of this model include wood cultivation technology, grid layout technology, and supporting technologies include pest control technology, thinning technology and so on.

Section2　Model and technology of agricultural resources saving and recycling system

I. Model and technology of saving agricultural resources

China's agriculture is resource-constrained, although the total amount of agricultural natural resources is large, the per capita ownership is small. The mountainous and hilly areas are widely distributed, and the ratio of the useless land is significant. The cultivated land area is insufficient. Water resources are in short, arid, semi-arid area are wide; Soil and water loss are serious, non-point source pollution is aggravating, and agricultural ecological environment is deteriorating day by day. The poor quality of the rural population and the problem of rural poverty have not been fundamentally solved, which have become serious constraints on agricultural development. This reality not only highlights the urgency of building conservation-oriented agriculture, but also becomes the starting point and basis for studying how to build conservation-oriented agriculture. For this, its significance is vigorously far-reaching to develop resource-saving agriculture .

The agricultural resource-saving model is an agricultural production model with improving the efficiency of resource utilization as the core and focusing on saving land, water, energy, fertilizer, pesticides, seeds and comprehensive recycling of agricultural resources. To save land, it is necessary to establish and popularize three-dimensional planting model and make full use of regional space to develop land-saving agriculture. To save water, we should popularize high-efficiency water-saving irrigation technologies such as pipeline water transmission, drip irrigation under film, and integration of water and fertilizer, build demonstration bases for dryland agriculture, and intensify efforts to popularize dryland water-saving agricultural technologies. To save energy, we will eliminate old agricultural machinery, promote the use of energy-saving agricultural machinery, promote the energy-saving transformation of pumping stations, and popularize energy-saving solar vegetable greenhouses. To save fertilizer, we should vigorously promote the technology of soil testing and formula fertilization, use chemical fertilizers in a scientific way, encourage farmers to increase the application of organic fertilizers, and reduce the unreasonable application of chemical fertilizers. To save pesticides, we should eliminate backward application machinery, promote the use of pesticides with high efficiency, low toxicity and low residue, and develop bases for organic agricultural products. To save seeds, we should popularize fine variety energetically, implement quantitative sowing. Due to the differences in natural agricultural conditions in different regions, we can develop distinctive models of saving agricultural resource according to the conditions of agricultural resources, natural economic and technological conditions, regional characteristics and development advantages of different regions. For example, the model of "saving land, saving time and saving water" in agriculture, the model of "saving grain - herbivory" in animal husbandry, the model of "saving bait - multi - layer" in fishery, and the model of "rapid growth - woody grain and oil - vertical shape" in forestry.

The basis of this model lies in relying on science and technology, highlighting economy and focusing on efficiency. The main technologies are related to land saving, water saving, energy saving, pesticide saving and seed saving.

II. Model and technology of planting, breeding, processing and marketing

The integrated ecological agricultural circular model of "planting, breeding, processing and marketing" is to grow crops in farmland, and the crops provide feed for the development of aquaculture industry. Crop straw and manure of breeding industry are used as biogas and returned to farmland after harmless treatment. Using methane as the energy, using planting products as raw materials, we would develop processing industry to produce clean vegetables, tofu, meat products, and characteristic produces, and the bean curb residues are returned to the breeding. Highly processed foods are provided for distribution center, into the sales channels. Their income returns to the planting, providing fund for reproduction. This model has realized the recycling of resources in the agricultural ecosystem, improved the conversion rate of resources, saved resources, protected the ecological environment, and constructed an agricultural production system with a virtuous cycle of high yield, high quality, high efficiency, ecology and safety.

The main technology is planting, breeding, processing technology, material recycling technology, marketing management technology, design technology of "planting - breeding - pr ocessing - marketing" integrated ecological agriculture model .

III. Model and technology of agricultural industrialization and low-carbon circular agriculture

In this model is in accordance with the symbiosis law of the material circulation in natural ecosystem to design the agricultural production system, based on the "leading enterprises" with green low carbon cycle (ecological) agriculture, based on the comprehensive development of local agricultural resources to determine the leading industry and leading products of agricultural industrialization, organically unite the the agricultural and sideline products production, processing and marketing, promote the transformation of green and low-carbon circular (ecological) agriculture into commercialized, specialized and modernized agriculture. In the process of agricultural industrialization, the biological chain and processing chain should be extended as far as possible to form a material cycle process of "natural resources-products-resources regeneration and utilization", so as to minimize the amount of natural resources invested, the discharge of waste, the degree of harm to the ecological environment and the maximum economic benefits.

The main technologies are the design of agricultural industry chain, agricultural product processing technology, resource recycling technology, clean production technology and so on.

Section 3 Model and technology of green low-carbon circular agriculture in courtyard

A courtyard refers to a yard in the countryside. The model of low carbon circular ecological agriculture in courtyar is that the farmers use their yard area to be engaged in the intensive production. It mainly plays the resource advantage of the yard, based on the principle of ecology, recycling economics, using system engineering method, to set up an ecological agriculture model of planting, breeding for comprehensive utilization. Some applied planting and breeding at the same time for comprehensive use; Some use limited space to develop three-dimensional planting and breeding industry. Due to the different size of the courtyard area, regional environment and resource conditions, the green low-carbon circular agriculture in the courtyard mainly has the following models.

Chapter 5 Development Model and Technology of Modern Eological Circular Agriculture

I. Model and technology of the three-in-one ecological cirlular agriculture of "toilet, pigsty and biogas digester" in courtyard

This model is to establish the trinity of "toilet, pigsty and biogas digester" ecological circular agriculture model, forming the virtuous cycle during which biogas promotes breeding, breeding promotes planting, planting promotes breeding, to achieve the unity of economic benefits, ecological benefits and social benefits of the courtyard.

This model takes biogas digester as a link, makes full use of the symbiosis and intergrowth of biological chains and organisms, and organically combines planting, breeding, energy and ecological environment protection through multi-layer energy transformation and material recycling to form a system. The biogas residue and biogas slurry provide fertilizer for planting industry, biogas provides energy for farmers, and planting industry provides feed for breeding industry, so as to realize the ecological circular agriculture model of overall coordination and material recycling and regeneration.

In terms of design and technology, an upflow biogas digester with floating cover is built according to the technical standard of biogas digester by choosing the appropriate position behind the house where the sewage of the pigsty and toilet is more convenient. The dark pipe is connected with the feed entrance of the biogas digester to input the human and animal waste and other polluting organic matter into the biogas digester for fermentation, so that the "toilet, pigsty and biogas digester" form a trinity. The outlet of digester is equipped with a discharging tank, which is convenient for planting industry to extract digester slag and slurry at any time. The biogas provides farmers for cooking and lighting.

The size of the three-in-one household biogas digester needs to cover an area of 30 to 42 square meters, one pigsty and one toilet need about 30 square meters, two pigsties and one toilet need 42 square meters. Its principle is that the structural layout is reasonable, with main body structure (be like housing) coordinated well, building ground level elevation should be 10 centimeters below the main room, with convenient transportation, and each department dimension is coordinated.

The digester building is arranged according to the "trinity". Try to build a tank in solid soil, low water table and higher terrain, sheltered from the wind and and exposed to the sun, taking comprehensive consideration of other supporting works and the convenience of the "trinity", as well as transport convenience. The appropriate volume is determined according to the gas demand of the household and the number of pigs. Take 14 cubic meters as an example, first prepare cement 1.7 tons, river sand (coarse) 4 tons, gravel 2 tons, red brick 800 pieces, diameter of 6 mm steel 15 kg, lime 30 kg, diameter of 300 mm concrete pipe or high quality hose 1, 1.5 meters long, as input pipe; Diameter 110 mm sewer quality hose 1.3 meters long, as outlet pipe. The biogas tank is a cylindrical one with the bottom being like a pot and a height of 2.05 meters. The earthwork is excavated strictly according to the standard, and the diameter error is not more than 5 cm.

The total height of the toilet should be 2.3 – 2.4 meters, and the net height of the internal space should be more than 2.2 meters, so as to double as a bathroom. Internal area of 2 – 3 square meters is appropriate; The height of the toilet door should be 1.85 meters, 0.8 meters wide; Toilets should be equipped with ventilation windows; The toilet floor is 10 cm higher than the pigsty floor; Toilet inside and outside should be whitewashed smooth, horizontal and vertical smoothy.

Pigsty's height and span should be consistent with the toilet. The interior and exterior walls should be painted smooth, horizontal and vertical smooth. The height of the pig pens is about 80 cm, and the area depends on the scale of the pig breeding and the area of the site.

Toilet and pigsty's roof can cover prefabricated board, can also be cast-in-place by supportting template . If cast-in-place, we should add wires, steel bars, bamboo bars or other reinforcement. The top and bottom are whitewashed smooth for seepage treatment.

When excavating digester pit, if the underground water level is high and the soil quality is quicksand, excavation shall be conducted according to a certain slope, and the excavated soil shall be stacked far away from the pit. Do not pile heavy objects around the pit to avoid cave-in. During the construction of the pool in the rainy season, the drainage ditch should be dug around the pit to avoid rain water to break down the pit wall; When casting the mold, the mold frame should be firm, stable, safe and reliable, easy to assemble and disassemble. Mold removal can not be too early, must wait for cement strength to be more than 70%. Be careful when removing the mold. No one can stand on the pool cover. For the first feeding, the inoculum should be 10% – 30% of the total fermentation liquid. Generally speaking, the greater the amount of inoculum, the faster the biogas fermentation starts. Heavy metal salts, pesticides and poisonous plants should be avoided in the use and management of biogas digesters.

The main technologies of the model are pig breeding technology, vegetable and fruit planting technology, biogas technology, system design of the model, etc.

II. Model and technology of planting and breeding ecological circular agriculture of "greenhouses, pig (chicken) houses and biogas digesters" in courtyard

This model is to use the limited courtyard reasonably and efficiently. In a courtyard we build a plastic greenhouse to plant vegetables in one end, on the other end build piggery or chicken house. We should feed the chickens with a perfect compound, and the chicken manure and the falling feed are added some concentrate and vegetables to feed pigs. The pig residues are put into the underground biogas pool, then added some weeds and straw, to produce biogas, which is used for lighting, cooking and heating. Biogas residue can also be used to raise earthworms. Earthworms can also be fed to chickens as protein feed, and the waste residue can be used as fertilizer for vegetables. This model makes use of the complementary relationship between the nutritional needs of vegetables and animals to form a virtuous cycle of biological food chain, maintain ecological balance, and produce safe green vegetables, meat and eggs.

In terms of design and technology, the courtyard construction is mainly divided into three parts: underground, ground and space, namely greenhouse in space, piggery on ground and digester under ground. It is the organic combination of pig, biogas and vegetable greenhouses, forming a virtuous cycle between each other. Plastic greenhouse can not only increase the temperature of vegetable growing environment, solve the problem that vegetables can not grow in winter, but also improve the humidity of pigsty and digester. Pig manure and urine into the pool to produce biogas, biogas can be used for lighting and cooking, biogas residue can be used as high quality fertilizer again, still can increase the surface temperature, advantageous to the pig bed to keep warm, the exhaled carbon dioxide by the pigs can meet the needs of the greenhouse vegetables, the vegetables exhale oxygen for pig use again, so that can keep the air fresh. The pigs in pigsty in winter greenhouses pigs grow fast with less disease.

The main technology of the model is pig and chicken breeding technology, vegetable planting technology, biogas technology, model system design and so on.

III. Model and technology of green ecological circular agriculture of "livestock - biogas - fish, duck, fruit and vegetable" in courtyard

In this model, farmers build pigsty to raise pigs in areas with water sources, and build biogas tank next to pig houses to produce biogas from pig manure and urine, which can be used by farmers as gas for cooking and lighting. At the same time, develop fish ponds, breeding fishes under the water, and ducks on the water. Use biogas slurry, biogas slag, duck manure to feed fish feed, or to plant fruit vegetables. This not only solves the problem of pig manure pollution, but also can produce biogas, as the energy needed by the people's life. It can also use biogas slag and biogas slurry as the development of pollution-free planting and breeding industry to provide green fertilizer and feed, forming a green ecological agricultural model of material recycling.

The main technology of the model is pig, chicken, duck and fish feeding technology, vegetable and fruit tree planting technology, biogas technology, model system design and so on.

IV. Model and technology of three-dimensional planting of fruits and vegetables (flowers, herbs, seedlings) in farmyard

The courtyard with a large area makes full use of land and light energy to plant fruit trees, such as pomegranate, apricot, jujube (date tree), papaya, persimmon, pear, grape, kiwi and so on. Interplant high-effectiveness cash crops between rows, such as strawberries, dwarf vegetables, fruit trees seedlings, landscape seedlings, flower seedlings, Chinese medicinal materials such as wolfberry, codonopsis pilosula (Dangshen), ligusticum Chuanxiong, and edible fungi, etc. Garden fruit trees require early fruits with high quality (good-looking, delicious, easy to sell), long economic life. This not only can beautify the living environment, but also can produce high-quality fruits, vegetables, medicinal materials, seedlings, etc., to increase the economic income of farmers.

This technology mainly includes: planting technology, post-planting management, pruning technology, pest control technology.

Section 4 Model and technology of green leisure agriculture and agricultural sci-tech (ecological) industrial parks

Green leisure agriculture and agricultural science and technology (ecological) park model is a new type of green low-carbon circular agriculture featuring the organic combination of serial development of agricultural production and agricultural leisure&tourism to enhance the added value and tourism value of agriculture. It is the use of agricultural production activities (process), pastoral and garden landscape, agricultural sci-tech parks, agricultural ecological parks, modern agricultural parks, facility agriculture, farming culture, farm life and other resources and environment. It provides tourists with leisure, sightseeing, vacation, entertainment, fitness, education and other tourism activities. It is a multi-functional agriculture integrating ecological functions, production functions, life functions, popular science education functions and tourism & entertainment functions. This combination of agricultural production and tourism industry has achieved mutual benefit and win-win situation. Leisure and sightseeing agriculture has driven the development of tourism industry, and tourism industry has promoted the development of leisure and sightseeing agriculture. At the same time, leisure and sightseeing agriculture promotes the innovation of agricultural production model, management model and people's consumption model. It is a new idea and new model for the future development of modern agriculture, with great market potential and broad development prospects.

Green leisure sightseeing agriculture technology mainly include all kinds of park planning and designing of landscape and function of the leisure sightseeing agriculture projects (travelism products), plant cultivation technology (vegetables, flowers, fruit trees and ornamental trees; three-dimentional, soiless cultivation, etc.), agricultural high-tech demonstration and display, farming culture heritage, tourism products design and development, etc.

Due to the differences in China's natural resources, human resources, agricultural resources and economic conditions, the development types and models of leisure and sightseeing agriculture show diversity. In terms of leisure content, sightseeing objects and activities, they can be divided into the following types.

I. Model and technology of green leisure and sightseeing agriculture

Green leisure and sightseeing agriculture is the use of agricultural production activities (process), pastoral landscape, garden landscape, agricultural facilities and other resources and environment, to provide tourists with activities such as leisure, sightseeing, experience, picking, tasting and purchasing melons, fruits, flowers etc, so as to enhance the value and efficiency of agriculture.

1. Model and technology of rural agricultural tourism

Rural agricultural sightseeing model is based on the agricultural production on field, through the facility agriculture, three-dimensional planting and ecological breeding, integration of planting-breeding-marketing, to combine organically with the leisure sightseeing function. It introduces good quality special varieties of vegetables, fruits, flowers and other ornamental plants, which are suitable for leisure sightseeing agriculture development, as well as advanced agricultural cultivation model and cultivation technology, which can improve the sci‐tech content, yield, quality and sales, as well as agricultural benefits. Based on the above, tourism activities will be developed such as appreciating rural scenery, watching agricultural production, tasting and purchasing green food, and learning agricultural technical knowledge etc, so that tourists can understand agriculture, experience agriculture, return to the nature and feel the ecology, benifical to health, both phsically and mentally.

The main technologies are facility agriculture, three‐dimensional planting, ecological breeding and other related technologies, as well as tourism management technologies.

2. Model and technology of landscape sightseeing

The landscape sightseeing model focuses on the production of orchard and forest garden, and builds the natural ecological sightseeing landscape of orchard and forest garden, so that the natural landscape, cultural landscape and agricultural garden landscape can be harmonious and unified, and the ecological environment can be a virtuous cycle. Relying on the natural and beautiful garden scenery, comfortable fresh air and green space of ecological environment, tourists can return to nature and enjoy the beauty of ecological nature. Green, ecological and natural orchards and forest gardens are used as carriers to provide tourists with sightseeing and scenery appreciation. Develop the tourism activities such as picking, scene‐watching, flower appreciation, spring hiking and fruit purchase, and make tourists watch the green landscape and experience the beautiful natural landscape of the garden.

The main techniques are the cultivation and management of orchards and gardens, as well as the planning and designing of them.

3. Model and technology of flower sightseeing

The flower sightseeing model takes various flower planting, cultivation, viewing and sales as the carrier, and develops tourist activities such as flower watching, flower appreciation and flower picking (flower buying) , so that tourists can watch the sea of flowers and experience colorful and beautiful flower scenery.

The main techniques are flower planting and management techniques, garden planning and designing techniques.

4. Model and technology of urban modern agriculture, leisure and sightseeing

Urban modern agriculture leisure sightseeing pattern is relying on modern agricultural facilities and modern agricultural cultivation engineering technology, through original cultivation of the facility crops and horticultural crops, to implement the productivity and appreciation of the facility cultivation, such as the use of walls, columns, multilayer cultivation bed equipment and engineering technology. The soilless cultivation technology of leaf vegetables is applied for cultivation of plane formation, to make full use of the variety and color of vegetables to form an ornamental point. Using nutrient solution tank and other equipment and engineering technology, the soilless cultivation technology for fruits, using the biological characteristics of fruits' (hot pepper, eggplant and tomato) unlimited growth, using the advantages of facilities' being environmentally controlled, we can fully extend the growth period of the hot peppers, eggplants and tomatoes, forming the tall tree shape, which is a point for appreciation. The facility cultivation of plant factory integrates modern agricultural cultivation engineering technology, intelligent control technology, sensing technology and environmental regulation technology into one, realizing seedling cultivation, vegetable cultivation and vegetable products, and forming a point for appreciation.

Urban modern agriculture leisure sightseeing model will integrate the crop cultivation techniques, ornamental plants and landscape artistics to form magnificent and colorful landscape, bring out the best between the urban facilities horticulture and sightseeing leisure agriculture, improve and enrich the connotation and denotation of urban facilities horticulture. So that the urban residents and tourists can meet the spiritual needs of leisure sightseeing in the urban modern agricultural parks, through viewing, picking and leisure walk etc. This not only provides a good leisure and sightseeing agricultural landscape for urban residents and tourists, but also provides rich agricultural product sales income and tourism income for producers and operators, and the comprehensive benefits are considerable. In recent years, in the suburbs of some big cities such as Beijing, Shanghai, Tianjin, Guangzhou and Shenzhen, urban modern agriculture has become a highlight of leisure tourism and a new economic growth point of the city.

The main technologies are plant factory facilities, modern agricultural cultivation engineering technology, intelligent control technology, sensing technology and environmental regulation technology, as well as urban modern agricultural parks planning and designing technology.

II. Model and technology of life-type leisure agriculture

The life-type leisure and sightseeing agriculture model enables tourists to enjoy the beauty of agriculture and rural ecological nature, the joy of rural customs and the taste of green and organic agricultural products by participating in agricultural production activities and experiencing agricultural life, such as "eating rural food, living in rural house, doing rural work and watching rural scenery". Thus agricultural efficiency is increased.

1. Model and technology of experiencing the agricultural life

Farming life experience model provides a new way of tourism and leisure environment for urban residents and students. They participate in agricultural production activities, eat, live and work with farmers, experience the actual agricultural production, farming culture and special local flavor. This model improves the urban residents' leisure life grade, and changes the students' cognitive way of society and agricultural science and technology, which is conducive to the harmonious and coordinated development of society.

The main technology is farmland production management technology, management of tourists' food, housing, travel.

2. Model and technology of farmhouse leisure (agritainment)

In the farmhouse leisure model, farmers make use of their own courtyard, their own agricultural products and the surrounding rural scenery and natural landscape to provide tourists with a comfortable, healthy and safe farmhouse living environment and delicious farmhouse characteristic dishes to experience the farmhouse life, as well as sightseeing, entertainment, leisure, shopping and other tourism activities. "Eating rural food, living in rural house, doing rural work and watching rural scenery" let tourists enjoy the beauty of rural ecology and nature, enjoy the joy of rural customs.

The main technology is the design of agricultural tourism variety, the scientific management of food, accommodation and transportation.

III. Model and technology of trade-type leisure agriculture

Agricultural trade model is a leisure and sightseeing agricultural model that combines leisure & sightseeing contents with agricultural economy & trade activities organically, by making use of various large and medium-sized agricultural and sideline product distribution markets, business exhibition centers and agricultural product processing parks to provide tourists with leisure & sightseeing, high-quality agricultural and sideline product procurement and other services, such as all kinds of farmers' trade fairs, agricultural products fairs and so on.

The main technology is business exhibition design, agricultural products processing park, agricultural products distribution market management.

IV. Model and technology of leisure agriculture in agricultural sci-tech parks

1. Sightseeing model and technology of agricultural sci-tech park

Agricultural sci-tech park leisure sightseeing model takes in agricultural sci-tech demonstration garden as its carrier, to carry out high-tech agricultural production, agricultural stereoscopic plantation and soilless cultivation, ecological agriculture, popular science education, which are integrated in the agricultural leisure tourism. It shows the visitors the modern high and new agricultural science and technology and new exotic plants, crops, agricultural products, the originality of agricultural high-tech sightseeing, and the charm of uniting the modern agriculture and agricultual recreational sightseeing agriculture. Visitors can learn about agriculture and agricultural science and technology through the exhibition of vegetable and flower cultivation, facility agriculture and ecological agriculture in the agricultural science and technology park.

The main technology is agricultural high-tech production, agricultural (flower) three-dimensional planting, soilless culture, ecological agriculture technology, park planning and designing technology, park management technology.

2. Leisure and sightseeing model and technology of agricultural ecological parks

The leisure and sightseeing model of agro-ecological park is to develop agricultural and sideline products and tourism products with regional characteristics centering on ecological agricultural production and utilizing pastoral landscape, natural ecology and environmental resources, so as to provide tourists with leisure activities such as sightseeing, touring, tasting, shopping, participating in farming, etc. It is a multi-functional new industrial park integrating ecological agriculture, science and technology demonstration, leisure and sightseeing agriculture, green organic agricultural production and popular science education, and realizing the unity of ecological benefits, economic benefits and social benefits.

The main technology is ecological agriculture production and management technology, ecological park planning and designing technology, park management technology.

V. Model and technology of leisure sightseeing of farming culture

The farming culture model is to carry out agricultural culture leisure and sightseeing activities by using farming skills, farming tools, farming seasons, farming displays, agricultural products processing and so on. That is to inherit the unique rural life culture, industry culture and many folk cultures through the development of leisure and sightseeing agriculture, and create a special style of rural culture. China's traditional farming civilization & culture has a long history, extensive and profound, and is closely related to leisure and entertainment ideology and culture, with new meanings. At the same time, China's rural conditions and customs, folk culture resources are rich. We should carry out folk culture leisure and sightseeing activities by using folk customs of housing, clothing, diet, etiquette, festival and recreation. We should develop rural culture leisure and sightseeing activities by using folk songs and dances, folk skills, folk dramas, folk performances, etc. We should develeop ethnic cultural leisure and sightseeing activities by using ethnic customs, ethnic habits, ethnic villages, ethnic songs and dances, ethnic festivals, and ethnic religions etc. We can also use the Naochunshe on February 2 for men to compete Tian drum in the field, drink spring wine for revelry. We can go for an outing before and after the Qingming Festival, can climb moutains and appreciate chrysanthemum on the September Ninth. The activites such as dragon boat race, chess competition, martial arts, lion dance, and new rural construction and other new features of the era can be combined into the agricultural culture leisure and sightseeing activities, making the original agricultural culture have a broad development prospect.

The main technology is the digging, collection and arrangement of farming culture and folk culture resources, as well as the development, design and management of leisure and sightseeing products.

VI. Model and technology of educational agricultural parks

Educational agricultural park is a special type of agricultural park, which combines agricultural production and popular science education, and is the base of new quality education and popular science education. Educational agricultural parks use crops planted, animals bred and equipped facilities in the agricultural parks, such as the special plants, tropical plants, farming facilited cultivation, traditional farming tools display, etc. to carry out agricultural science and technology demonstration, ecological agriculture demonstration, and agricultural knowledge educaiton. It integrates science and technology, popular science, education, leisure and sightseeing, and has strong local cultural characteristics. It is a good form to carry out agricultural popular science education. The educational farm has the following functions:

① Science popula rization education function. The beautiful and unique plants, animals, natural landscape become educational resources, become a popular science education base and comprehensive practice education base. It is a good place for students' spring outing, autumn outing, summer camp, military training, laboring, sketching, writing, photography and other extracurricular education.

② Production and living function. All kinds of trees, flowers, melons and fruits, vegetables, poultry and livestock bred in the garden can provide fresh, healthy and safe food for visitors. In addition, soilless cultivation, three-dimensional cultivation, grafting, hybridization, transgenic technology and other applications can provide tourists with more new, excellent, special food and flowers, such as colorful bell pepper, banana, zucchini, magic beans, pets, all kinds of flowers and so on.

③ Sightseeing, leisure and entertainment function. Green garden such as trees garden, flowers garden, vegetables garden, fruits garden, garden landscape, architecture, water body, farming culture, folk culture and so on also formed a marvelous rural landscape, which can attract visitors to tour, taste, experience andentertain, and they held various activities, such as meetings, family gatherings, elderly leisure vacations.

At present, educational parks have been developed in many parts of the world. Since the 1990s, many countries in Europe and the United States have developed a large number of educational parks. Most are private ones to increase income, integrate traditional production and living resources, and provide urban residents with agricultural and rural life experience. There are also educational agricultural parks set up by government agencies, churches and other social organizations from the perspective of social welfare or ecological education. Such agricultural parks are not for profit, mainly to promote agricultural technology and stimulate people's love for the natural environment and biology. The Japanese government invests a large amount of money to organize and implement educational activities for teenagers every year. The "School Park" established by the Japanese government is a demonstration base for moral education, mental health education and environmental protection education for teenagers.

China's educational agricultural park (agricultural science popularization education park) is to provide tourists with educational activities to understand agricultural history, learn agricultural technology and increase agricultural knowledge through agricultural sightseeing park and agricultural science and technology park, such as the use of agricultural research testing ground, which integrete the agricultural production, science and technology demonstration, science and technology education into a whole, to provide agricultural science and technology education to farmers and students. Make use of the resources and environment of the agricultural parks, modern agricultural facilities, agricultural operation activities, agricultural production process, high-quality agricultural products, etc., to carry out agricultural tourism, participation and experience, DIY education activities. Make use of local agricultural cultivation, animal husbandry, feeding, farming culture, agricultural technology, etc., to let students participate in leisure agricultural activities, receive agricultural technical knowledge education.

The development and construction of educational agricultural parks can provide people with the importance and significance of understanding agriculture, understanding agricultural production process, experiencing rural life, understanding rural culture and ecological environment through popular science activities. Through the demonstration of agricultural science and technology and ecological agriculture, people can cherish the ecological environment and natural resources of agriculture and rural areas more, and increase the consciousness of protecting nature and environment. Above all, the purpose is to make people understand that agricutural and sidelire food are hard-won, thus cherish the food move.

The main technology of educational agricultural park is the planning and designing of educational agricultural park, the forms of popular science education activities, the development, design and management of new leisure and sightseeing products.

It has been discussed above that the development of green leisure and sightseeing agriculture is not only an important means to adjust the agricultural industrial structure and promote the income increase of agricultural farmers, but also an effective way to improve the quality of life of urban and rural residents. China's leisure and sightseeing agricultural resources are unique and rich, with elegant taste, distinct regional characteristics and diverse ethnic styles. According to the characteristics of local resources, local conditions, overall planning, scientific layout, continuous innovation, the development of leisure and sightseeing agriculture should be suitable for local development, to promote the development of local leisure sightseeing agriculture. With the growth of the consumption demand for leisure tourism of urban and rural residents, the requirements for the quality of leisure activities will also improve. In the development of leisure agriculture, we should not only pay attention to the scale and quantity of leisure agriculture products supply, but also put the development of differentiated products and continuous improvement of product quality in an important position. Give full play to the government guidance, enterprise participation and market driving mechanism, reasonably adjust and grasp the development direction of travel demand, develop a variety of leisure sightseeing agriculture projects with modern agriculture and high-tech features, rich rural scenery, simple customs, and rich rural culture to make the green leisure sightseeing agriculture become bigger and stronger.

Chapter 6 Basic Theory of Modern Ecological Circular Agricultural Park Planning

Section 1 The concept, types and significance of modern ecological circular agricultural parks

I. The concept of modern ecological circular agricultural parks

Modern ecological circular agricultural park based on circular economy theory to carry out the inner layout and agricultural production, with resource efficiency and recycling as the core, with low consumption, low emissions, high efficiency for this feature. In a certain regional ecological system, we carry on the agricultural production activities, cultural activities, leisure, entertainment, etc. , and through material recycle and energy flow, information transmission, we can make the park activities reach the state of being the highly adapted, coordinated and united. The resources, reduction and recycling technology are optimized and combined in the industrial chain model. It is the integration and display of basic principle, technology and model of circular economy applied to the park space, to achieve the purpose of energy conservation and emissions reduction and income increase, promote the sustainable development of modern agricultural park.

Generally speaking, the circular modern ecological agricultural park is an agricultural production model that uses the principle of material recycling and regeneration and material multi-level utilization technology to achieve less waste production and improve the utilization of resources. It is an important way to transform the agricultural increase model, improve the utilization efficiency of resources and improve the rural ecological environment. The modern circular agricultural ecological park is characterized by intensive assembly of agricultural clean production and comprehensive utilization of waste technology, becoming an important demonstration carrier for the transformation of sci-tech achievements of circular agriculture and modern circular agricultural production. As a kind of environment-friendly agricultural method, circular agriculture has a wide range of social, economic and ecological benefits. Only by constantly inputting technology, information and funds into a dynamic system engineering can we better promote the recycling of rural resources and the sustainable development of modern agriculture.

Modern ecological circular agriculture park is a new type of agriculture that is ecologically self-sustaining, with low input and strong vitality in economy. It is acceptable in environment, ethics and aesthetics. In practice, it tries to overcome the crisis brought about by "petroleum agriculture". Its purpose is to establish agriculture on the basis of ecology rather than on the petroleum chemistry. In fact, the ecological agriculture is the result of ecological thought on the agricultural production, requires people to make full use of local natural resources, use the interdependence among the animals, plants, microorganisms, use modern science and technology, realize productions with no waste and no pollution, provide as much as possible of cleaning products, meet the needs of the people's production and living, romote the development of scale economy and create a beautiful ecological environment.

II. The main types of circular agricultural ecological parks

Circular agricultural ecological park is the trend of large-scale, modernized and standardized agriculture in the future. The planning and construction of circular agricultural ecological park is a technology-intensive work to integrate and coordinate the local regional agricultural ecological system, so that regional agricultural production construction and development can give full play to the advantages of local leading industry and product characteristics. Fully explore the local culture, greening and beautifying the environment. Give full play to the financial advantages of enterprises and individuals, vigorously strengthen the construction of rural production infrastructure, and establish links to normalize the transformation of agricultural scientific and technological achievements. Give full play to the advanced role of scientific and technological personnel and modern technological equipment. The circular agricultural ecological park has more demonstration, guidance and promotion significance to the development of agricultural industry economy in the region.

According to the fact of agricultural development in China, and the influence of regional transportation, economy and environment, the development ways of circular agricultural ecological park in China are as follows.

① According to the leading industries, it can be divided into planting leading type, breeding leading type, processing leading type and tourism leading type;

② According to the leading function: agricultural science and technology park, agricultural tourism and leisure park, leisure farm;

③ According to the regional types: suburb of the urban city, surrounding area near the scenic region, characteristic villages;

④ According to the organization type, it can be divided into community collective management, enterprise independent management and farmers independent management.

Different types of circular agricultural ecological parks are all development planning models based on local conditions, such as regional realities and site characteristics. Each park is based on good environmental, economic, technological and industrial advantages, which is more conducive to the construction and development of circular agricultural ecological parks.

III. The significance of building ecological circulation agricultural parks

1. To improve comprehensive production capacity

The construction of the ecological circular agricultural park emphatically mobilizes and exerts overall function of agricultural ecosystem, setting about to large-scale planting and breeding, in accordance with the principles of overall consideration, overall coordination, recycling, and sustainable development. We should comprehensively coordinate and plan, adjust and optimize the agricultural structure to the greatest extent, make the involved agricultural form coordinated with the comprehensive development of industries, so that the various industries can support each other, cooperate with each other, so as to achieve the purpose of improving the comprehensive production capacity.

2. To give play to diverse industrial advantages

To construct ecological circular park, we should consider the actual situation of agricultural production in our country, because our country is large country with many differences between the natural conditions, resources foundation, economic and social development level. Ecological circular agriculture should fully absorb the essence of traditional agriculture, combine with modern science and technology, apply the variety model of ecology, ecological engineering construction and practical technology to agricultural production. Through continuous and in-depth analysis and comparison, we explore the advantages of agricultural production in different regions, transform backward production modes, and then integrate different and diverse development models and industries, highlighting the diversity of development.

3. To improve the industrial competitiveness

After China's accession to the WTO, it brought more opportunities than challenges to the agriculture, and the impact was huge. It made the chronic diseases in China's agriculture stand vividly revealed on the paper, such as backward infrastructure, small scale, single structure, low efficiency and lack of competition. So how to meet WTO's challenge has become an important issue for our agriculture. To solve the series of problems we should establish ecological circular agricultural parks to raise the level of modern agricultural development in our country, optimize the agricultural structure. By large-scale, industrialization management, we should integrate some low benefit industries together, set up the a circular agriculture platform, realize the high connection between production and market demand, at the same time protect the agricultural environment, in order to enhance the competitiveness of industries.

4. To promote sustainable development of benefits

Construction of ecological circular agriculture park can promote the scientific development of ecological circular agriculture, can effectively protect agricultural ecological environment, prevent the harm chemical fertilizers, pesticides and other harmful substances on land, crops, livestock, agricultural products. This not only maintains the ecological balance, but also improves the quality and safety of agricultural products. Environmental protection, agricultural production and economic development are closely combined to ensure the supply of agricultural products and meet people's consumption demand, to make our modern agriculture develop to a more stable, harmonious, sustainable direction.

5. To play an exemplary role

In the western mountains, farmers are dispersed, transportation is inconvenient, communication technology is backward, information sources are not fast, farmers can not master more agricultural practical technology, utilization of resources is low, increasing farmers' income is difficult. All these result in weak markets for agricultural products, planting and breeding of low level repeated. It is urgent to establish a collection park with outstanding industrial advantages, new technologies, new varieties, new management and new ideas. First, ecological circular agricultural park can effectively introduce new agricultural varieties for cultivation, promote the development of agricultural high and new technology, vigorously promote advanced and applicable technology, and promote the progress of agricultural science and technology. Second, it is conducive to organically combining scientific research units, scientific and technological personnel, industrial entities and farmers, collecting various strengths, and giving full play to comprehensive advantages. Third, it can explore a development road suitable for regional rural economy, easy to operate, and being scientific, efficient, high quality and pollution-free. Fourth, we can train farmers to improve their overall quality of farming.

Section 2 Theory and functional industry orientation of modern ecological circular agricultural parks

I. The theory of modern ecological circular agricultural parks

In essence, modern ecological circular agricultural park is a new agricultural organization form integrating production, research & demonstration, leisure & sightseeing, science popularization & education to promote the development of local economy. Therefore, the responsibility of modern ecological circular agriculture park is relatively significant, which is not only related to the economic interests of the government, enterprises, rural collectives and farmers, but also related to the prospect of regional agricultural development and the optimization and upgrading of agricultural structure. Modern ecological circulation agricultural park planning's overall policy is to serve the people, and serve the production. It is necessary to follow the principle of adjusting measures to local conditions and reasonable layout from the actual point of view, which not only conforms to the level of national, provincial and municipal agricultural development, but also creates a guiding and demonstrative modern ecological circular agricultural park system with local characteristics and innovative spirit.

The planning concept is the general and clause limitation and guidance of the whole planning of the modern ecological circular agricultural park, which directly affects the development orientation and strategic target of the park, the selection of key projects and the future development direction of the park, etc. , which are some basic concepts that must be followed in the planning process. The planning concept is the dominant idea established by the planning unit and the planning staff in the process of compiling the planning scheme of modern ecological circular agricultural park, which endows the scheme with unique cultural connotation and characteristics. The scientific planning concept of measuring to the local conditions plays an important role in the construction of modern ecological circular agricultural park. It is not only the essence of planning, but also can make the planning scheme more local and professional. At present, the overall planning of modern park mainly includes the following planning concepts:

1. Ecological concept

Ecological concept applied in the modern ecological agriculture park, starts from the planning point of view, bases on the characteristics of the urban and rural ecological environment to balance the development and ecological environment, pays attention to the environmental capacity and the limits of land development and utilization, emphasizes the overall environment coordination, comprehensive utilization of resources, maximizes the protection and improvement of ecological environment to achieve the symbiosis, mutual integration and mutual reflection between architecture and environment, so that the modern ecological circular agricultural park presents a harmonious scene of the symbiosis and mutual integration of ecology, production and life, and becomes an urban back garden with strong local cultural characteristics and ecological landscape texture.

2. Energy-saving concept

The planning and layout of modern ecological circular agricultural parks should fully reflect the concept of energy conservation. Buildings such as office and production areas should ensure sufficient sunshine spacing, and use exterior wall materials with good thermal insulation performance, so as to ensure the comfort of work, production and life, also reflect the characteristics of modern energy conservation. Agricultural facilities such as connected greenhouse, solar greenhouse and arch shed are designed according to the local sun height angle to design the most economical lighting requirements, and the spacing between the buildings is designed scientifically. At the same time, try to make the space formed between the buildings closely combined with the surrounding landscape, water, sunlight, natural ventilation conditions, to achieve the harmony of human and nature. In addition, in the production, we should make full use of the energy saving production process, optimize the artificial building environment, save artificial energy, and avoid energy waste.

3. Water-saving concept

The planning and layout of the modern ecological circular agricultural park will fully reflect the concept of saving water, and the sewage in the production, life and processing area can reach the discharge standard and be used for farmland irrigation after treatment. The low-lying area and the indoor area of the park shall have rainwater collection and utilization engineering facilities as far as possible. To avoid the loss of rainwater, set up stormwater drains in the management area and processing area to make rainwater flow to the reservoir for backup irrigation of farmland. Drip irrigation is used as far as possible in open field cultivation, and under membrane drip irrigation is used in greenhouse irrigation.

4. Sustainable development concept

The planning of modern eco-circular agricultural parks should be integrated with recent projects and developments, and a flexible long-term development framework should be established to delineate a clear outline of the future of the parks. Due to the rapid development of modern science and technology, high-tech industry, production procedure reconstruction and renewal, as well as the pace of production scale expansion, all these requires that the base construction projects must be flexible, architectural design must leave flexible space. At the same time, the concept of sustainable development should also be reflected in the comprehensive utilization of land, water, power distribution and other resources, and the regional space for future development should be reserved to meet the needs of future technological, social and economic development.

5. Humanized design concept

When the people-oriented concept is more and more prominent, material production and social and economic development more and more reflect the human requirements, and the planning and designing of work, life, production environment needs to have good efficiency and comfort. Therefore, the planning of modern eco-circular agricultural parks should deal with the harmony between human beings and buildings, human beings and facilities, and human beings and the environment. No matter it is demonstration area, production area, scientific research area or sightseeing and leisure area, it should reflect the humanized design concept of people-oriented.

6. Personalized design concept

Modern ecological circular agricultural park planning should have a main tone in the whole architectural style, but in the specific design they can have their own characteristics. The layout of each functional area of the park should not only meet the basic requirements of the overall pattern, but also reflect its characteristics. It should have a distinct personality in the combination of the architectural form and the natural space, and the logo of the enterprise, so as to enrich the visual effect and cultural environment of the whole project area.

7. Landscape design concept

On the premise of satisfying the demonstration of high and new agricultural technology and industrial development, the construction of the park strives to be planned and laid out in accordance with the idea of lanscape. Arrange the display projects of famous, special and excellent melons, fruits and vegetables, flowers, seedlings, grassed and so on, specifically design the facility engineering, ecological sightseeing park, different architectural styles, and elaborately design and plan the supporting projects such as roads, landscape, water surface, bridges, ditches and green corridors. All these gradually form a high-quality modern ecological circular agricultural park, which is suitable for the surrounding climate, environment and cultural landscape.

8. Intelligent design concept

With the development of modern science and technology, the construction and production system needs the characteristics of intelligence, sociality and structure clarity, which fully embodies the characteristics of modern ecological circular agricultural parks and intelligent buildings. On the premise of meeting the production and living functions, in the planning and design of the park, priority should also be given to the automatic management of safety, fire protection, energy saving and other aspects between the independent functional areas and the buildings of the units, as well as the functional settings of power, communication and network services, and the introduction of advanced facilities and equipment, so as to reflect the intelligent characteristics of the modern ecological circular agricultural park.

II. The functional industry orientation and functional area analysis of the modern ecological circular agriculture parks

The functional orientation of modern ecological circular agricultural park must be determined based on the actual development of local industry, so that it will have better development advantages and competitive advantages in the later development process. Generally, the analysis data is based on the local leading industry, and the industrial positioning should be carried out according to the advantages of the regional leading industry and its own characteristics to highlight the advantages, tap the potential and form characteristics. Through the function orientation, we can solve the product structure model, and the low technology content of the production operation mode. Using the principle of regional economics, industry economics, on the basis of regional resource advantage and comparative advantage, we should study in layers on the function orientation, area and time sequence of industry development orientation. The functional orientation is the foundation of the park planning, the starting point and the main idea of the park planning. The industrial planning, product planning and project general plan of the parks are carried out based on the functional orientation, and meet the standard requirements of the "Construction of Agricultural Standardization Demonstration Zone".

1. Technology innovation and technology transformation function

Modern ecological agriculture park is different from the past and other modern ecological agriculture park. It must take science and technology as the guide, the project as a carrier, the enterprise as the backing, the talent as the base and efficiency as the goal, to form the good model of ecological sustainable development. It is a comprehensive platform to promote the transformation of agricultural science and technology into real productivity under the condition of socialist market economy. It can effectively improve the market competitiveness of agricultural products, not only ensure national food security, but also promote the agricultural industry to go out for the international development, and become the growth pole of regional agricultural economy. Modern ecological circular agricultural park has the characteristics of good ecological benefits, high scientific and technological content, high conversion rate of scientific and technological achievements, and high comprehensive economic benefits. It centers on the new revolution of science and technology, and depends on the units of agricultural scientific research, education and technology promotion, with science and technology innovation as the core, and with the system and mechanism innovation as the power. Its main content is science and technology development and demonstration, popularization and promotion, which realized the effective combination of the modern science & technology with the ecological cycle of agriculture industry, and promoted the integration between product origins and cities, intregration between the parks and towns, and the integration of the three industries.

2. Enterprise cultivation function

Modern ecological circular agricultural park is an important base for the integration and innovation of new agricultural technologies. For one thing, the park can attract universities and research institutes to establish scientific research bases and R&D centers in the park, carry out technological innovation and experiment, develop and cultivate new varieties with high technological content, high market demand and high added value, and develop fresh preservation, deep processing and related supporting technologies of agricultural products. For another, the transformation of scientific and technological innovation achievements is a key link to support the development of modern agriculture, so the park also needs to play the role of "incubator", promote the transformation of scientific research achievements, and constantly incubate modern agricultural high-tech enterprises.

3. Agglomeration function of agricultural science and technology talents

The construction of agricultural science and technology parks can attract agricultural science and technology enterprises and scientific research institutions to gather, promote information exchange and knowledge sharing, and accelerate product and technology innovation. In addition, the agglomeration of various production factors enables enterprises to complement their partners in resources and reduce transaction costs, gradually forming an industrial network, attracting a large number of external talents, technologies and funds, and further promoting the agglomeration of agricultural science and technology personnel.

4. Production and processing function

The modern eco-circular agricultural park is an economic entity in nature, which undertakes the important task of applying the achievements of modern agricultural science and technology, and the production and processing of high-tech products is its basic function. By attracting relevant agricultural science and technology enterprises to settle in the park, large-scale production and industrial management can be realized. At the same time, as the key carrier of agricultural production and processing, enterprises in the park can combine agricultural science and technology with production and processing, and improve the added value of products and enhance market competitiveness through intensive processing.

5. Ecological and tourism sightseeing function

Focus on characteristics, we fully display modern agricultural science and technology and regional traditional agricultural culture, through tourism to bring a huge flow of people, logistics and information, promote the development of modern ecological circular agricultural park. In the process of park construction, in addition to maintaining the natural attribute of agriculture, we should build a modern agricultural sightseeing park being scientific, artistic, and cultural, being united with the nature, by a new type of agricultural facilities and display of the high and new technology, combined with the integral design of the garden and the production and demonstration of rare and special fruits, vegetables, flowers, and fishes.

6. Popularization drive function

The introduction of new varieties, popularizaiton and demonstration of technology and leading the surrounding economic development are the main task of the agricultural science and technology park. Through the display, demonstration and popularization of these technological achievements, it is helpful to transform and upgrade traditional agriculture, promote agricultural modernization, and promote the transformation of agricultural growth mode from "resource-dependent" to "sci-tech dependent". As the "growth pole" of agricultural economic development, modern ecological circular agricultural park can promote the economic development and the improvement of scientific and technological innovation ability in the surrounding areas, and help further improve the enthusiasm of local farmers for breeding and planting and their willingness to start businesses, so as to increase farmers' income.

7. Educational demonstration function

In recent years, land-lost farmers have become an invisible vulnerable group in the process of national reform & opening and urbanization. It is one of the effective ways to solve the re-employment problem of land-lost farmers by carrying out vocational skills training and improving their comprehensive quality and employment ability. Modern eco-circular agricultural parks can provide education and training for technicians and local farmers based on the advantages of human resources and scientific and technological resources, such as improving their knowledge and skills through technical training and improving their entrepreneurial ability through entrepreneurship education. The establishment of a sound agricultural technology popularization system and a sound training mechanism in the park can effectively improve the training personnel's cultural quality, technical level and employment and entrepreneurship ability, so as to make them adapt to the needs of agricultural development and rural revitalization.

Section 3 Planning principles and contents of modern ecological circular agricultural parks

The planning and designing of circular agricultural ecological park is a systematic project, which integrates the basic principles and methods of circular agriculture, ecological engineering, landscape architecture, agricultural management and other aspects from the perspective of discipline, and belongs to the interdisciplinary field. The principles of circular agriculture and ecological engineering include engineering technologies such as reduction, resource, reuse, residues reused as feed, residues as materials, residues as energy and fertilizer. The principles of modern agricultural science and technology include facility agriculture, modern agricultural equipment, animal husbandry, agricultural industrialization and information technology. Principles of landscape science include landscape ecology and landscape engineering technology. Agricultural management science includes organization management, product marketing, property management and logistics management, agricultural product quality management and so on.

The relevant theories of planning and designing include the theory of agricultural location optimization, the theory of reasonable matching of resources, the theory of compound ecological agriculture, the theory of agricultural transformation and upgrading, the theory of agricultural engineering system, the theory of high and new technology industry, the theory of agricultural environmental protection, and the theory of agricultural resource utilization. The code framework includes agricultural system code, planning and building system code and related policies and laws.

I. The planning background and guiding ideology of modern ecological circular agricultural parks

1. Planning background

In recent years, our country pays more and more attention to and regulate the development of modern agriculture industrial park. In October 2016, the State Council issued the National Agricultural Modernization Program (2016 – 2020) which proposed that we should innovate the mechanism of the first, second and third industrial convergence. The development of agricultural industrialization should base on industry. We should build a batch of rural forerunner area of the three industries and agricultural industrialization demonstration bases, to promote the clustering of farmer cooperatives, family farms, leading enterprises and supporting service organizations. In December 2016, the Central Rural Work Conference proposed that "modern agricultural industrial park is an important carrier to optimize the structure of agricultural industry and promote the deep integration of three industries". In 2017, the first central document proposed for the first time, we should promote the rapid development of modern agricultural industry park in China, based on large‐scale farming base, relying on the leading enterprises of agricultural industrialization, gathering modern production factors, and constructing modern agricultural industry park of "production, processing and technology". In March 2017, the Ministry of Agriculture jointly the Ministry of Finance issued "Notice on Constructing the National Modern Agricultural Industrial Park", which required by creating a national modern agriculture industrial park to make it a leading agricultural supply‐side structural reform of the new platform, focusing on the functions of industrial convergence of modern agriculture industrial park, farmers leading, technology integration, income increasing. We will explore new experiences to foster new drivers of agricultural and rural economic development, open up new ways to increase farmers' incomes, and provide a new carrier for accelerating agricultural modernization. During April to may, 2017, the Ministry of Agriculture and the Ministry of Finance has successively held the "National Conference of Centralized Deployment of Modern Agricultural Industrial Park", held training course for Planning and Perfecting of National Modern Agricultural Industrial Park, jointly submitted "Plans to Selecting the National Modern Agricultural Industrial Parks", which promoted the creation of the national modern agriculture industrial parks. In 2017, the Ministry of Finance and the Ministry of Agriculture announced the first batch of 11 national modern agricultural industrial parks, and in 2018, the Ministry of Agriculture and Rural Affairs and the Ministry of Finance jointly issued the "Notice on the Establishment of National Modern Agricultural Industrial Parks in 2018" to standardize the conditions and tasks for the establishment of national modern agricultural industrial parks.

2. Guiding ideology

The planning should be in accordance with the requirements for high quality development, implement the strategy of rejuvenating the country, in order to promote agricultural supply-side structural reform as the main line. It based on the advantageous industries, depends on the construction of large-scale farming and breeding bases, the leading industrialized enterprises, and the union of modern production elements, and the production, processing and technology of modern agriculture industry cluster. It will promote the development of the three industries convergence, innovate the benefit coupling mechanism of increasing farmers' income, cultivate the agricultural rural economic development momentum, build high starting point, high standards of modern agriculture construction sample area and rural prosperous industry leading area. It will lead the construction of provincial (autonomous regions and municipalities directly under the central government), city and county parks with ranks, to provide strong support for the agricultural modernization and rural revitalization.

II. The planning principles of modern ecological circular agricultural parks

The planning principles are guided scientifically and reasonably by the principles of foresight, demonstration, marketability, peculiarity, adaptation to local conditions, operability and sustainable development.

1. Principle of government guidance and market dominance

The remarkable characteristic of the development of modern ecological circular agricultural park is that it is closely combined with the market development. The planning should be guided by the market while the government regulates, grasp the latest market demand, adjust the agricultural industrial structure, combine the market segmentation and the development trend of the market demand, and make scientific and reasonable planning. Therefore, modern ecological agriculture park should be planned under the government guidance, accurately and timely grasp the market situation, analyze the modern agriculture industrialization development trend, research on the agricultural production conditions and production potential, market capacity and the outlook of agricultural industry, determine the target market of park, and give full play to the market entities' leading role in the industry development, investment, construction, and product marketing.

2. Principle of adjusting measures to local conditions

China is vast, with big span between east and west, south and north, and different in topography, soil, water, vegetation conditions, traffic condition, climate condition, and energy supply condition. National modern ecological circular agriculture parks, due to the different location conditions and geographical environment, have different resource conditions, different climate conditions and different economic development. So, there are different ways of planning and construction. Therefore, in the process of making the planning & designing and the choice of characteristic industry projects we should combine with regional characteristics, adjust measures to local conditions to consider local resources condition and ecological elements, as well as industry advantage, economic base, production conditions, and scientifically select appropriate leading industry and agriculture production project. We should focus on the local leading industries, develop the famous and excellent products to ensure the continuous growth of grain and other bulk agricultural products, and develop and utilize the agricultural resources in diverse ways, so as to make better comprehensive use of regional resources, reduce construction costs, and maximize the benefits of the park.

3. Principle of multi-objective synergy for sustainable development

We should adhere to the unity of economic benefits, social benefits and ecological benefits and take the road of sustainable development. Pay attention to the cooperation of industrial structure and landscape ecological elements in the park, and build a long-term mechanism for green, low-carbon and circular development. In the process of planning and designing, it is necessary not only to meet the functional requirements of the basic operation of the park industry, but also to enhance the beauty of the landscape to the greatest extent. It's important to set short-term goals and take long-term interests into account. Strive to protect and improve the ecological environment within the maximum range allowed by land development and utilization and environmental capacity, and promote the sustainable development of the park. In the economically developed areas with high level of science & technology, we should establish modern ecological circular agricultural parks with a high starting point, combined with high standard of science & technology and intensive capital. For the regions with abundant natural resources and lagging economic development level, we should first establish modern ecological circular agricultural parks with leading science and technology, intensive labor and resource strength. With the development of economy, the means of production and conditions should be improved gradually, so that the construction of modern ecological circular agricultural parks will present a multi-level, multi-type and multi-form pattern. The establishment of multiple chains from the industrial level is conducive to making full use of resources and promoting, complementing and symbiosis among industries in the park.

4. Principle of putting people first

The planning and designing of the park introduces the principle of "being human-oriented". The goal of modern ecological circular agricultural park construction is to solve the problems of "agriculture, rural areas and farmers", increase farmers' income and improve farmers' living standards. People-oriented considerations are hierarchical, and the focus should not be one-sided consideration of individuals, but comprehensive consideration of different groups, combined with the region, culture, society, benefits, etc. , to reflect the people-oriented planning concept. Therefore, the planning of modern ecological circular agricultural park should not only meet the production needs, but also meet the material and cultural needs of consumers and tourists. In the process of planning, it is necessary to carry out comprehensive humanized design from macro to micro, from whole to part, and pay attention to the coordination and benefit sharing among government, enterprises, scientific research and service institutions around industrial development and construction.

5. Principle of being forward-looking and operable

The planning of circular agricultural ecological park must be combined with the development goal of the park and the scope of consumer groups, considering the market supply and demand and future market saturation, and the project should be set in a forward-looking way in terms of technology and product production. In terms of production mode and technology, the principle of advanced & practical and moderately look-ahead should be followed to avoid the phenomenon that planning measures cannot be implemented into concrete construction steps. The planning of modern ecological circular agricultural park must be feasible, conform to the local agricultural production conditions, economic and social conditions and national policies, and avoid the abstract and visual planning that can not be implemented into the actual work. Operability and prospectivity are relative, which need to be adjusted timely in the dynamic development to ensure a relatively reasonable state. In general, on one hand the content and aspects of the planning should meet the requirements of being leading for a considerable period of time. On the other hand, the existing planning and designing specifications should be referred to, and the planning should highlight the openness, scientificity and ecology for the long-term development of the local area.

III. The planning content of modern ecological circular agriculture parks

1. Collection and analysis of relevant data

The first step in the planning and designing of modern ecological circular agricultural park should be to analyze the basic data of the site conditions, including the collection, analysis and field research of relevant data.

We should collect the region, base and policy conditions of the planning park, the results of the macro-layout planning of agriculture in the area where the park is located, and the research results of similar modern ecological circular agricultural park. Regional research is to study the macro location conditions of the region where the planned park is located. Including geographical environment, natural conditions (including environmental pollution severity), climate characteristics, resources distribution, social conditions and current situation of the development of social economy, the situation of agricultural development (productivity level, technical level, the main industry, etc.), agricultural industry (leading industry, characteristic industry, potential industry) structure development direction and market demand (including regional, national and international market), etc. Base research is to study the geographical location, traffic location conditions, land use regulations, infrastructure construction, planting types, enterprise construction status and so on. If there is an upper planning, it should also be analyzed. Policy research is to study the guidelines, policies and norms for the planning and construction of modern ecological circular agricultural parks issued by the national and local governments to ensure that the planning and designing meet the national regulations.

Through the analysis of basic data, the situation of agricultural resources and the overall level of agricultural production in the location of the agriculture park can be determined, which provides a basis for the positioning of the park. An overall grasp of the development direction of the park is helpful to determine the guiding ideology and strategic goals of the park, which mainly includes the nature and scale of the park, the main function and development direction of the park, the development stage of the park and the development goals of each stage.

We should investigate the local agricultural conditions, agricultural overall level, economic and social development level, and the development status of agricultural characteristic industries, etc. We should also investigate the current agricultural conditions of the park, including geographical location, natural conditions, and the severity of environmental pollution, etc.

2. Determination of the guiding ideology and strategic objectives

The guiding ideology of modern eco-circular agricultural park planning is the fundamental basis and action guide for the park development, as well as the theoretical basis for establishing the development orientation, construction target and project plan. The guiding ideology and goal of the planning should be based on the actual situation of the park, from the national and provincial conditions, to the village and household conditions. According to local conditions, according to the specific environmental conditions, we choose the content of planning and construction, guide the industrial park to develop advantageous agriculture, and achieve the strategic goal. The main contents are: to determine the possible way to achieve the park target, to find out the core factors to improve the competitiveness of agricultural products, and to formulate the development strategy of the park.

3. Function zones and layout

The basic functions of the national modern ecological circular agricultural park include agricultural production, agricultural product processing and sales, scientific and technological research and development, demonstrational popularizaiton, sightseeing and recreation, supporting services, leisure tourism and ecological protection, etc. These basic functions are integrated into the park planning, and on the basis of making the park profitable, also with social welfare functions such as scientific and technological innovation, mechanism innovation, demonstration popularization, serving farmers, driving industrial development, as well as a significant ecological protection effect. Functional layout is an important part of national modern ecological circular agricultural park planning. There are many factors affecting the functional layout of the park, but considering the industrial planning as the core, the main factors include land use status and benefit level, industrial correlation degree, functional similarity, overall planning requirements and the qualitative positioning of the park, etc. According to the industrial layout, we would determine the functional zoning and economic and landscape axes, delimit the core area, demonstration area and popularization area. The functional zoning should be planned according to the guiding ideology, development goals and planning principles of the park planning, adhere to the principle of highlighting the key points and embodiments the characteristics, and create a functional zoning layout form with complete functions, close spatial connections, reasonable and efficient. The functional zoning of the park is combined with the industrial layout, and the industrial layout is suitable for the project scale. It is necessary to consider both the overall planning and local coordination, and fully ensure the advanced, scientific and operable nature of the project scale and technology selection introduced in each region. The agricultural industry plays an important fundamental role in the park. The industrial setting and layout need a relatively specific and detailed project construction plan. Several functional areas or industrial belts should be established according to the target positioning and the nature of resources in the park. Establish the scope of core area, demonstration area and popularization area. It is necessary to give full play to the role of the third industry such as tourism and popular science education in the planning of the park while centering on the first and second industries such as planting, breeding and agricultural product processing.

The core area concentrates the park administration, science and technology development, industrial institutions, which is the core and key of the park production and construction, and is the agricultural high-tech scientific research and industrial development base. The core area has complete boundaries and well-defined ranges. The core area is the main body of the park, where the administrative institutions, most of the economic entities and organizations, the biotechnology tissue culture center, the new variety introduction area, the facility horticulture area, the scientific research and development center, the training and exhibition center and the information center are gathered. The main functions of the core area include: to introduce and apply the modern high and new agricultural technologies and scientific and technological achievements, for the experiment and demonstration of agricultural sci-tech achievements; to cultivate leading enterprises in advantageous industries, to form leading industries and enterprises or enterprise groups with market competitiveness, to carry out the development of high-tech agriculture and facility agriculture projects, to transform traditional agriculture with high and new technology, to carry out agricultural industrialization management, and to improve the market competitiveness of agricultural products in the park; to manage, direct and coordinate the whole park, to provide sevices for agricultural industry development and agricultural production such as technical services, transport and marketing services for production materials, transport, and marketing & processing services for agricultural products; to introduce experts of agricultural science and technology from home and abroad to tackle key problems and consult on the development of agricultural science and technology industry in the park; by agricultural science and technology training, to popularize and promote the advanced agricultural technology for farmers and agricultural enterprises in the region, and improve the quality of agricultural workers.

The demonstration area is the agricultural production base and the test base of agricultural science and technology achievements. It is also the driving base of agricultural industrialization in the core area, the main demonstration platform of modern ecological circular agriculture park and the direct object of the core area. The demonstration area has a complete boundary and clear scope. Generally, the demonstration area is close to the core area, and the area is 3-5 times of the core area. Demonstration area absorbs the core of the new technology and new varieties of the core area, guided by the technology, varieties, talents, funds, training of teh core area, through the operation of the "company plus farmers" form, carries out the standardization of agricultural production and demonstration, explores and develops modern agriculture industrialization management pattern. And through the technical support of the core area, the demonstration area focuses on the exhibition of biotechnology, saving and efficient cultivation techniques, modern facility agriculture technology, water saving irrigation technology, intensive farming, livestock products processing technology, information management technology, agricultural products processing technology and achievements of agricultural industrialization, showing the combination of artificial and natural environment improvement, showing the renovation and construction results of modern country, in order to pull and drive the development of agricultural economy of surrounding area thereby.

The popularization area is far away from the core area, and it is not within the boundary of the modern ecological circular agricultural park. There is no complete boundary and clear scope, so it can only have a rough radius. The popularization area is the surrounding agricultural production and rural economic area involved and affected by the leading industry in the core area of modern ecological circular agriculture park, or the same or similar type of agricultural area with similar geographical environment, resource characteristics, production and economic characteristics. The popularization area is generally close to the demonstration area, the area being 10 – 15 times the size of the demonstration area. According to the theory of "development poles", advanced technologies, new varieties and new products of modern ecological circular agricultural parks exert diffusion and traction effects on farmers in surrounding areas through the demonstration of planting enterprises in the demonstration areas. With the national modern ecological circular agricultural park as the demonstration base, the advanced productive factors gathered in the park are organically integrated, and the intensive production process of modern agricultural technology enterprises is transformed into competitive agricultural products, and profits are obtained through market sales.

4. Industrial planning

Industrial planning is the core content and design focus of the modern ecological circular agricultural park planning, through the selection and analysis of regional appropriate superior agricultural industries, to determine the industrial development content of the park, which is an important measure to achieve the industrial development of the park, obtaining economic, social and ecological coordination. Industry determines the strategic development direction of the park, and its function orientation should be based on the actual conditions of the local society and economy, and appropriate construction content and technical route should be chosen according to local conditions. National modern ecological circular agriculture park should analyze the technical support ability and the level of agricultural capital services business before the industry planning, and fully tap the potential industry development potential. After clear information before the planning and layout, with the first industry as the basis, second industry as the pillar, third industry as the links, we can realize the integration of agricultural industries.

The first industry, namely agricultural production, is the foundation of the existence and development of the park, covering the largest area. The spatial layout should be spread out in pieces or rings around the core area of the park, and the planned layout should be on agricultural land with good basic strength of agricultural production, reasonable distribution of rural population and distinctive agricultural production characteristics. The secondary industry is an industrial chain integrating the processing, storage and sales of agricultural products. Its spatial layout should be based on the principle of land intensification and industrial cluster, and should be concentrated into blocks and planned on construction land or reserve land for long-term construction. The third industry is an industrial form with the main functions of agricultural high-tech research and development, science and education publicity, training and promotion, sightseeing, leisure and entertainment, with diversified spatial planning and layout forms. The third industry can be combined with the spatial planning forms of first and second industries to carry out recreational activities such as agricultural production experience, leisure sightseeing and picking in industrial parks. It can be combined with agricultural culture, folk culture to plan and construct agricultural culture theme park; It can also display high and new agricultural technology, and independently build science and technology research and development center, technology extension center, agricultural science and technology tourism greenhouse, ecological greenhouse leisure projects and other development of popular science and entertainment combined construction projects.

The integration of three industries in rural areas can be divided into five types: one is industrial integration, such as the combination of planting and breeding; Second, the industrial chain extension type, that is, with agriculture as the center to extend forward and back; Third, industrial crossover, such as agriculture and tourism integration, agriculture and cultural integration; Fourth, technology penetration type, for example the rapid promotion and application of information technology, makes it possible to network marketing, online rental hosting and so on; Fifth, comprehensive type, that is, we comprehensively use modern engineering technology, biotechnology, information technology and other technological achievements, and to the maximum extent get rid of the constraints of natural conditions on agricultural production and management activities. Under relatively controllable environmental conditions, we achieve annual, all-weather, out-of-season agricultural enterprise production.

The project planning of modern ecological circular agricultural park mainly refers to the industrial planning and other project design closely related to the industrial planning content. Industry planning need to be clarified by selecting suitable advantageous agricultural industry to realize the industrialization, to realize the economic, social and ecological harmony, mainly including the choice of pillar industries, industrial scale, industrial chain development thought, industrial organization, involved in planting, breeding, processing, sales, research and development, logistics and tourism, and other fields.

① Planting industry. Mainly involved in vegetables, fruit trees, tea, field crops, ornamental plant planting and seed production and so on.

② Breeding industry. According to regional resource advantages, regional advantages and industrial advantages, we would develop beef cattle, dairy cows, sheep, pigs, poultry and other conventional livestock breeding; plan the breeding in water areas according to local conditions, promote environment-friendly breeding according to regional living resources and ecological environment; according to the contents of the park planning and regional characteristics, carry on the rare animals breeding. The main advantages areas of breeding products are: Northeast, central and southwest of pig producing; central China, Northeast, northwest, southwest of beef producing; the middle China, central and eastern China, northwest, southwest of mutton producing; Northeast, Inner Mongolia, north China, northwest, south and big city suburbs of dairy producing, north China, the middle and lower reaches of the Yangtze river, south China, southwest, northeast of meat and poultry superior region; East China, North China, Central China, South China and Southwest China of poultry eggs. In coastal areas, we will actively protect the ecological environment of mud flat areas, plan and develop ecological aquaculture, deep-water cage aquaculture and factory recycling aquaculture, and develop marine ranches. In coastal areas and the middle and lower reaches of the Yangtze River, we can build high-quality freshwater products processing industrial belt.

③ Agricultural products. The science and technology innovation platform of the processing park focuses on building an environmentally friendly, resource-intensive and market-oriented agricultural products processing zone by relying on the leading enterprises to focus on the leading industry and characteristic agriculture of the region, and effectively integrates the first and second agricultural industries. It mainly includes primary processing of agricultural products, intensive processing of grain and oil, processing of fruits, vegetables and tea, processing of livestock and poultry, processing of aquatic products, processing of other projects, and comprehensive utilization of by-products of agricultural products.

In the processing zone of modern ecological agriculture park, we should develop processing enterprises complemented each other, contacted closely, related to agricultural production. On the need of industrial park development vision and planning, we choose to develop processing industries such as industrial feed, livestock and poultry residue resource utilization, organic fertilizer production, agricultural machinery production, greenhouse equipment manufacturing.

④ Agricultural leisure tourism, catering industry. Basing on the agricultural elements and the local culture as the background, with the ecological landscape as a starting point, according to the situation of site factors, we adopt the group layout, applying the principles of activity zoning, using the drainage or garden road for function zoning, combing the integrated functions of protecting ecology, leisure, sightseeing, experience and education demonstration and other functions, to meet people's desire to be close to nature, learn agricultural knowledge, feel folk customs, experience farming fun and enjoy farm life. When planning we should pay attention to the combination of dynamic tour and static viewing, effectively protect and develop resources, protect agricultural environment, and try to grasp the balance between production and tourism. In addition to maintain the natural property of agriculture, we can also use the new type of agricultural facilities and new high-tech display, coupled with the integral design of the garden and perennial production and demonstration of fruits, vegetables, flowers and special rare birds, ornamental animals, forming the modern agricultural leisure tourist attractions combing science, arts, culture as a whole, with the unity of nature and human being.

5. Project planning

The project planning of modern ecological circular agricultural park is mainly aimed at the design of supporting and service projects of the park industry, which mainly includes the construction purpose, project scale, key technologies, technological processes, safeguard measures, risk assessment, benefit analysis and other contents. The project categories are usually set according to the functions of the park, such as facility vegetable cultivation project, coarse grain processing project, infrastructure construction project and so on. Park project planning must be coordinated with land use index, population development index and ecological cycle planning in the process of development.

6. Planning of the operation model

Determining the fundamental position of the advantageous industry in the park is only one aspect of planning. The normal operation and development of the industry cannot be separated from a set of operation mechanism and management mode. After analyzing the resources, investment, technology, management system and industrial characteristics of the park, the operation mode suitable for the development and construction of the park is an important link in the park planning. The other key points of the planning is the park project planning. Specifically, the planning should make arrangements for the organization and management mode, fund raising mechanism, land transfer mechanism, science and technology research and development and application mechanism, park operation mechanism and risk guarantee mechanism, etc.

7. Benefit analysis

The infrastructure construction of the park is mainly financed by the government, and the industrial development projects adopt the market operation mode and are invested and constructed by enterprises. No matter what channel the capital input comes from, it should be the first task to drive farmers to increase production and income and obtain direct economic benefits. At the same time, indirect benefits should focus on social and ecological benefits.

(1) Investment estimate and source of funds

The investment estimate of modern agricultural industrial park consists of two parts: fixed asset investment and current asset investment. The investment in fixed assets mainly includes the fixed assets expenses and other expenses used for construction of factories, buildings and equipment purchase, such as land collection fee, survey and design fee, site leveling fee, construction project fee, public infrastructure fee and personnel training fee, etc. Current capital mainly includes the cost of purchasing raw materials, fuel and power needed for agricultural production, as well as the cost of paying staff wages, equipment maintenance and operating activities in the park. In addition, it also includes administrative costs and unforeseen expenses.

The sources of funds mainly include policy funds, social financing, enterprise investment, foreign investment and farmer investment. In the process of planning and construction of the park, all sectors of society should be encouraged and attracted to participate in the construction of the park to broaden the funding channels, give full play to the government's leading role in the investment and construction of the park, and encourage various production factors to be added to the construction of the park. At the same time, we will vigorously introduce foreign capital, strengthen international cooperation in science and technology, and introduce foreign advanced production equipment and management concepts to ensure better construction of the park. In addition, in order to reduce the financial risk, we should not blindly expand the borrowed funds, but should make the borrower funds to maintain an appropriate proportion, so that the project operation always maintain a certain solvency and good financial condition.

(2) Analysis and risk assessment

The benefit of national modern agricultural industrial park mainly involves three aspects: economic benefit, social benefit and ecological benefit. The economic benefit mainly refers to the direct economic benefit and indirect one produced by the park and its popularization area. Social benefit refers to the social influence of park construction on promoting employment, improving environment, increasing fiscal revenue and so on. Ecological benefit refers to the benefit brought by improving the production efficiency of the park, reducing the pollution of the surrounding environment and improving the ecological environment.

The main risks of national modern eco-circular agricultural park include economic policy risk, market risk, scientific research and technology risk, engineering risk, financial risk, investment estimation risk, environmental risk and social impact risk. The risk assessment of the park is based on the prediction and assessment made in advance when the deviation of market situation, time connection, technical arrangement and project management causes the benefit change of the park and threatens the investment security. The risk assessment system of modern ecological circular agricultural parks is generally carried out by expert survey, analytic hierarchy process, CIM method and Monte Carlo simulation method. Therefore, in the period of planning and construction, we should conscientiously use the risk assessment method to analyze the potential risks of park construction, put forward corresponding countermeasures and suggestions according to the specific risk factors, and take practical methods to reduce the investment risk in the implementation process.

8. Other special plans

Road traffic planning, landscape planning, building engineering planning, agricultural engineering planning, service facilities and municipal engineering facilities planning are serving for the park industrial planning and project planning landing. The details are not covered in this section.

IV. The planning results of modern ecological circular agriculture park

The basic planning procedure of modern eco-circular agricultural park is: commissioned planning → basic data collection and analysis → planning; designing period → scheme optimization → scheme implementation; planning results; basic data compilation → feasibility analysis → planning description → planning atlas.

The design results of the overall planning of modern ecological circular agricultural park include planning text and planning drawings.

(I) The planning text

The planning text of modern ecological circular agricultural park (also known as planning description, planning report) should comprehensively reflect the content of planning and designing, and make a comprehensive introduction and explanation of planning conditions, concepts, basis, construction scheme, organization scheme, operation scheme, possible benefits, etc. In general urban and rural planning, it is required that the text of the master plan should adopt the form of provisions in the format, with standard words, accurate and positive language, and some technical arguments and explanations should be expressed in the manual. The planning results of modern eco-circular agricultural parks can draw lessons from the requirements of urban and rural general planning, but they need not be restricted to form. In most cases, it is allowed to merge the text and the specification and express it in the form of the plan text. Some texts will be accompanied by a research report, which will provide a more in-depth analysis of the client's concerns, and the results will be attached to the planning text in the last part of the text and atlas.

1. Text preparation requirements

The planning text outline of modern ecological circular agricultural park generally includes the following contents: project outline, policy background, planning basis, current situation analysis, the basic principles of park construction, guiding ideology, development goals (strategic objectives and specific indicators), park location, space layout and functional zoning, industry planning, the construction of the park content, park's organizational structure and management system, the operation mode and mechanisms of the park, park investment estimation and financing, analysis of the effect of park construction and suggestions on measures to ensure the implementation of planning.

The content of the planning text should be related, including the consistency of the data and contents; the basic logic is clear, and the purpose of each chapter is crystal clear. In addition to the main text, it is also necessary to indicate the names of both parties, the names of the planning projects and the members of the compilation.

2. Frame of the planning text

The text frame of the overall planning of modern ecological circular agricultural park

Chapter 1　　Planning Background
1.1　Planning background
1.2　Main planning basis
1.3　Policy planning stage
1.4　Planning Scope and Term

Chapter 2　　Basic Analysis
2.1　Location and transportation
2.2　Conditions of natural resources
2.3　Conditions of historical and humanistic resources
2.4　Social and economic conditions
2.5　Industrial status and trend
2.6　Case Reference and Analysis / SWOT Analysis

Chapter 3　　Target Positioning
3.1　Basic principles
3.2　Guiding ideology
3.3　Function positioning

3.4 Development objectives

3.5 Spatial structure/Functional partition

3.6 Plane layout/Industrial deployment

Chapter 4 Content of Zoning Construction

4.1 ×××× area

Including construction location, construction scale, construction content.

4.2 ×××× area

4.3 ………

Chapter 5 Infrastructure Engineering System Planning

5.1 Road system planning

5.2 Farmland water conservancy planning

5.3 Water supply and drainage planning

5.4 Power planning

5.5 Heating planning

5.6 Communication planning

5.7 Environmental sanitation project planning

5.8 Comprehensive disaster prevention planning

Chapter 6 Special Planning Guidelines

6.1 Agricultural science and technology services

6.2 Landscape recreation system

6.3 Brand agriculture cultivation

6.4 Comprehensive ecological environment

6.5 Investment promotion

Chapter 7 Organization Operation and Investment Estimation

7.1 Organization operation

Including organizational structure, operation mode, management mechanism.

7.2 Investment estimation

Including development timing, investment estimates.

7.3 Comprehensive benefit analysis

Including economic benefits, social benefits, ecological benefits.

Chapter 8 Guarantee Measures

8.1 Policy guarantee

8.2 Talent guarantee

8.3 Fund guarantee

8.4 Technical guarantee

(II) Planning drawings

The planning and designing drawing of modern ecological circular agricultural park is an important expression form of the planning results. The design drawing includes the analysis diagram of current situation and the planning diagram. The analysis diagram of current situation mainly expresses the location, traffic layout, water resources, comprehensive landscape of the site, and land use situation. Planning drawings include spatial layout drawings such as spatial structure, general plane, functional zoning and aerial view, special planning drawings such as land use, road, water system and power, and special planning drawings such as landscape, recreation and environmental protection.

1. Requirements for drawing

The drawing ratio of the planning drawings of the modern eco-circular agricultural park is generally 1 : (5000 – 10 000), and that of the key planning area is 1 : (1000 – 3000). The drawings should be printed and submitted in A3 format, in color, attached to the planning text or bound into a book separately. The format of planning and designing drawings are JPG, DWG and so on. The general requirements of each drawing are as follows, but not limited to this

(1) Preliminary analysis diagram

① Location map shows the geographical location of the park in its own level and the upper level, as well as the situation of its subregions, and can superimpose geographical elements such as roads, water system, railways and airports.

② The status map uses color block photo symbols to show the land use status quo, topography, vegetation, infrastructure status quo, artificial buildings and other spatial information.

(2) Part of the planning diagram

① Functional partition map   It outlines the boundaries of different functional areas in the way of color blocks, and explains in the form of words and tables.

② General plane map   It uses graphics, images, pictures, symbols, color blocks, lines and other ways to reflect various planning elements in a map.

③ Area plane map   The general plane is divided according to different functions, partly showing the planning elements, and accompanied with captions, pictures, intention maps, etc.

④ Special planning map   It is used for roads, farmland water conservancy, forest network and tourism, leisure, tourism, processing, circulation and others.

⑤ Effect drawing   Using dimensional modeling software or image processing software to bird view the plane map, showing the effect after the implementation of planning in three-dimensional and other ways to show.

2. The contents of planning drawing

The work content of the modern ecological circular agricultural park includes: landscape conceptual planning and designing, scheme design, preliminary design, construction drawing design, late docking site and so on.

According to the project conditions, the work of modern ecological circular agricultural park can be divided into four stages as follows.

Stage 1: Conceptual planning and designing

1.1 Site survey, project related data collection and sorting; (to design task book, current situation map, planning map, geological survey report; to collect meteorological data, hydrogeological data, local cultural and historical data; to survey on the spot and take photography, etc.)

1.2 Communication with the owner to determine the theme concept; (Customer requirements)

1.3 Site space concept analysis, landscape scheme general plane, landscape axis analysis diagram, traffic analysis diagram and zoning analysis diagram, etc.; (to show and analyze the solution)

1.4 Perspective intention of main scenic spots; (draft or schematic)

1.5 Concept design description;

1.6 Design cycle according to the project situation;

1.7 Stage achievements delivered.

Stage 2: Scheme design

2.1 Further improvement of the conceptual design, and re-communication with the owner to reach a consensus on the conception, which becomes the final landscape plan after expert review; (Customer demands analysis)

2.2 Deepening the conceptual design, improving the design concept and chief plane map. Add vertical, facilities, landscape, pavement and other analysis pictures, and display effect pictures such as overall bird's eye view, perspective effect pictures of important nodes, etc. ; (Deepening plan)

2.3 Special design: vertical (overall topographic change treatment scheme, including the main slope map and elevation map) ; Special plants design (plant greening principles, framework, zoning, characteristics, variety intention, etc.) ; Sponge city project (principles, analysis of rainwater collection and application, LID facility analysis, etc.) ; Special pavement; Facility services, etc.; (added according to project requirements)

2.4 Scheme design description;

2.5 Landscape Engineering estimate;

2.6 Design cycle according to the project situation;

2.7 Stage results delivered.

Stage 3: Preliminary design

Communicating with the owner again based on the results of the scheme, reaching a consensus on the scheme of modern ecological circular agricultural park, and carrying out the preliminary design of the deeper scheme.

The catalog of the overall planning of modern ecological circular agriculture park is as follows:

3.1 Location analysis chart

3.2 Road traffic status diagram

3.3 Land use status/planning map

3.4 Comprehensive landscape map of land use status

3.5 Spatial structure and industrial layout

3.6 Function Partition Diagram

3.7 Master/chief Plane map

3.8 An aerial view

3.9 Construction sequence diagram of key projects

3.10 Function Partition Intention Diagram

3.11 Road traffic system planning diagram

3.12 Landscape system planning diagram

3.13 Recreation system planning diagram

3.14 Irrigation planning diagram

3.15 Drainage Planning diagram

Stage 4: Construction

4.1 Construction drawing design is completed fully before the real construction;

4.2 Confirm sealing samples for the main paving materials selected by the construction unit to ensure that the landscape project is constructed according to the drawing;

4.3 Coordinate major construction nodes to the site;

4.4 Assist the developer to solve technical problems in the construction, answer design problems of the construction unit, and be responsible for design changes;

4.5 Participate in the acceptance inspection of all works related to the garden and issue the certificate of completion.

Section 4 Comprehensive evaluation system of modern ecological circular agricultural parks

The present situation of modern ecological circular agricultural park construction is reflected in two aspects, establishing demonstration park with village as the main body and enterprises as the main body. Based on the countryside and enterprises as the main body of the demonstration park, characteristics of diversification should be shown in the agricultural development. Among them, sightseeing agriculture and tourism agriculture have become new economic growth points. In addition, the active application of modern production technology and equipment, so that agricultural production is more standardized, on the basis of improving the output and quality of agricultural products, effectively improve the utilization rate of agricultural resources, to achieve agricultural ecological cycle development, obtain good economic benefits.

I. Research on standardization paths of construction&evaluation of modern ecological circular agricultural parks

(I) Present construction situation of our modern ecological circular agricultural parks

1. Establishment of demonstration parks with village as the main body

Many areas regarded countryside as the main body, vigorously developed modern ecological circular agricultural parks, adopted cooperative models, constantly strengthened the main position of rural village in the construction of modern ecological circular agricultural parks, and established agricultural enterprises in rural areas, so that agricultural enterprises and farmers establish cooperative relationship to accelerate the development of rural ecological circular agricultural parks. The advantages of this model lie in that it is based on the rural relationship and can quickly reach cooperation. Land can be used together, and the maximum rights and interests of farmers can be guaranteed through the form of cooperatives.

2. Establishment of demonstration parks with enterprises as the main body

Many enterprises participated in the modern ecological agriculture park construction and took the enterprise as the main body to establish the modern ecological agriculture parks. Enterprises and farmers established relations of cooperation, made the main goal of "one controll, two reduction, three basics", by using the combined model of planting and breeding, putting the fertilizer produced in breeding into planting, which helped to achieve development goals. Establishing demonstration parks with enterprises as the main body can organically integrate the advanced management experience of enterprises with agricultural production resources, and improve the deep processing of agricultural products and enhance the added value of agricultural products by means of enterprise industrialization.

3. Establishment of ecological parks dominated by counties, cities, and provinces

The modern eco-circular agriculture parks are established in various regions. There are more than 100 national parks, more than 500 provincial parks, and more than 2000 eco-agricultural demonstration sites. These modern ecological circular agricultural parks are characterized by large scale, absorbing more enterprises, being able to fully integrate all kinds of resources, and maintaining a high level of infrastructure such as water, electricity, roads and factories, so as to achieve scale effect.

(II) Establishment path of ecological circular agricultural parks in China

1. To transform the traditional agricultural production mode and actively apply the concept of modernization to build an ecological cycle

Modern ecological agricultural parks based on the modern concept, build a management system, production system in the zone, and introduce ecological tourism project. We can use the elaborating management mode to deal with water management, and we can recycle the resources, with the help of modern technology on water disposal. We can make the agriculture and tourism industry mutually integrated at the same time, promote the rural third industry's rapid development, and expand the scale of rural economic development. We should do well in the overall construction planning of the parks, reasonably arrange the layout of agricultural production space, especially the underground pipe network construction, and pay attention to the secondary utilization of resources. In addition, we should pay attention to the protection of the surrounding environment, prevent the pollution of fertilizer and water to the environment, and develop healthy and green agriculture. In agricultural tourism, we should pay attention to the brand construction and publicity of agricultural products to enhance the brand value and market recognition of agricultural products.

2. To improve the effective utilization of resources and promote the rapid development of modern ecological circular agriculture

In the development of modern ecological circular agricultural parks, we should improve the effective utilization of resources, follow the four-R principles, which are the reduction principle, recycling principle, reuse principle and the regulation principle. When following the reduction principle, we should enlarge the application scope of water-saving technology in the modern eco-circular agricultural park, use low-toxic pesticides and reduce pesticide residues on crops. Following the principle of recycle and reuse, we should make full use of livestock and poultry residues and carry on the innocent treatment of crop residue at the same time. The straw disposal is an important part, tranforming the straw-farmland single planting patterns to the hybrid planting patterns with fresh water breeding , which can make full use of agricultural residue and improve the efficiency of resource utilization. When following the principle of regulation, we should establish agricultural production system, deep processing system, logistics system and sales system based on the agricultural production content, making the produced crops transferred from the field to the table of users in a short period of time, so that consumers can quickly eat fresh and nutritious crops. At the same time, we should cooperate with universities and research institutes to carry out ecological circular agricultural research, improve the applicability and pertinence of scientific research results, and realize the rapid landing and promotion of scientific research results. In addition, we should actively learn modern agricultural production models from foreign countries to effectively reduce agricultural production costs and improve the quality of agricultural products.

3. To establish a standardized management development model of ecological circulation

Application of standardized management model in the modern ecological agricultural park, requires that the management content must be concise, which can enable farmers catch the main points of the standardized management, and carry on standardization production strictly in the process of agricultural production, conductive to promoting standardization of modern ecological agriculture park development. With the improvement of Chinese farmers' knowledge level and cultivation skills, farmers have strong internal need to improve agricultural economic benefits through standardized management. In the future, farmers can be taught standardized management methods through pictures and videos on the Internet, so as to help farmers improve their awareness of standardized production and improve the quality of agricultural products and the level of agricultural production standardization.

II. Comprehensive evaluation system of modern ecological circular agricultural parks

Systematic evaluation is for newly developed or modified systems. According to the predetermined system target, with the method of system analysis, we should evaluate and select the various programs of the system design from the technical, economic, social, ecological aspects, in order to determine the best or second best or satisfactory system program. The construction of comprehensive evaluation system for modern ecological circular agricultural parks is a very important technical link for park planning and designing. Scientific evaluation and understanding of the characteristics of ecological functional areas in which regions or projects are located are the basis for planners and managers to formulate relevant policies. It is of great significance to promote regional ecological function, maintain regional ecological security, support planning and regional economic and social sustainable development. On this basis, the comprehensive evaluation system of modern parks was constructed by using frequency analysis method, expert consultation method and analytic hierarchy process.

(I) Comprehensive evaluation steps of modern ecological circular agricultural parks

The evaluation steps generally include: to clarify the objective system and constraint conditions of the planning scheme of modern ecological circular agricultural parks; to determine the relevant index system of the evaluation project of the park; to collect relevant data and work out the evaluation methods; to research on the feasibility of the comprehensive evaluation system of the park; to evaluate the technics and economy of the park; to evaluate comprehensively.

The above is the general procedure of the comprehensive evaluation system based on the systematic evaluation. According to the stage of the park planning system, the system evaluation of modern eco-circular agricultural park can be divided into pre-planning evaluation, in-planning evaluation, post-planning evaluation and follow-up evaluation.

① Pre-planning evaluation. This is the evaluation in the planning stage. Since the construction system of the park is not clear and the target plan is still in the stage of search, exploration and demonstration, the comprehensive evaluation system of the park can only be based on the area where the park is located, local economic conditions, industrial conditions, technical conditions, traffic conditions, policies, etc. By referring to existing materials or using virtual simulation, we can predict and evaluate the development planning scheme of the park, and invite relevant experts and scholars to discuss the scheme, so as to determine the development planning direction of the park for evaluation.

② Planning evaluation. It is the evaluation in park planning implementation stage. It is to determine whether the key argument and the inspection are in accordance with the previous basis, at the same time keep up with the frontier of industrial development. On the basis of the original scheme, we should implant the innovation elements of the era to make seamless connection with the social development, such as using plan coordination technique to evaluate the park construction schedule.

③ Post-planning evaluation. It is the evaluation conducted after the implementation of the park planning system, to evaluate whether the system has reached the expected goals of the park planning in the early stage. It is easier to evaluate because the park plan has already been completed. The qualitative evaluation of the social, economic, ecological, technological and other factors of the park planning system should be carried out through the participation of relevant experts and scholars.

④ Follow-up evaluation of park development. It is the evaluation of the impact on other aspects after the park planning system is put into operation. For example, the impact of straw returning to the field for reuse on the ecological benefits of the park.

(II) Systematic evaluation method

There are four types of systematic evaluation methods as follows.

① Expert evaluation, which is conducted by experts based on their own knowledge and experience. Delphi method, scoring method, voting method and checklist method are commonly used.

② Technical and economic evaluation, which calculates the benefits of the system in various forms of value to achieve the purpose of evaluation. Such as net present value method (NPV method), profit index method (PI method), internal rate of return method (IRR method) and Sobelman method, etc.

③ Model evaluation, which uses mathematical models to simulate on the computer for evaluation. Such as system dynamics mode, input-output mode, econometric mode and economic cybernetics mode and other mathematical modes can be used.

④ System analysis, which makes quantitative and qualitative analysis on all the aspects of the system to evaluate. Such as cost-benefit analysis, decision analysis, risk analysis, sensitivity analysis, feasibility analysis and reliability analysis. One of the most common methods used in system evaluation is the correlation tree method. The correlation tree can represent the whole target system. Correlation trees can be used to analyze the relative importance of each factor at the same level to the subordinative factors at the upper level. The relative importance of each factor to the overall goal is known by working up level by level until to the zero level. Quantitative assessments of importance are generally obtained by consulting the experts. Experts usually estimate the relative importance of each factor and then average the estimates. At the end of the evaluation of one level, the next level is recurred, each time considering the relative importance of the combination.

(III) Comprehensive evaluation system of modern ecological circular agricultural parks

The comprehensive evaluation system of modern ecological circular agriculture parks is based on the above research, in view of the specified evaluation method, conducive to the healthy development of the park, to establish the park planning and comprehensive evaluation system, which consists of three evaluation systems such as "economic evaluation+social benefit evaluation, ecological evaluation".

① Economic benefit evaluation system. The economic benefit evaluation system is composed of seven subsystems, including science and technology subsystem, capital subsystem, talent subsystem, product scale subsystem, production management subsystem and product subsystem. The total index and single index are used to comprehensively evaluate the science and technology content of the park, achievements transformation, capital benefits, production scale, product design, market operation, personnel allocation, mobility, per capita net income of farmers, per capita net income growth rate of farmers, etc.

② Social benefit evaluation system. The social benefit evaluation system is mainly composed of qualitative indicators to evaluate whether the demonstration and driving effect of the park is obvious; whether it drives the region's agriculture on the development road of good variety, product quality, large-scale production, rationalization of structure layout, service socialization, economic characteristics, marketization, high efficiency produciton; whether it can further enhance the farmers' consciousness of "developing agriculture through science and technology"; Whether it has the display and appreciation function, for the excellent modern ecological circular agricultural park is the "window" of agricultural development; whether it can display high and new technology and further promote the transformation of scientific and technological achievements; whether it can improve the ecological and living environment, provide sightseeing and tourism places for residents, and meet the needs of people's spiritual life; whether it can help relieve the pressure of social employment.

③ Ecological benefit evaluation system. The ecological benefit evaluation system is mainly composed of technical indicators such as agricultural efficiency and environmental protection technology to evaluate whether the park has implemented pollution-free crop planting, water-saving irrigation, biological control and other high-tech technologies; whether the production base of green food and organic food has been established and how about its scale; whether the green coverage rate has been improved, whether it can reach the national or international advanced level; whether the soil erosion is effectively controlled; whether it caused less pollution to the environment and so on.

④ Continuously updated development evaluation. According to the demand of the era and industrial development, we should make timely anticipation and the connection. The regional leading industry and agriculture technology innovation will directly influence the development of the direction, the index will directly affect the core area of the park of the industry layout, market planning and the application of modern agricultural technology and equipment. The continuous improvement of social environment affects the layout and development of tourism function, service function and other important agriculture-related industries in the park, as well as the planning of landscape system and infrastructure of the region or project. Relevant norms and legal confirmation system ensure the safety of project planning and construction and scientific planning and designing. The continuous development and application of the expert consultation and evaluation system can broaden the scope of knowledge and ideas of designers and managers, and provide a strong guarantee for the forward-looking, scientific and sustainable planning and designin-g of the park.

III. Measures for establishing the standardized evaluation system for modern ecological circular agricultural parks in China

(I) To determine the evaluation index

In the process of determining evaluation index, the principles of setting evaluation index must be determined first. In the principles, we should ensure that they have scientific and practical characteristics. The scientific principle is the basis of determining the evaluation index of the modern ecological circular agricultural park. Following the scientific principle, the evaluation index will be more comprehensive and the evaluation process will be closer to the real development of the modern ecological circular agricultural park. In the process of determining the evaluation index and setting principles, we should consider comprehensively the economy and environment of the area where the modern ecological circular agricultural park is located, and the evaluation index conducive to the healthy and stable development of the modern ecological circular agricultural park. In addition, the evaluation indexes should reflect the characteristics of dynamic and static integration. The dynamic evaluation indexes can be determined to evaluate the dynamic development of the modern ecological cycle agricultural park. The static evaluation index can be used to evaluate the internal and external factors that affect the development of modern ecological circular agricultural park.

(II) To determine the evaluation content

In the developmentof modern ecological circular agricultural parks, it is necessary to continuously plan the park to ensure that the resources can be fully utilized, and the development of the park is conducive to creating more benefits for the society. Based on the park planning, resource utilization and social benefits, the evaluation content of modern eco-circular agricultural park can be determined. After determining the evaluation contents of the park, the park can clarify its development goals, and at the same time, the park will take improving the utilization rate of resources as the development core, and the park will be built on the recycling resources, which will help the park planning to be more standardized and reasonable, and the resource utilization rate of the park will be continuously improved. In the process of creating more benefits for the society, economic benefits and ecological contribution are important components of benefits. Taking economic benefits and ecological contributions as evaluation criteria, we will continue to expand the scope of influence of modern ecological circular agricultural park in local economic construction.

(III) To establish an evaluation index system

Planning index, resource index and social index are considered as important components in the process of establishing standardized evaluation index system of modern ecological circular agricultural park. In the planning index, the emphasis is on the evaluation of industrial planning, fund guarantee and team construction. Industrial planning refers to the strategic planning of the industrial development and the type and scale of the industry. Fund guarantee refers to the special funds invested annually in production facilities and equipment and the measures for fund management. Team building refers to the organizational structure and rules and regulations. In the resource index, the emphasis is on the evaluation of resource recycling and resource reduction. Resource recycling refers to the effective utilization coefficient of chemical fertilizer, the resource utilization rate of livestock and poultry residues and the comprehensive utilization rate of straw. Resource reduction refers to the level of pesticide use, the level of agricultural film use and the intensity of agricultural mechanization use. In the social index, it focuses on the evaluation of economic contribution, social contribution and ecological contribution. Economic contribution refers to the output value of agricultural GDP per unit area and the growth rate of total agricultural output value. Social contribution refers to the growth rate of per capita annual income, the number of standardized trainees and the index of science and technology extension. Ecological contribution refers to the growth rate of the improved soil area and the annual increase of standardized production base.

(IV) To set the weight of index

In the process of setting the index weight, the index weight should have the characterisitics of being comprehensive, because the modern ecological cycle agricultural park involves a variety of industries. At present, in the process of setting the index weight, the index score will be set, and the relationship between the index score and the index weight will be established. When the index score is changing, the index weight will also change. Under the background of modernization, the original index weight is optimized and improved, and the matrix variance calculation formula is applied in the index weight calculation, so that the index weight obtained is more scientific, and the index weight produced provides a reference for the development of modern ecological cycle agricultural park.

1. Ordering the index weights

When the index weight order is arranged, a comparison level is generally set to reflect the importance of the index weight according to the change of the level. Many modern ecological circular agricultural parks will use Bi and Bj respectively when setting the index weight, and Bij represents the quantified value of importance. Modern ecological gricultural parks will set the index weight in four levels, and the quantified value will be arranged in order from high to low. Bi and Bj represent equal importance in the first level quantization values, and in the fourth level quantization values, Bi is more important than Bj. Most modern ecological cycle agricultural parks use first-level index to compare the importance of Bi and Bj.

2. Calculating the index weight

In the process of calculating the index weight, a modern ecological agricultural park was taken as an example. YAAHP1.0 software was used to calculate the planning index, resource index and social index in the park. Taking the planning index as an example, the calculated weight value is 0.5714, among which the weight value of industrial planning is 0.3143, that of the demonstrated industrial development strategy planning is 0.2095, and the value of industry type and scale is 0.1048. The weight value of fund guarantee is 0.1373, that of annual special funds invested in production facilities and equipment is 0.0915, that of fund management method is 0.0458, that of team construction is 0.1199, that of organizational structure is 0.0799, and that of rules and regulations is 0.0400. Among the resource indexes, the weight of resource index of the modern ecological cycle agricultural park is 0.2857, that of resource recycling is 0.1905, that of effective utilization coefficient of chemical fertilizer is 0.0952, that of resource utilization rate of livestock and poultry residues is 0.0476, the comprehensive reuse rate of straw is 0.0476. The weight value of resource reduction is 0.0952, that of pesticide use level is 0.0381, that of agricultural film use level is 0.0190, that of using medical mechenics is 0.0381.

(V) Measures and suggestions

1. Adhere to the standardized evaluation system and promote the sustainable development of modern ecological circular agricultural parks

Modern ecological agriculture parks should adhere to the standardization development path, and in the process of standardization development, establish a standardized evaluation system. Through optimizing the evaluation and adjusting the development pattern of the modern ecological agriculture parks, we should make the park establish contact with the external market, at the same time make the park continuously improve product quality, effectively expand the impact in the market environment. Modern ecological agricultural parks should pay enough attention to the establishment of standardized evaluation system, through the standardized evaluation system, we can obtain the real operation information of the park, and according to the information we grasp the development of the park. Using the standardized evaluation system, we can accumulate continuously the development experience, avoid repetitive problems, and guarantee the healthy and stable development of the park.

2. Adhere to the standardized model and strengthen the management of ecological parks

The establishment of standardized production model can continuously improve the production quality of the park and promote the modernization of the park. In standardization production model, the park should strictly carry out and implement national and industry standards, make every production link get effective management, such as selecting seeds, breeding to planting. During the planting period, the application of water-saving irrigation techniques should be all in standardization management model, which can facilitate the farmers to operate in accordance with the process operation. The operation process is not only intuitive, which can reduce the difficulty of production operation, but also it can make farmers have unified thinking, which is contribute to the manage work of the modern ecological circular agricultural park.

3. Adhere to the standardized long-term mechanism to accelerate the development of eco-circular park economy

Establishing standardized long-term mechanism is the key to accelerating the development of eco-circular park economy. During the establishment of standardized long-term mechanism, the park should actively train local personnel and establish a talent management system suitable for local development, so that the provisions of the mechanism can be implemented into practical work. In addition, establishing standardized publicity mechanism, this mechanism is used to guide farmers to form the correct consciousness, actively participate in the ecological cycle in the establishment and development of modern ecological agriculture park. Farmers get standardized agricultural production technology through publicity, thus the traditional farmers are transformed into professional farmers, which can improve the development level of modern ecological agricultural parks. Modern ecological agricultural parks should establish a standardized reward mechanism, local governments should increase the investment of policies and funds, and constantly strengthen the application value of standardized long-term mechanism.

In the development of modern ecological agriculture parks, we should base on modern management concept, apply modern philosophy in the park construction and development, give enough attention to establishing standardized evaluation system, and determine the evaluation index, the content &weight. Through the evaluation system we can improve the development level of modern ecological agriculture park. At the same time, regional authorities should give enough help, make the ecological agriculture park to expand unceasingly, to introduce a variety of economic production model actively. The park should improve their economic development ability, and bring more farmers to participate in the construction and development of modern ecological circular agriculture parks, which can create more benefit for the society.

Section 5 Main problems and strategies in the planning and development of modern ecological circular agricultural parks

In recent years, with the continuous development of economy in our country, since 2006 the central No. 1 file explicitly has put forward to the development of circular agriculture, and from 2007 to 2016, the No. 1 file continued to emphasize and encourage the development of circular agriculture. Circular agriculture, as a new agricultural economy and ecological agriculture development direction, has more and more got the attention of the government and society at all levels in our country. Some circular agricultural demonstration parks also appear in various parts of China. Modern ecological circular agricultural park is a kind of modern park which takes ecological agriculture and circular economy as the development concept, constructs the recycling development industrial chain of the planting, breeding and processing, and pays attention to the introduction and application of ecological technology and new energy. In the actual development process, they have opportunities and challenges, so how to turn the existing unfavorable factors into favorable conditions according to the market development demand, is the top priority of the planning and development of the modern ecological circular agricultural parks.

I. The problems in the planning and designing of modern ecological circular agriculture parks

1. Lacking of scientific planning and designing management

The main goal of the planning and designing of the modern ecological circular agriculture parks is to carry out ecological demonstration and expand the functions of ecological agriculture and sightseeing agriculture. But some circular parks' planning and designing lack of sufficient investigation on the raw material, lack of scientific planning on the leading industries, product scale and the operation mode. They didn't fully explore the rich resources of the modern park, making the core functional areas lack of efficient agricultural technical contents. The functional partition is not clear, with low resource utilization. In addition, the construction of the park blindly pursues the scale of land use but has low utilization rate of capital, which directly leads to the slow construction speed of the park and the limited space of production capacity. This seriously undermines the confidence of investors. Moreover, due to the gap in understanding, the marketization of park management and the standardization of agricultural production have not been paid attention to, and the demonstration and driving effect is not obvious, and even has a counter-effect.

2. Inaccurate positioning

The development of modern ecological circular agricultural parks takes modern agricultural production as the core, so it needs reasonable overall positioning and scientific planning, increasing investment in the transformation of scientific and technological achievements, improving production capacity, expanding market, and forming leisure tourism in combination with rural ecological environment. At present, some agricultural ecological parks are only limited to the use of climate environment, green resources, etc., unilaterally improving the tourism and catering functions of the park, and the scale of the resort is becoming larger and larger, far from the theme of modern ecological circular agricultural park. There is no concept of "cultivating high-quality products", lack of fist products and featured products. Therefore, the image of the whole ecological park is not obvious and the brand is not prominent, which eventually leads to that the production capacity is not increasing, the tourism projects are monotonous and unattractive, and the in-depth development meets the policy and market restrictions.

3. Weak demonstration effect

The planning of modern ecological circular agriculture parks is to adopt the ecological sustainable model for the layout and production of agriculture in the park. Based on the high-tech content of ecological agriculture, it embodies the connotation of "ecological culture" and embodies the harmonious unity of production, product, operation, sightseeing and catering models. Some parks focus on "high-quality agriculture", such as high-tech and facility production, and are keen on introducing foreign equipment and facilities, which are divorced from the actual needs of local agriculture and rural economic development, and do not have strong driving capacity for regional development. Some ecological parks simply pursue profit development, without adopting the standard model of ecological agriculture to plan and design, the product market and product quality can not be guaranteed, and do not fully follow the production mode of organic agriculture. This kind of eco-park economy lacking cultural connotation, its investment value and development potential will be greatly reduced. Therefore, it is difficult to play the corresponding ecological agriculture demonstration role, but also does not have the ability to carry out green food production through organic agriculture, it is difficult to achieve the unification of economic, social and ecological benefits.

4. Weak basic conditions

Modern ecological circular agricultural park is essentially a model for the development of agricultural science and technology in China or a regional base of agricultural science and technology innovation, but its status is not fully affirmed. There is still a lack of clear and stable support policies and support channels for governments at all levels and sci-tech departments. Land transfer is a sensitive and difficult problem in the construction of most agricultural science parks. For the needs of industrial development, the shortage of scientific and technological talents and achievements in modern ecological circular agricultural parks is still prominent. The unbalanced development of parks in different regions is quite prominent, and the phenomenon of similar projects and mutual imitation among parks still exists. As a result, the planning of modern ecological circular agricultural park is not serious or there is no project, lacking of scientific demonstration by experts.

II. The solution strategy of planning, designing and constructing of modern ecological circular agricultural parks

1. To strengthen the circular agriculture and modern standardization

Popularization of production management knowledge and the household contract responsibility system liberated our agriculture productivity after the reform and opening-up, agricultural production relations have been adjusted more actively, and farmers have been on a rich road. But from the perspective of the requirements of modern agricultural standardization, land contract brought scattered small-scale production and management, the lack of unified plan and instruction on the planting and breeding at the beginning, middle and later period, thus there is no guarantee that the products are strictly in accordance with the requirements of the normalization, standardization production. So in the new situation, the modern ecological agriculture park planning must be in line with national industrial policy, to explore and create new production system, to change the small agricultural production mode by the household contract responsibility system to gradually develop moderate scale management, to promote the new form of agricultural production with the combination of specialization and scale management, to form a big agricultural structure as soon as possible with standardization, large-sacle and ecological industrialization, to smoothly realize the planning and construction of circular agriculture and ecological park, and to lay a solid institutional foundation for standardizing agricultural production.

2. To strengthen the construction of agricultural engineering planning team

On July 6, 2011, the Organization Department of the CPC Central Committee and the Ministry of Human Resources and Social Security issued the Medium and Long-term Plan for the Construction of Highly Skilled Personnel (2010-2020), which is China's first medium and long-term plan for the construction of highly skilled personnel. Firstly strengthening the construction of agricultural engineering planning team conforms to the requirements of national talent planning. In addition the training, technical integration and recruitment of senior talents for the existing agricultural engineering planning team is of great significance to the planning and construction of the park.

3. To improve the comprehensive evaluation system of circular agricultural ecological park construction projects

The comprehensive evaluation of the construction project of circular agricultural ecological park is actually a key link to test whether the plan can be implemented. The rational evaluation of the indicators in the comprehensive evaluation system is actually a qualitative and quantitative investigation of the optimization design, resource allocation, industrial layout and product scale, etc. We should comrehensively evaluate whether the functional positioning, industrial positioning, layout of functional areas and other infrastructure, landscape and tourism of circular agricultural ecological parks are scientific and reasonable, whether the degree of product marketization can be realized, and whether the prediction of payback period will meet some unreasonable and imperfect links in the process of comprehensive evaluation system. Before the park has not been formally invested in the construction, try to find the problem, solve the problem, optimize the planning scheme, which is conducive to saving the cost of the early stage for the circular agricultural ecological park construction project, and to contribute the comprehensive evaluation system to the later investment.

4. To improve the investment risk management of circular agricultural ecological park construction projects

As an important part of project management, project risk management is a set of programs and methods to deal with various problems caused by uncertain factors. The circular ecological park planning should introduce investment risk evaluation system, which is a very effective early stage management system to investors and investment security. Planners, managers and administrative departments should improve and perfect the investment risk management system of the agricultural ecological park construction, in line with serving the people and contributing to the agricultural production.

Developing circular agriculture is an effective means to realize agricultural clean production and sustainable utilization of agricultural resources, and it is also an inevitable choice to solve the dilemma of modern agricultural development. As an environmentally friendly farming method, circular agriculture has good social, economic and ecological benefits. At present, the development of circular agriculture in China follows the "5R" principle, namely "the principle of reduction, reuse, recycling, reproduction of resources, and replacement of resources", which better promotes the recycling utilization of agricultural resources and sustainable development of modern agriculture in our country.

China has a good development base of agricultural circular economy, with rich contents of circular economy, including some specific development models such as the coupling of matter and energy within agricultural production departments, the coupling of matter and energy between the departments of agricultural industry, the coupling of matter and energy between agriculture and industry, and the coupling of matter and energy between urban and rural resources use, etc. Ecological agriculture is an important form and carrier of the development of agricultural circular economy, but to achieve the goals and requirements of circular economy, we must realize the renewal and improvement toward industrialization, scale, standardization, marketization and functional diversification. At the same time, the corresponding countermeasures must be taken, including strengthening the legal system construction of agricultural circular economy, the construction of ecological agriculture industrialization, the treatment and utilization of agricultural residues and the construction of modern ecological circular agriculture park, etc.

Chapter 6 Basic Theory of Modern Ecological Circular Agricultural Park Planning

Section 6 Management and operation of modern ecological circular agricultural parks

American scholar Fred R. David defines strategic management as "the art and science of making, implementing, and evaluating that enable an organization to achieve its goals and can make cross-functional decisions." He believes that strategic management is committed to the comprehensive management of the marketing, financial management, production operations, research & development and computer information system, in order to achieve the successful development of enterprises. The purpose of strategic management is to create and exploit new and different opportunities for tomorrow's business.

The key word of strategic management is not strategy, but dynamic management, which is a new management thought and management way. The key to the formulation and implementation of strategy lies in the analysis of the changes in the external environment of the enterprise and the audit of the internal conditions and quality of the enterprise, and on this premise to determine the strategic objectives of the enterprise, so that the three can reach dynamic balance. The task of strategic management is to achieve the strategic goal of the enterprise under the condition of maintaining the dynamic balance through strategy formulation, strategy implementation and daily management.

As a result, strategic management can be defined as a dynamic management process during which the enterprise determines its mission, makes its strategic objectives according to the specific external environment and internal conditions, plans in order to ensure the correct implementation and implementation, relys on the enterprises' internal ability to put planning and decision into action, and controls in the process of implementing.

I. Strategic management of modern ecological circular agricultural parks

The strategic management of modern ecological circular agricultural parks is a systematic decision-making and implementation process based on the full possession of information, which must follow a certain logic system and contain a number of necessary links, thus forming a complete system.

(I) Establishment of the strategic guiding ideology of modern ecological circular agricultural parks

The strategic guiding ideology of modern ecological circular agricultural parks is a series of ideas that guide the management activities of modern park under certain social and economic conditions. It is the fundamental starting point of strategic research and the operational criterion of modern eco-circular agricultural parks. It points out the basic tasks of the park's survival and development as well as the goals and norms of behavior that the park should achieve.

The strategic guiding ideology of modern ecological parks must conform to the social and economic development level, national policies, laws & regulations, and the relations between people, people and society, people and parks, and parks and society. These constraints are basically the same for all modern eco-circular agricultural parks, but in fact, the strategic guiding ideology of different parks is different. Even for the same park, its strategic guiding ideology may change in different management periods.

(II) Analysis of the internal and external environment

Because the organization of modern ecological circular agriculture parks is an open system, the external environment is the important information source of the park organization, so for the strategic management we must analyze the environment more carefully. Through a large number of research and investigation, we understand the current situation of the park, and on this basis we should judge the challenges facing the park and the development opportunities. External environment analysis mainly includes all the real and potential factors which may affect the behavior of modern park such as politics, economy, science & technology, culture and social environment, etc. We should master its change rule and developing trend, and determine if these changes will bring modern parks more opportunity or more threats. The focus is to study the nature of the market structure and the strengths and weaknesses of competitors and their strategies. The analysis of internal conditions should focus on understanding the relative position of the agricultural park itself, its resources and strategic capabilities, and also need to understand the interests of stakeholders related to the modern park.

(III) Formulation of strategic objectives

According to the strategic guiding ideology, we should determine the strategic direction and goal of the management of the park. Generally speaking, the strategic objectives of modern park mainly include: ① Market objectives. It mainly refers to the expected market position of modern park, which requires careful analysis of customers, target markets, products or services, sales channels and so on. ② Innovation goal. It mainly refers to the degree of technological innovation, institutional innovation and management innovation in modern parks in the society with intensified environmental change and fierce market competition. ③ Profit target. The profit goal of modern parks depends on the resource allocation and utilization efficiency, including the input-output goal of human resources, production resources and capital resources. ④ Social goals. The social objectives of modern eco-circular agricultural parks reflect their contribution to society, such as environmental protection, energy saving, employment, cultural inheritance, science popularization education, etc.

(IV) Evaluation and selection of the strategic programs

In the field of future business, we can have a variety of ways and methods to achieve the strategic target, so we may have many possible strategic plan. We should have specific demenstration on these plans, mainly to check if it is technically advanced and economically reasonable, so as to make comprehensive evaluation, and by comparing the pros and cons of each solution, we choose a satisfactory, practical and feasible solutions. Hiroyuki Itami (Yidan Jingzhi), a Japanese strategist, believes that an excellent strategy is an adaptive strategy, which requires strategy to adapt to external environmental factors, including technology, competitors and customers. At the same time, the strategy of modern parks should also adapt to the internal resources of the park, such as the state of natural resources, the level of economic and social development, human capital and so on. In addition, the strategy of modern parks should also adapt to its organizational structure. Specifically speaking, we should mainly judge from the following aspects: ① The strategy should have differentiation, which is different from that of your competitors. ② The strategy should be concentrated. The resource allocation of modern ecological parks should be concentrated to ensure the realization of strategic objectives. ③ To make a strategy we should grasp the opportunity. The strategy formulation of modern parks should follow closely the development trend of similar parks at home and abroad and be launched in a favorable policy window period. ④ The strategy should be able to take advantage of its core competitiveness. Modern parks should be good at utilizing their own advantages and transforming them into core competitiveness. ⑤ The strategies can inspire employee morale. ⑥ The strategy should be combined skillfully with other elements. The strategy of modern ecological park should be able to combine all the elements of the park skillfully and make all the elements produce synergistic effect.

(V) Implement of the strategy

The intention strategy determined in the strategy formulation stage should be transformed into concrete organizational actions to ensure that the strategy achieves the predetermined goal. The implementation of a new strategy often requires an organization to make corresponding changes and take corresponding actions in the organizational structure, business process, capacity building, resource allocation, corporate culture, incentive system, governance mechanism and other aspects. In the strategy implementation practice of modern parks, relevant departments should submit the operation strategy report to the management layer of the park. The management made the strategy and announced to all departments, according to the internal and external situation and characteristics of the park, and then decomposed, implemented and supervised the strategy in each work to ensure its progress. Strategy implementation needs to be achieved in the continuous cycle of "analysis‐decision‐execution‐feedback‐reanalysis‐decision‐execution".

(VI) Strategic control and evaluation

Strategic control and evaluation is an important link in the process of strategic management. Strategic control mainly refers to establishing and improving the control system in the process of the implementation of the strategy, which is to compare the strategy implementation results of every stage, every level, every aspect with the anticipated target, in order to find the strategic gap in time, analyze the reasons of the deviation, correct the deviation, and ensure the completion of all the strategic plan. Strategic evaluation is to check the scientificity and effectiveness of the strategy by evaluating the business performance of the enterprise, which is the last stage of strategic management. This mainly includes reviewing external and internal factors, measuring performance, and taking corrective actions.

The control and the evaluation of strategy implementation are both different and related. To carry out the control, the evaluation must be carried out. Only through evaluation can control be achieved, and the evaluation itself is a means rather than the end. To discover problems and to control them is our purpose. Strategic control focuses on the process of strategy implementation while strategic evaluation focuses on the evaluation of the process and the results of strategy implementation.

II. The operation of modern ecological circular agricultural parks

Modern ecological agriculture park, as a high‐level form and production organization way in the process of modern agricultural development, caused the reform and innovation of agricultural organization, resource allocation methods, agricultural operation ways and management system. Its construction and development is of great practical significance to accelerate the transformation and application of agricultural science and technology achievements, to promote agricultural technology progress and to enhance the quality and efficiency of agricultural development. The construction of the park should not only have high science and technology content, focus on the combination of advancement, applicablility and popularization, but also have a reasonable organizational form and a coordinated, flexible, efficient and perfect operation mechanism to ensure the sustainable survival and development of the park.

(I) Operational objectives

There is a famous saying in management that "if you cannot measure it, you cannot manage it". The operation of modern ecological parks is in a constantly changing environment, which requires setting clear and measurable goals to guide the park to develop in the direction of normalization, standardization and modernization. Operational objective is the assumption of the expected effect of operational activities, and is the reflection and embodiment of the purpose of operational activities. The determination of operation objectives is an important basis and logical starting point for the operation of parks, an important guide to determine the future development direction of parks, and can also be used to measure the operation performance, efficiency and quality of the parks. As an industrial model with relatively high input, modern park correspondingly requires its output level, scientific and technological content, social and economic functions to be consistent with the input level, so as to improve its market competitiveness through good operation. In general, to create a modern park with good economic benefits, strong social service function and high ecological value is the expected effect of the operation of modern park. Therefore, this section will introduce the operation objectives of modern park in detail from three aspects: economic benefit, social benefit and ecological benefit.

1. Economic benefit target

In the traditional sense, economic benefit refers to the proportional relationship between input and output in economic activities, which is a comprehensive index to measure the results obtained by economic activities. As an independent accounting economic organization, the basic function of modern park is to achieve maximum economic benefits through the industrialization of scientific and technological achievements under the guidance of the market, with the goal of developing beneficial agriculture. The realization degree of the goal of economic benefit reflects the management performance of modern parks in using resource elements and creating value in the market, and represents the development speed and potential of the parks in the future. At the same time, as a kind of operation of investment behavior, modern ecological agriculture parks must take economic benefits as the center, especially for the less developed areas, through the reasonable operation of modern parks, they not only can increase the economic income, but also can improve the local agricultural structure and promote regional economic development. The maximization of economic benefit is the basis for the survival and development of modern parks, and it is the direct embodiment of the advanced nature and competitiveness of modern park. The modern park whose economic benefit is difficult to guarantee is also difficult to exert social and ecological benefits. Therefore, the operation of modern parks should first pay attention to economic benefits, and maximize the economic benefits of survival and development on the basis of obtaining economic benefits, which is the most basic requirement for the operation of modern park.

In the process of operation of modern park, we should adhere to being market-oriented, take benefit as the center, in line with "the introduction, integration, demonstration, popularization, and service" principle, in accordance with the market economy rule. While carrying out the cost control on the management and daily operation, we should actively explore the way of improving agricultural efficiency and farmers' income. Through increasing investment, expanding scale, strengthening management and other measures, we can effectively improve the economic benefits of the park. Under the premise of giving full play to the functions of the modern park, the economic benefits of the park should be constantly improved, so as to further promote the sustainable operation of the park.

Generally speaking, the quantitative analysis of economic benefits can be measured by relative indexes and absolute indexes. Relative index means that economic benefit is equal to the ratio of labor fruits and labor consumption & resource occupation.

Economic benefit M= labor achievement W/labor consumption and resource occupation K

The larger the calculated value of economic benefit in this formula, the better the economic benefit.

Measured by absolute indicators, economic benefits are equal to the difference between labor achievement and labor consumption & resource occupation.

Economic benefit M= labor achievement W - labor consumption & resource occupation K

If the calculation result of economic benefit in this formula is positive number, it means there is surplus, and the larger the surplus number, the better the economic benefit. In practice, the commonly used evaluation indicators to measure the realization of the economic benefit goal of the operation of modern parks include the total output value of the parks, the annual profit rate of the parks, the benefit-cost ratio of the parks, the labor productivity of the parks, and the land productivity of the parks.

The total output value of the park refers to the annual total output value during the construction and normal operation of the modern park.

The annual profit rate of the park refers to the ratio between the total profit and the gross annual production value. This index can be used to measure the profit level of park operation.

The benefit-cost ratio of the park refers to the ratio of the average annual benefit to the average annual cost in the normal operation years. Among them, benefit refers to the sales income of products in the park; costs refer to the park expenses, including annual operating expenses and taxes. This index can reflect the utilization of park funds.

The labor productivity refers to the ratio of park production value to the number of workers in the park. This measure can be used to measure the average productivity of employees in the park.

The land productivity of the park refers to the ratio of the time production value of to the total land area needed to consume.

The income level of producers in the park refers to the average income level of workers engaged in agricultural production in the park.

2. Social benefit objectives

Any human organization is social. As an organ of society, the organization has a direct impact on the society, so we must always pay attention to the impact of the organization's activities on the society. It bears due social responsibility, and modern park is no exception. The modern park is not only a profit-making economic organization, but also must always focus on the impact of the organization's activities on the society and the due social responsibility. Social benefit refers to the indirect benefit and auxiliary benefit produced by the construction and development of modern park. As a new type of organization for the introduction, integration, demonstration, popularization and service of new varieties, new technologies, new equipment and new models of modern agriculture, the modern park has attracted extensive attention from all walks of life and assumed a series of social functions since its emergence.

Specifically, social functions of modern parks are mainly embodied in the following aspects: first, a successful modern park built, sustained, steadily operated can create or provide lots of jobs, thereby increasing regional labour accommodation, which can, to a certain extent, help solve the problem of the rural employment of the park and the surrounding areas. Second, a basic task of modern ecological park is to transform traditional agriculture with high and new technology, and create a new mechanism and environment for the transformation of sci-tech achievements through the demonstration project of agricultural high and new technology, so as to quickly transform fully "matured" sci-tech achievements into real productive forces. Therefore, in the process of continuous operation, the modern park should be able to promote and drive the progress of agricultural science and technology and agricultural comprehensive development in the surrounding areas. Third, at present the most modern ecological agriculture park will take the agricultural high and new technical training as one of the main functions of the park. Relying on the park's mechanism of being resource intensive, high-tech talent intensive and service efficient, it carries out various forms of agricultural skills training activities on a large scale. Through training, it cultivates a new type of agricultural talents, to enhance the overall quality and development of regional agriculture to provide human resources guarantee.

In this regard, according to the above social functions of the modern park, we can set the social benefit targets, and convert it into quantifiable indicators, which can be used to measure the contribution and benefit created by the modern park to the society. Based on this, it is evaluated whether the operation has given full play to its social function. These indicators to measure the realization of social benefits include the following:

The number of employment can reflect the ability of modern park to provide employment opportunities for society. Specifically, it refers to the number of employment opportunities created for the surrounding areas or the number of labor force employment during the construction and normal operation of the modern park.

The popularizaiton extension area of the park can reflect the guiding, demonstrating and driving effect of the modern park. In particular, it refers to the total area of production of new varieties and new technologies introduced into the surrounding areas of the park in each year.

The annual training number of farmers can reflect the contribution of modern parks in cultivating agricultural talents. It specifically refers to the number of farmers trained for the surrounding areas every year during the construction and normal operation of the park.

The number of farmers to get rich reflected the park's driving role. Specifically, it refers to the number of farmers actually driven to become rich by the park in the year, and the measurement standard is the number of farmers in the extension area of the park whose annual net income is greater than the annual net income of local farmers per capita.

The proportion of popularization and transformation of achievements refers specifically to the ratio of the number of the achievements popularized to the total number of research achievements in the year of the park.

(II) Objectives of ecological benefits

Modern park is not only the interior enterprises' production operation place, and as a "window" and "model" of agricultural development, it should also be the enterprises' cooperation and exchange business district, a hotbed of attracting investment, and the daily leisure sightseeing park for the surrounding residents. All these rely on the good natural ecological environment in the park. Therefore, a good ecological environment is the guarantee for the sustainable survival and development of modern parks, and is also an important value standard for the success of the operation of modern parks.

Ecological benefit refers to the beneficial influence and effect of modern parks on human production, living and environmental conditions. Its ecological benefits are mainly embodied in the characteristics of being clear, clean, beautiful and green, create a picturesque ecological landscape, improve natural environment and maintain ecological balance and improve the quality of living environment, act as urban green belt, prevent and control the urban environmental pollution to maintain pure, fresh, and quiet life environment, and prevent urban over-expansion, etc.

General Secretary Xi Jinping pointed out in the report of the 19th National Congress of the Communist Party of China to speed up the reform of the ecological civilization system and build a beautiful China. This highlights that in the context of economic development has entered the new normal era, the concept of ecology and environmental protection should be integrated into the operation process of agricultural parks, tranforming from the past "benefit from the scale" to "benefit from ecology". Continuing operations of the modern parks should be conducive to the sustainable and efficient utilization of resources and the improvement of ecological environment of the parks and surrounding areas, and observe the ideas of "harmless, low emissions, zero damage, ecological, sustainable and environmental". Through the efficient and rational use of resources, we can improve efficiency, reduce costs, and protect the environment. By encouraging planting farmers to apply biological control, formulated fertilization and other high and new agricultural technologies, we can lead them to reduce environmental pollution caused by pesticides and fertilizers in the process of planting, maintain ecological balance and promote sustainable agricultural development. In the processing and production of the extension of the agricultural chain, we should give priority to clean production, clearly limit the environmental pollution generated in the production process, reduce as far as possible the environmental pollution of the agricultural processing industry, and force enterprises in the park to take effective measures to reduce the impact of product residues on the environment. By these, we can establish a long-term stable ecological security pattern for the area where the modern park is located. We should carry out harmless treatment to the processing residues and sewage, effectively prevent the pollution to agricultural products, soil, water, etc. , and ensure the quality safety and ecological safety of agricultural products. At the same time, we should optimize the design and configuration of sightseeing agriculture such as famous, excellent and special vegetables, flowers and fruits, strengthen the construction of landscape and water conservation forest in and around the modern park, form a beautiful landscape ecological pattern, and protect and improve the regional ecological environment.

The realization degree of ecological benefit goal in the operation process of modern park can be measured by the following indicators:

① Green area ratio of the park refers to the ratio of the current green area to the total area of the park.

② The proportion of pollution-free products in the park reflects the agricultural environmental protection and the safety of agricultural products in the park. Specifically, it refers to the proportion of the output value of green and pollution-free agricultural products in all products in each year during the normal operation of the park.

③ The utilization of clean energy in the park reflects the utilization of clean energy for production during the construction and normal operation of the park.

④ The development level of circular economy in the park reflects the development condition of the circular economy during the construction and normal operation of the park, including the waste disposal and sewage discharge in the park.

III. Operation mechanism

Mechanism is the relationship of mutual restriction, mutual influence or contradictory unity of various components in an organism or organization. Operation mechanism is the combination, linkage and circulation of various functions which are possessed by the system, and it is the operation mode of mutual connection among various elements of the system. The operation mechanism of modern park refers to the functional system that promotes, regulates and restricts the normal operation of various elements in the modern park in the process of its construction and development, so as to achieve the development goal of modern park. The construction and healthy development of modern park cannot be separated from the efficient operation mechanism. The improvement of the operation mechanism can not only solve the problems in the development of modern park, but also improve the operation efficiency of the park, and ensure the play of the park function and the realization of the operation goal.

Through the operation of modern park, it is necessary to integrate the advantage resources inside and outside the park and build a platform for the development of the park industry. In order to expand the service scope of the park, the various elements in the whole agricultural industry chain should be integrated organically. Thus the modern park can maintain the core competitiveness in the market, and we constantly shorten the production process, play the scale effect of modern park, and pursue sustainable profit economic goals.

The operation mechanism of modern park can be divided into investment and financing mechanism, management mechanism, technology selection, docking and diffusion mechanism, production mechanism, demonstration mechanism, brand marketing mechanism and risk defense mechanism. The healthy development of modern park is not a simple integration and accumulation of various mechanisms, but a complex systematic engineering, with different mechanisms interacting and influencing each other. They gradually unified and coordinated in the continuous development, and once the operation mechanism formed it has a relative stability.

(I) Investment and financing mechanism

The construction and sustainable operation and development of modern park cannot be separated from capital, and the accumulation of capital needs good investment and financing mechanism. If lacking of capital, it will affect the scale construction, benefit, and the promotion & development of the advanced technology. Investment and financing mechanism refers to the general term of organization form, investment and financing method and management mode of investment and financing activities in the operation process of modern park. The establishment of multi-level, multi-channel and diversified investment and financing mechanism is the premise to ensure the sustainable material foundation and financial support & development of modern parks.

The main financing channels of modern parks include government policy investment, social subject investment, bank loan and enterprise stock market financing, etc. However, whatever financing channels are adopted, they should follow the principle of "who invests, who gains" to carry out market-oriented operation. By forming more and more wide financing channels, we can make the amount of financing more and more large, and gradually form a stable financing mechanism, so as to more effectively promote the operation and development of modern park.

As for the government's policy investment, all levels of government are as the entities input a certain number of start and guide funds, which are mainly used for infrastructure construction of modern park, important service facilities and projects with indirect benefit and long-term benefit, and it often is a "catalyst" for the follow-up investment promotion and capital introduction. The government's policy investment funds are mainly through the integration of agricultural projects and social development projects, such as the integration of funds, the integration of land, agricultural comprehensive development, returning farmland to forest, rural transportation, poverty alleviation etc. And at the same time, they activate a large amount of social capital into the construction and daily operation of modern parks. On the basis of full integration of various types of agriculture-related funds, we may invest the various types of agriculture-related funds used for the development of poor households into the modern park, to make the poor households enjoy the dividends. This financing mechanism can not only make the poor households get stable income, but also make the operation of modern park get effective financial support.

It is necessary to rely on the government's policy investment for the operation and development of modern parks. However, the financial support from the government is limited and relatively scattered. It is difficult to achieve sustainable development solely relying on government investment, and it will bring great pressure to the finance. Therefore, government investment can only be used as the introduction and catalyst for the initial construction of modern parks, and the fundamental way of financing is to rely on social legal persons, individuals and foreign investors in accordance with the law of market economy. Modern parks should optimize the investment environment in the park, formulate supporting preferential policies, in the form of contracting, leasing, auction, stock cooperation, allowing land, technology, and management as a shareholder, or give the owners of technical achievements with a certain property rights to attract the main social investment from all walks of life. To strengthen investment attraction, it is necessary to attract all kinds of enterprises and social funds to participate in the construction with preferential policies and good infrastructure facilities, and promote the operation and development of the park.

The operation of modern park is inseparable from the development and reform of financial system, so it is necessary to actively seek support from banks. At present, many banks, including the Agricultural Development Bank of China, have launched loans to meet the capital needs of the construction and operation of various modern parks, but they have made certain restrictions and requirements on the registered capital, credit status, management experience, qualifications and other aspects.

Capital market is a place for medium and long-term financing, and a variety of financial instruments in the capital markets offers a variety of channels in different ways and time limit for the continuing operations of investment and financing. They effectively reduce the government's financial burden, reduces the credit risks with much bank loan, and realizes the risk-sharing of all kinds of investors, the efficient allocation of capital elements. Therefore, agricultural enterprises entering the capital market should become the main financing channel of modern parks. For the agricultural high-tech enterprises in the park which meet the requirements of financing in the capital market, we should actively guide them to issue stocks or bonds in the securities market to get finance, and accelerate and improve the subsequent construction and operation of the park in the form of shareholding system.

(II) Management mechanism

Institutional economics holds that the main reason for economic development is not capital, manpower and natural resources, but organizational system, which itself determines the operation and development state of the organization, and the system is also the productive force. The management mechanism of modern parks, in a certain sense, is also the product of the system construction of modern parks. It is the sum of the organizational forms and management methods of the construction, operation and management of modern parks. The scientific and effective management mechanism is the guarantee for the development of modern parks, which is a special production organization with advanced form. Modern enterprise management mechanism emphasizes the integration of internal resources and the establishment of order. It runs through the whole operation process and is the core of the operation mechanism of modern parks. The operation of modern parks has a variety of modes, and different operation modes involve different subjects participating in the construction of modern parks. The organizational structure should be set according to the actual situation of modern parks, but no matter which operation mode is adopted, in the management process we should follow the principle of "independent-operating, self-financing, self-disciplining and self-developing, gradually establish the modern enterprise system with a "clear property rights, clear responsibility, separation of enterprise from administration, and scientific management", keep continuously perfecting the modern enterprise management system with the organic combination of market orientation and the technological innovation, and ensure the orderly operation of modern parks.

① The management committee system. It refers to the comprehensive management of the development, construction and daily operation of the modern parks by the agricultural park management committee. The management committee of modern parks can be an administrative government management mode, which is combined with the local government. It can also be a governing body sent by a local government. Generally the management committee have management authority of land requisition, planning, project examination & approval and personnel, responsible for the overall development & construction, transformation of sci-tech achievements, sci-tech enterprises' incubation, cultivation of sci-tech talents, new technology training and guidance, which has demonstration and radiation effect to the agricultural industrial structure adjustment. With the economy as the link, the surrounding farmers should be organized to participate in the production of planting and breeding industry, and gradually form an enterprise-oriented production and circulation system with the park as the leader, so as to promote and drive the development of rural economy.

② Corporate system. For modern parks established through the joint model of agricultural companies and multiple parties, the management mechanism in the operation process mostly adopts the corporate system. That is, the general corporation shall comprehensively manage and operate the modern park. If a joint stock company or a limited liability company is established, the general meeting of shareholders, board of directors, board of supervisors and general manager shall be established accordingly to regulate the operation. The advantage of this management mechanism is that it is easy to establish a modern enterprise system and directly carry out the enterprise operation. The disadvantage is that the head office has no administrative management function and no administrative authority in land requisition, planning, project examination & approval and personnel, which restricts the operation and development of the park in many aspects.

③ Contract management system. The contract management system is generally adopted for sci-tech contracting type of modern parks. After the government or the collective economic organization has built the public facilities in the park, the facilities and projects for production and management in the park will be put into the hands of farmers and individuals through the contract system or the rental system. The two-tiered management system is implemented, that is family management is the basic, and centralization is combined with decentralization. Farmers are both the investment and management subjects. And adopt unified planning and construction, unified varieties, unified technology, unified brand, unified sales, to realize the production, processing and marketing links. This will not only help farmers closely connect with the park development, mobilize the enthusiasm of farmers, but also help farmers to manage carefully and operate dedicatedly.

(III) Technology selection, docking and diffusion mechanism

Agricultural technology is one of the most important support elements in modern agricultural productivity. Representative technology selection & application and the continuously docking & diffusion is the foundation and key to the development of modern ecological agriculture parks. It is also the inevitalbe choice to speed up the transformation of agricultural sci-tech achievements, and to promote the development of agricultural industrialization. In general, the modern park should choose the corresponding technical system according to the functional orientation set at the beginning of the planning and establishment, and constantly carry out research, experiment, application, innovation and demonstration of the selected agricultural technical system, so as to highlight the characteristics of the park itself. On this basis, we should further give full play to the radiating and driving function of the park, and guide the farmers of the surrounding areas to use new varieties, new technology and new agricultural techniques. By standardization and intensive production, we realize the transformation of agricultural technology from potential productivity into real productivity, so as to promote the agricultural development of regional economy and promoting the development of modern agriculture.

Generally speaking, the technologies selected in modern parks can be divided into three categories: high and new agricultural technologies, practical technologies with great popularization value and applicable technologies with demonstration and driving effect. When choosing technology, we should follow the principle of composite multivariate, smart principle, high output & quality principle and resource-saving principle, and choose appropriate technical system, according to the function orientation in the planning period and the characteristics of the area. And we should equip a complete set with good quality, strong function, and high cost performance.

The technology docking of modern park refers to a series of processes in which the technology is gradually transmitted from top to bottom through the technology docking carrier between the subjects of agricultural technology. The main body of technology docking includes three parts: experts, enterprises and farmers. Among them, experts are the creators and providers of technology, and enterprises and farmers are the recipients and demanders of technology. The technology docking carrier is composed of government, market and intermediary organization, which together constitute the link and bridge between the subjects of technology docking. There are a lot of modern parks with different kinds in China, and due to the differences in natural, social and economic conditions, the choice of technical docking model suitable for the development of this region is the premise of realizing the effective extension of technology. At present, the typical models of agricultural technology docking mechanism in China include "expert + farmer household", "expert + agricultural intermediary organization + farmer household", "expert + leading enterprise + farmer household", "expert + market + farmer household", and so on. Among them, the technology docking model of "experts + farmers" belongs to the process spreading technology to farmers from top to bottom by the government and experts. The technology docking model of "experts+agricultural intermediary organizations+farmers" is that the experts introduce new technologies and culitvate new varieties, based on the local industrial structure and resource characteristics, in accordance with the commission and requirements of agricultural intermediary organizations, and then conducts technology docking with farmers through agricultural intermediary organizations. In the technology docking model of "experts+leading enterprises+farmers", leading enterprises aim to obtain high-quality primary products and production raw materials, and sign supply and marketing agreements with farmers under the guidance of the experts and the project execution agency. Experts carry out technology research, introduction and guidance according to the needs of enterprises. Leading enterprises provide pre-production, mid-production and post-production services for farmers. Farmers are responsible for agricultural production and provide agricultural products to enterprises for further processing, and finally put the products to the market. "Expert+market+peasant household" technology docking model is that the modern park management committee and experts build large professional market close to scientific research institutes and areas with intensive experts. Experts bring the new agricultural varieties and all kinds of practical new technologies to the market, the surrounding farmers come to the market to buy technology and seeds of the expert. At the same time, the management committee establishes a strict supervision team and technical feedback mechanism.

The technology diffusion or popularization of modern parks refers to a new technology, new products or new way of agricultural management that is widely popularized, with the help of information carrier, through different channels, by one person or a minority of people, a minority areas to more people, or greater scope for popularization. Generally speaking, the technology diffusion of modern parks has three main bodies for the agricultural high-tech popularization, which are formed by the technology supply institutes, the leading enterprises and farmers, through technology publicity, typical demonstration, education and training follow-up services, with the technology transfer and shares as the link. In addition, the agricultural sci-tech demonstration & popularization model based on the university is also a kind of specific technology diffusion mechanism, which is a kind of new form of popularizing agricultural science and technology for the agricultural university to serve the agriculture and rural economy. It is guided and promoted by the government, market-oriented, relying on the sci-tech, talents, information and other resources advantages, linked with projects. It integrates and makes use of resources for agricultural sci-tech popularization in relevant research institutions and grassroots communities. It demonstrates and popularizes advanced and applicable new agricultural technologies and transforms the agricultural high-tech achievements, with farmers' demonstration households as the starting point, and with the rural economic cooperation organizations and agriculture-related enterprises as the integration point.

(IV) Production mechanism

The production mechanism of modern parks is the general term of the production system of modern parks. The goal of the modern park production is to produce high quality products with high efficiency, low energy consumption, flexibility & punctuality, provide satisfactory service, achieve less input and more output, and finally achieve the best economic, social and ecological benefits. The production mechanism of modern parks includes the following aspects:

Above all, it can arrange production and organize work from macroscopical level. This work is mainly to determine the production content of modern parks, accordingly implement labor quota and labor organization, and set up production management system. Through the arrangement of production organization work, according to the requirements of the enterprise goals, it can reasonably arrange time and space for each of the component part in the production process, set up a production system which is technically feasible, economically reasonable, materially and environmentally allowable, and can make the production activities of modern parks efficient smoothly.

Secondly, it can formulate the production plan of modern parks. Production planning is the basis for production management. The production tasks of modern parks is to make overall arrangement, prescribe the indicators such product type, quantity, quality and schedule of the enterprise in a certain period of time, which is the action plan for achieving production targets in the planning period, and is also a basis for the modern parks to make other plans. The production plan of the modern parks should be based on the principle of balanced production, in accordance with the full exploitation of production technology and the full utilization of resources.

Finally, it can effectively control all aspects of the production process of modern parks, including production schedule control, equipment maintenance, inventory control, quality control, cost control and so on. Among them, the production schedule should be controlled in order to ensure the stipulated output of products and the products delivered on schedule. All kinds of machines and equipment will be used in the production of the park. In order to avoid the failure of the equipment and delay the production schedule, it is necessary to timely maintain and repair all kinds of equipment and facilities in the park to reduce the possibility of equipment failure. In order to ensure the normal operation of the parks, it is necessary to set a reasonable inventory level and adopt effective control methods to keep the inventory quantity, cost and occupied capital at a minimum. Quality control is to ensure that every product produced from the park meets the requirements of quality standards, which is the key for the modern parks to be selected and trusted by the market. In addition, it is necessary to control the material costs, inventory occupancy expenses and labor costs in the production process, so as to ensure the profits in the operation of the parks.

(V) Demonstration and driving mechanism

The modern eco-circular parks should be able to show the latest agricultural scientific and technological achievements, the most advanced agricultural management means, the most dynamic agricultural management mode, and predict the development direction of the future agriculture, which requires the demonstration and driving role to be fully played in the operation of the modern parks. The demonstration driving mechanism is a demonstration model of modern agricultural industry development path by introducing, absorbing and integrating modern agricultural factor conditions and innovating modern agricultural management system and mechanism. It represents advanced agricultural productivity and has a strong expansion effect. It leads the surrounding areas mainly through policy demonstration, linkage of industries universities and research institutes, industrial chain drive, and information service; promotes the agricultural productive relations of wider regions to further adapt to the development of agricultural productivity; promotes the construction of modern agricultural industry system; increases farmers' income; promote the development of urbanization; makes it the growth pole of the rural development and the regional development; radiatively drives the development of surrounding areas and realizes the virtuous circle of regional economic development.

The demonstration role in the operation process of the modern parks relies mainly on the parks' own technical characteristics, talent support, industry developmentadvantages. Through policy guidance, industry-university-research cooperation, the whole industry chain drive, etc., we can promote the leading role of the modern parks on the regional agriculture, increasing farmers' income the boot. First, in terms of policy guidance, construction and operation of modern parks was carried out as first experiment and experience in such aspects as capital raising and use, the land circulation and scale management, talent introduction and training as well as the production organization pattern, which create a good investment environment and comfortable environment. These policy directions are worth learning and promoting in other regions. Second, in the aspect of industry-university-research cooperation, as a platform for technology research and achievement transformation, the modern park has established cooperative relations with domestic and foreign scientific research institutions, sci-tech enterprises and many other units, attracting various talents and research teams to settle in. Through the combination of industry, university and research, various resources are integrated to the maximum extent, which enhanced and strengthened the park's functions of introduction, integration, application, demonstration, promotion of new varieties, new technologies and new facilities. Third, in terms of the whole industry chain drive, modern parks combine with the present regional development situation and industry characteristics, break through the traditional mode of production and operation, plan and integrate the resources and elements inside and outside the park, develop the characteristic industry with great effort. Based on industry, we should speed up the development of quality agriculture, green agriculture and characteristic agriculture, realize the optimization and upgrading of industrial structure, and maximize economic benefits. At the same time, we should strengthen the close connection of production, harvest, packaging, transportation and sales to form a complete and tight industrial chain, which will continue to promote the overall level of competitiveness of our agriculture.

(VI) Brand marketing mechanism

A good brand can produce "premium" brand marketing, which is a marketing strategy and process in which enterprises make use of consumers' demand for products, and then use the product's quality, culture and uniqueness to create a brand's value recognition in the mind of users, and finally form brand benefits. For the modern eco-circular agricultural parks, the brand means the representative technology of the modern parks and the reputation of the products, and it is the symbol of quality assurance and safety in the market and consumers. Brand marketing is to get the modern park products like industrial products processing and management, in the process of continuously building the brand, to establish and maintain mutually beneficial exchange relationship with the target market, to continuously improve the added value of agricultural products, to gain market trust, and to promote the competitiveness of the park, and to aquire a satisfactory comprehensive benefits. Brand marketing in the operation of modern parks should be the organic combination of hierarchical strategies, diversified strategies and branding strategies.

Today's society is a symbolic one. For the brand marketing, the differentiated personality which can deep influence consumers' core value of brand is a special symbol of modern park. It can let the consumer clearly identify and remember the selling point and personality, which is the main force that drives consumers to identify and love a brand. In the process of brand building, the modern parks should first ensure the quality of products and win the recognition of consumers according to its own market positioning. On this basis, we should rely on the novel packaging, unique design and symbolic name of the product to further attract more consumers. In addition, in the sales process and after-sales service we should also let consumers have a good experience. Through the above points, we can create the brand which can most reflect the characteristics of the park.

In the process of building a good brand and marketing to the outside world, we should make full use of the resources of the information age to popularize and promote the park brand in an all-round and multi-angle way to maximize and optimize the publicity effect. We should use advertising, business promotion, public relations, personal promotion and other methods to promote the target market customers. For the local market we can rely on the guidance and support of the government and relevant departments, combined with local news media publicity; we can also come to the organs, organizations, social institutions to publicize, and give a certain business discount, quantity discount, cash discount, seasonal discount and other preferential measures. For one thing, it can reduce the burden of capital and storage costs, and for another, it can gradually form a group with high head-turning rate and stable consumption. In view of the foreign market, we can set up offices in the central city to expand business, and through multi-channels to participate in the trade fairs and trade shows held in the central city and neighboring cities to vigorously promote the parks' products. We should make full use of opportunities such as traditional holidays, large-scale trade fairs and others to actively carry out brand marketing.

(VII) Risk defense mechanism

The basic meaning of risk is the uncertainty of loss, representing a negative adverse consequence. With the accelerated process of agricultural modernization, the pace of agricultural centralized production and management is also moving forward, which makes agriculture in greater uncertainty while advancing, and facing more and more complex risks. As a kind of advanced form and production organization mode in the development of modern agriculture, modern ecological parks will face many risks in the operation process. In order to avoid the risk of production and management, to reduce the degree of loss caused by the risk, and to ensure the realization of the operation goal of the park, it is necessary to clarify the types of risks that may be faced in the operation of the modern parks and build the risk defense mechanism.

The risks faced by modern parks include system risk, decision risk, production risk, scientific research risk, information risk and market risk. ① System risk. The operation of the modern park is carried out under the norms of the country's political system, financial system, tax system, land system and so on, which may have an impact on the operation of the parks.

For example, the existing systems may have a certain degree of incompleteness, and because of this incompleteness, the deviation between the expected return and the actual return will be caused. The deviation degree is in direct proportion to the final risk. The risk will arise when the insufficient system supply, the system changes or the system innovation process bring the uncertain influence to the benefit of the modern agricultural parks, especially when the influences change or develop in the unhelpful direction. ② The decision risk. Due to the decision-makers' ability level, constraints of the conditions for decision, or failure to grasp completely all the conditions before making operating decisions, all these lead to deviation in the process of operation with the expected benefit, thus risks may arise. ③ Production risk refers to the risks that may be encountered in various links of the production of modern parks, including the purchase, installation & commissioning of agricultural machinery and equipment or facilities, cultivation & selection of varieties, production management and other aspects of the risk. ④ Scientific research risk refers to the risk caused when modern parks carried out the varieties's choice & breeding, introduction, improvement, development, the technology research & promotion of all kinds of production methods and links, due to the fact that the technical innovation cycle is long and is impacted by the regional environment, and the technological tranfer and promotion is restricted by the operation scale of farmers. ⑤ Information risk means that the operation of modern parks must fully grasp social, economic and other kinds of information. If the operation of the park is carried out under the condition of incomplete information collection or serious information asymmetry among various departments, the operation results may deviate from the actual development needs of the society, and the expected returns cannot be obtained, which may lead to risks. ⑥ Market risk refers to whether the products produced in modern parks can be accepted by the market and consumers, such as whether the products are marketable, whether the sales channels are smooth, whether the funds can be timely returned to the risks and challenges.

For the various types of risks that may occur during the process of modern park operations, we should take a series of risk management measures, methods and means to build a modern park operating risk defense mechanism, avoid park operating risk, reduce the degree of risk loss, ensure the implementation of park operation objectives. ① We should establish a complete set of modern park operational risk assessment system, which can carry on the comprehensive identification and evaluation of risk, to make the risk management of the whole process more professional, systematic and organized. We should establish risk management organizational structure and organizational relationship, strengthen the relationship between various departments, and ensure the completeness and reliability of the information. We should organize relevant personnel to participate in risk management training to improve the skills of risk identification, assessment, prevention and treatment. ② We should establish the risk liability system for the operation of modern parks. The responsibility should be specific to the people involved in this process of the various work. If one link in risks caused by human error, we should investigate this error and find out who is to blame. This management measures is condictive to improving the risk responsibility consciousness of the staff in modern parks, avoid risks caused by carelessness. ③ We will strengthen the support and popularization of agricultural insurance. The risks faced in the operation of modern parks, especially some non-human risks, can be prevented through agricultural insurance.

IV. Operation model

For the organization of modern parks, it is required to have the concept of operation in its management process to pursue the maximization of economic, social and ecological benefits. The pattern can be regarded as the combination of relevant elements and the paradigm of operation flow within or between systems. The operation mode of modern parks can be understood as a general description of the characteristic operation mode and operation. As a complex system, modern eco-circular parks have many elements, complex structural relations, diverse operating mechanisms and easy to be affected by the surrounding environment, which form the diversity of the operation mode of modern parks. In general, the operation of the modern parks should be combined with the regional characteristics, and the operation mode suitable for the local area should be established, so as to realize the coordination between the modern parks and the surrounding economy and the sustainable development of the system. Modern ecological agriculture parks, after years of development, gradually formed a number of representative operation pattern. The operation models described in this book are mainly classified according to the properties of the operator, that is the government dominating operation model, enterprise dominating operation model, "government + enterprises" mixed operation model, the scientific research universities leading operation model.

(I) Government dominating operation model

The government dominating operation model emphasizes the important role of the government in the whole operation process, including the relevant national departments, provincial offices, municipal offices and county offices and other levels, and the investment subject is the governments at all levels or their functional departments. The modern ecological parks led by the government have high grade, large scale and full functions, which play an important role in promoting the national economic and social development. Therefore, on the one hand, the goal of its operation is to focus on the overall development of regional economy, which requires the operation of modern parks to improve the level of agriculture, promote the increase of farmers' income, and play a leading role in the construction of modern agriculture and the adjustment of agricultural structure. On the other hand, it also emphasizes the social benefits of the operation of the park, and requires the operation of the park to be able to promote agricultural technology to the surrounding areas and form a radiating driving effect. In form the government takes the lead in organizing relevant functional departments to form a leading group, which is responsible for formulating park planning, determining operation objectives, providing preferential policies and infrastructure construction. The management committee established in the zone, as the resident agencies of the government, has great economic management authority and the corresponding administrative function, has a certain control over the modern parks' ecological cycle, operation and the development, and is mainly responsible for the examination and approval of the high-tech enterprises' entry, organize the implementation of park planning and other administrative functions authorized by the government. In this model of operation, the government has provided a large amount of funds, land and professionals for the sustainable operation and development of the modern parks, and formulated relevant policies to attract enterprises and scientific and technological personnel to participate in the in-depth construction of the parks.

The characteristics of the government dominating operation model are as follows: ① The main investment body is relatively single, focusing on the public welfare of the park operation. ② The government is directly involved in the construction and operation management of the park. ③ The modern parks invested by the government have a strong guiding and demonstration role. The construction, management and operation of the modern parks are carried out by the government through the park management committee, which actually exercises part of the government's functions. Therefore, the centralized power of park management committee is conducive to the reform and innovation of operation system.

In the actual operation process of the modern parks invested and operated by the government, the government undertakes the function of agricultural technology promotion, so that the modern parks can obtain good social benefits. However, due to the complex government management system, low market sensitivity, inflexible operation mechanism and other shortcomings, the modern parks have some problems, such as low efficiency of staff, insufficient operational effect, lack of incentive mechanism, and relatively difficult to achieve substantial improvement of economic benefits, which limits the development of the park. According to the actual situation of the region, the government should formulate the rules and regulations in the operation process of the park, strengthen the operation supervision and improve the evaluation system, so as to make the construction and operation of the park follow the rules, so as to maximize the benefits.

(II) Enterprise dominating operation model

The enterprise dominating operation model refers to a complete system, during which the construction and operation of management organization is formed by single or multiple enterprises, under the guidance of the government, in order to realize the economic benefit as the main purpose, the integrated use of market means and business enterprise principle of management for agricultural technology research and development and technical results transformation. In this operating model, the government does not establish a local agency, but the development company is a legal person in economy, which is responsible for the modern parks' planning and designing, the construction of infrastructure, project selection, investment, financing, and organization of the operating activities in the area of the modern parks, and take some part of the government's functions. The investment enterprises follow the principle of being "independently-operating, self-financing, self-disciplining and self-developing", and they adopt the enterprise-style management. They allocate the factors of production such as personnel, finance, resources, focusing on the benefit maximization goal, and design and layout the technology selection, product positioning, market direction and so on. They make full use of its capital management concept & information, network and other resource advantages to construct and operate the park, form the industrial "diffusion effect", and drive the rapid development of the surrounding regional economy, so as to realize the effective combination of capital and resources. Generally speaking, modern parks led by enterprises are established by enterprises of different nature, such as township enterprises, private enterprises, joint ventures and so on. The main investors are enterprises, especially private enterprises, which account for a large proportion.

The characteristics of enterprise dominating operation model are as follows: ① It adopts the operation model of company system. The whole modern park is a standardized joint venture to manage capital as a legal person, to organize farmers to produce, to obtain economic benefits. ② It is closely connected with the market, pursuing economic benefits with a flexible operation mechanism. Because of the avoidance of administrative intervention, the park can be operated according to the law of marketization and the characteristics of corporate management. The enterprise will take the most profitable project as the main direction of development, and combine with market demand to carry out key research and development and technological innovation, and improve the efficiency of operation.

With the continuous acceleration of the process of marketization, the enterprise dominating operation model has become the choice of more and more modern parks due to its development potential and market vitality. However, under this model of operation, enterprises mostly aim to maximize their own economic benefits, and it is often difficult to obtain sustainable benefits in terms of social and ecological benefits in the process of operation.

(III) The hybrid operation model of "government+enterprises"

The mixed management mode of "government+enterprises" refers to the establishment of both the park management committee, which can exercise certain economic management authority, and an investment and development company in the park. The management committee is responsible for the government's administrative function, and the investment and development company is responsible for the enterprise operation function. Among them, the management committee is a dispatched agency of the government, which exercises the management power of the government, intervenes in the business activities of the enterprise with administrative power, and mainly plays the role of supervision and coordination. In this operation model, the government selected the appropriate area, prepare the park's overall planning and investment guide, invest on the construction of the infrastructure, set up industrial park management committee, make some related management measures and preferential policies, create a good investment environment and provide a foundation platform in the park for the enterprise to enter. After being examined and approved, the park's construction project shall be invested by the enterprise for independent management. In the operation of modern parks, the government and enterprises will be "mixed", the industrial development of the parks can be wider, the corresponding technology, investment subjects and financing channels will become wider, which is conducive to fully driving the enthusiasm of all aspects.

This park construction operation model embodies the principle of "small government, big business". It is conducive to giving full play to the government's administrative function. In addition, it can make up for the shortcomings of modern park itself, with the help of abundant capital, advanced technology and management experience, to give full play to the economic functions of enterprises in line with the market rules. It can give full play to the macro-control role of the government and the professional management ability of enterprises, and the mutual promotion and cooperation of the two are conducive to the construction, management and sustainable operation of modern ecological circular agricultural parks.

(IV) The scientific research institutions leading operation model

In order to speed up the innovation and development of advanced agricultural technology and promote the transformation of high-tech achievements, scientific research institutes and universities set up modern parks. These parks generally have strong scientific and technological ability and high technological level, and the leading role of industrial science and technology development is obvious, which is conducive to promoting industrial innovation. The leading operation model of scientific research institutions can be generally divided into two types: one is the modern parks established by the scientific research institutions, focusing on the research and development of agricultural high and new technology, and promoting the close combination of production, education and research. Its characteristics is to integrate these factors in one such as talent training, knowledge innovation, achievement transformation, technical consultation, enterprise incubation and industrialization development. The other is the modern parks jointly established by scientific research institutions and foreign investments. The research institutes implement the engineering-centered guiding strategy, guided by the market and aimsing at the engineering and industrialization of agricultural technological achievements, and attract the investment through new technologies, new achievements and new products.

However, in the actual operation process, the venture capital fund of the modern parks led by scientific research universities is relatively weak, difficult to reach the requirement, and the incubator is not sound enough. Therefore, scientific research institutions can establish venture capital funds with the appropriate power of the government to encourage and support the sustainable operation and development of the parks.

Chapter 7 Planning & Designing and Case of Modern Ecological Circular Agricultural Parks

Section 1 Application of ecological circular concept in agricultural parks

In the face of the problems and relevant countermeasures of the development of circular agriculture in China, the planning and construction of modern ecological circular agricultural parks under the concept of ecological circulation are analyzed by taking the "modern agriculture circular economy industrial parks under the background of rural revitalization" planned by an enterprise in 2020 as an example.

I. The project background

The project is located in the warm temperate subhumid monsoon climate zone, rich in natural resources, beautiful soil, fertilizer and water, rich in food oil, medicine, grain, pigs and other agricultural and sidecar products. The project area has convenient transportation and is located in the central service area of the urban economic circle, with great potential for industrial development.

II. The construction objective and principle

1. Construction objective

To build a comprehensive park with circular economy as the main function and demonstration training and exhibition and sightseeing as the auxiliary functions.

2. Planning principles

Measures adjusted to the local conditions, ecology priority, recycling use, energy conservation emission reduction, industry optimization, scientific promotion, leading projects, cycling development.

III. The functional area planning

The project area is divided into four parts: modern agricultural circulation demonstration area (core area), efficient grain and oil production area, facilited fruit and vegetable planting area and ecological circular breeding area.

1. Demonstration area for modern circular agriculture (core area)

The core area focuses on the construction of circular economy, and develop the planting, breeding and processing recycling industry chain, with both sightseeing display and science & technology training functions. The core area covers an area of about 32,000 mu, which is divided into seven zones: ecological energy-saving new rural community, green facilited fruits and vegetables demonstration area, high-standard grain and oil production area, high-quality flowers and seedlings demonstration garden, livestock and poultry breeding demonstration area, circular economy exhibition area and agricultural products processing and logistics center.

2. Production area for efficient grain and oil

Centralized planning and construction of farmland around the project area. Relying on a good ecological environment, the use of advanced farming technology, we will achieve large-scale, standardized, modern farming of grain and oil, and improve the economic benefits, social benefits, ecological benefits of the crops in the project area.

3. Planting areas for facilited fruits and vegetables

We will vigorously develop the fruit and vegetable industry, mainly for the core regional market in the urban, adjust the internal structure of leafy vegetables, solanaceous fruits, melon vegetables, strengthen the production, processing and marketing of fruits and vegetables with pollution-free green facilities, and extend the industrial chain. At the same time, it will combine with other industries in the project area to develop circular economy and build a provincial green standard fruit and vegetable production and supply base.

4. Breeding area for ecological circulation

Through the development of ecological recycling breeding, the agricultural industrial structure of the surrounding towns can be adjusted, resources can be utilized comprehensively and efficiently, agricultural benefits can be improved, and farmers' incomes can be increased continuously. Using ecological and environmental protection, the latest animal husbandry technology and the principle of circular economy, through the effective use of straw, poultry manure, site space and other resources, reduce the production costs of planting and breeding industry in the project area, protect the environment while maximizing economic and social benefits.

IV. The ecological circular design of key projects

1. Main models of agricultural circular economy in industrial parks

Through the combination with the livestock breeding industry and processing & logistics industry in the park, the comprehensive utilization ecological model of biogas as the link of clean energy will be developed. The specific manifestation of this model is: breeding (pig, cow, chicken) - biogas - planting (grain, oil, vegetables) and so on. Model principle is as follows: This model is based on land, with biogas application as the link, combined with planting and animal breeding, to form a supporting development and production model of planting for livestocks, livestocks for biogas, biogas for agriculture, planting and raising combined.

2. Ecological utilization model of livestock manure in breeding industry

(1) Comprehensive utilization model of pig manure

For example, the core area stores 100 000 pigs and produces 300 tons of pig manure every day. The pig manure can be used for biogas production, and the biogas produced is used for domestic gas. The biogas slag and biogas slurry produced by the biogas project can be processed into biological organic fertilizer and resource reuse.

(2) Comprehensive utilization model of cow manure

The core area has 10 000 cattle, and produces 200 tons of cow manure every day. The cow manure can be used for the domestic gas consumption of residents and the earthworm breeding and agaricus bisporus planting in the circular economy exhibition area, and the fresh earthworms can be partially supplied to the eel breeding in the core area. Through earthworm breeding and agaricus bisporus planting, 4058 tons of cow manure can be used, and the produced earthworms can meet the requirements of 16 mu of eel breeding, and the remaining cow manure can be used for composting to form the recycling of resources.

(3) Sewage treatment and recycling

The sewage treatment tank is established according to the amount of sewage produced by different industries. After the sewage enters the treatment tank, the treatment water discharged into the reuse tank meets the standards of COD_{Cr}, BOD_5, SS and pH, which can be used for :

① cleaning the production and processing places of fruit trees; ② providing production of the seedlings in seedling cultivation center and irrigating the new variety demonstration area; ③ watering the green plants in the factory; ④ flushing the washroom, etc.

The excess treated water in the reuse pool can also be discharged through channels for farmland irrigation, forming a recycling chain of "industrial wastewater - sewage treatment - secondary utilization".

Under the concept of ecological circulation, the construction planning of modern ecological agricultural parks is a complex work to regulate and coordinate the planned regional agricultural ecosystem. The planning and construction of the modern agricultural circular economy industrial park explained in this chapter solved the problems such as weak industrial chain correlation, unreasonable resources utilization, serious agricultural non - point source pollution, and disorderly piling of livestock manure in the process of agricultural development, and promoted the smooth and efficient construction and development. So that the park has more demonstration, guidance and promotion significance for the development of circular agriculture industry economy in the region.

The green agricultural ecological model of "livestock and poultry manure - biogas - electricity - heat - organic fertilizer - crops - feed - breeding" has set out a road of green development with enterprise as the core, industry as the link and ecological and environmental protection as the goal. Park planning follows the "reduction, reuse, recycle, resource, resource regeneration of alternative" concept of development, links with the modernization of biogas project, integrate the planting, breeding, processing, tourism and the living in the new rural residential building, and form the energy chain and industry chain of planting, breeding and processing integration, the three industries mutual promotion The introduction of various circulation cultivation modes makes the plant and animal resources in the park fully utilized. At the same time, the application of new technology provides an effective technical guarantee for the circular development of the park, and achieves the unity of social benefits, economic benefits and ecological benefits.

Section 2　Designing of Zhaori Lvyuan (Sunrise & Green Source) ecological circular farm

Zhaori Lvyuan Farm, located in Muyudian town, Laiyang city, Shandong province, was established in 2006 as a joint venture by Asahi Brewery Co., LTD. (73%), Sumitomo Chemical Co., LTD. (17%), and Itochu Corporation (10%). The leased land is 1500 mu, five villages (Wujiatuan, Daming, Nanwang, Xiaodian, Zhongwang), 1000 farmers, 1000 yuan per mu (the rent is very high, at that time, the general rent is 580 yuan per mu), the lease period is 20 years. It is an agricultural company to produce, process and market the high value-added agricultural products such as vegetables, fruits, and milk etc.

Zhaori Lvyuan company introduces the advanced agricultural technology in Japan, effectively use circular agriculture, energy-saving equipment, IT, and implement and build food system and strict quality managementfrom the cultivation of crops to the food logistics & sales, providing domestic secure, safe, delicious and high value-added agricultural products. It aims to contribute to the improvement of China's food life through cooperation with Shandong Provincial government, realize the new agricultural operation demonstration model, cultivate the next generation of leading technical talents in the agricultural field, and help solve China's agricultural problems as the goal.

At the beginning of the establishment of the farm, Zhaori Lvyuan Farm proposed that "profit is not our goal, we are to establish an ecological circular agriculture demonstration project, producing safe, secure and high-quality products is our only goal at present. This is the first wholly foreign-owned farm in China and the first farm in China to follow 'circular agricultural production', and Zhaori Lvyuan is ready to take a gamble."

I. The circular agricultural model of Zhaori Lvyuan Farm

Zhaori Lvyuan adopts the circular agriculture model, investing 3 million yuan to build a composting plant, using the manure of farm cows and organic matter to produce organic fertilizer to improve the soil, so as to improve the quality of crops, and explore circular agricultural production. Zhaori Lvyuan also spent $2 million to bring in wind power and solar power equipment to power offices and farms.

1. Maintainance of the land

The land maintainance has become the first step in the development of ecological agriculture and organic agriculture in Japan, Germany and other European and American countries. "The health of land" directly determines the quality of agricultural products.

As early as December 2003, Zhaori Lvyuan group began to organize agricultural experts to carry out feasibility research, and has carried out strict investigation on water, soil and other environmental indicators of 8 agricultural lands in Zibo, Weifang, Jiaonan, Zhangqiu and Laiyang cities of Shandong province for more than two years.

However, Sun Yinghao, a supervisor assigned to the company by Zhaori Group, said the test results showed that the soil of the eight agricultural fields did not meet the requirements for growing green fruits and vegetables, and the final choice of Laiyang Muyudian town can not meet the requirement, so Zhaori Farm spent many year in maintaining the land.

Zhaori Lvyuan Farm hopes to improve the health of the soil and organic matter by maintaining the land, thereby affecting the quality of agricultural products. On the surface, the Japanese are making a fool of themselves by leasing land that produces only half as much as the locals, but in reality the Japanese way of farming represents the future of world agriculture, and there is much to learn. No fertilizer, no pesticide, no herbicide farming method is actually the hope of agriculture to return to nature. The most important feature of modern agriculture is the application of fertilizers and pesticides with the aim of increasing production, but the long-term consequences and bad result of such practices are becoming increasingly apparent. It is true that the use of chemical fertilizers year after year can increase the production of agricultural products in the short term, but the side effect is that the soil is hardened and salinized, leading to soil degradation. Moreover, the quality of agricultural products is also decreased because the application of nitrogen fertilizers, such as urea, increases the nitrate content of crops and deteriorates their quality.

Similarly, a large number of pesticides can eliminate pests and weeds, and increase the yield of crops. However, the widespread use of pesticides leads to the ecological deteriorating and because of increasing resistance of pests and weeds in pesticide, resulting in a vicious cycle of "virtue is one foot tall, the devil ten foot". This not only increases the dose and the cost of agricultural production, but also can kill other insects and weeds beneficial to humans and the environment, a great damage to biodiversity. In this sense, the Japanese way of farming without chemical fertilizers, pesticides and herbicides is not "destroying the land", but protecting the land. Zhaor Lvyuan's vice president Keiji Maejima believes that their concept of farming is in accordance with the ancient Japanese saying, "before planting, first make soil, before making soil, first cultivate people". Although Laiyang's land is fertile, the long-term use and infiltration of chemical fertilizers and pesticides have degraded the land. What they have to do is to put a lot of effort and money into restoring the soil in the first few years.

Of course, the Japanese can lose money by farming without chemical fertilizers, pesticides or herbicides. But this is only a small loss, in the long run they are taking a big advantage. For one thing, they can make up for their losses through the earnings of different agricultural products. Zhaori's milk, for example, is priced at 22 yuan per liter, 1.5 times more than the domestic price. Their strawberries are priced at 120 yuan per kilogram. Five years ago, Zhaori Lvyuan's strawberries hit record prices in Shanghai.

2. Awesome attitude

Farming is not just about obtaining agricultural resources, it is about respecting agricultural resources, so that it is possible to respect the laws of nature. The Japanese certainly deserve praise for their "reverence for agriculture or nature". Zhaori Lvyuan Farm, for example, has a strict management system. Staff are not allowed to touch cows or shout at them. If a cow died, the staff should collectively mourn it. After giving birth, the cows are fed miso soup (a Japanese dish made of snapper, red and white radish, fish bones and miso) to stimulate their appetite.

3. Positioning the high-end market

Zhaori's green products are aimed at the high-end market in China and overseas, and will be accepted in the low-end market in the future, because their products are green and pollution-free. Instead of fertilizer, they compost with cow manure. Remove the grass without herbicide, but hand pulling and hoeing. They rarely used pesticide, but occasionally, also need to be guided by experts. Soil is regularly tested to ensure nutrient balance. This planting model conforms to the law of nature and ecology.

Such products are not only high quality, safe and acceptable to all, but ultimately sustainable because they protect the environment and preserve biodiversity. Such planting concepts and methods are not only in terms of economic benefits, social benefits and long-term development, but also in line with the traditional Chinese concept of the unity of nature and man. It is also a way of living according to the laws of nature.

4. Circular agricultural model

The core of Japan's circular agriculture model is the establishment of ecological value chain. Through the establishment of ecological value chain, Japanese agriculture maintains the relationship and balance of various ecological chains.

Likewise, the biggest feature of Zhaori Lvyuan Farm is the practice of circular agriculture in Japan. To put it simply, basically no pesticide or chemical fertilizer is applied. The corn stalk can be used as feed for cows, and the manure of cows can be used as fertilizer for crops, so as to improve the quality of soil and thus the quality of agricultural products. The whole production process does not use any pesticide, fertilizer.

Planting crops in a way that is in keeping with nature is not just a Japanese wisdom. Farmers in China have already implemented it and achieved success. If the concept and action of green planting can be promoted, it will benefit not only the people of today, but also the future generations.

In addition to the above four points, the modern management concept and precise market position of Japanese agriculture are also worth learning for Chinese agricultural managers.

II. Analysis of the reasons for failure of circular agriculture model of Zhaori Lvyuan Farm

Zhaori Lvyuan Farm is a fully designed business model and management model, but why did it fail? There are three reasons for this.

1. Product pricing being higher than consumption level

Zhaori Lvyuan Farm's cost of maintaining land is calculated into the price of its products, and the farm hopes to reduce its losses and inputs with the help of high pricing.

For example, the products of Zhaori were first introduced into stores and supermarkets in first-tier cities. Its main products are priced at 320 yuan per Jin (500 g) for strawberries, each sweet corn is 8 yuan and the milk is more than 20 yuan per liter, all of which are several times higher than the price of the most high-end products in China. Such a high price leads to that Zhaori Farm sales can not keep up, and even the products are not sold.

2. The shame of scale

The yield of ecological agriculture has been low. This is an indisputable fact of the industry.

Due to the dispersive nature of China's land, coupled with China's "pesticide" based agriculture and many other reasons, Zhaori Farm has been unable to increase in scale.

"Zhaori Farm has been losing money, largely because it has not broken through the bottleneck of scale." Agriculture is typical of economies of scale, said Mr. Sun, a supervisor at Zhaori. It's hard to make a profit on the scale of production you have. To expand the scale of cultivation, land is the primary constraint factor.

3. Single industry form

From the Zhaori Farm's plan, the farm mainly raises cows, supplemented by agricultural farming. In other words, the farm's produce is basically internally digested. It is well known that one of the agricultural premium comes from the processing end. But Zhaori Farm has not been involved in the processing end, but with "high price agricultural products" to the market. Therefore, the single industrial form caused the short circuit of the pure premium part of Zhaori Farm.

III. Reference from the circular agricultural model of Zhaori Lvyuan Farm

What lessons can be learned from the failure of Zhaori Lvyuan Farm?

1. To find out the premium points

The circular agriculture model is only compared with the traditional agriculture model, but the key of circular agriculture is to rely on the industry derivative, especially the premium point of circular agriculture industry. For example, by adopting the duck and rice symbiosis model, the benefits of the industry can not only produce high-quality rice, but also derive a new interest chain based on ducks, so as to reduce the low-yield effect brought by ecological agriculture and improve the yield benefit per mu by other industries.

2. To have scale effect

On the whole, agricultural profits come from the scale, which can reduce costs and improve overall agricultural profits. In fact, there are two ways to increase the profit of agriculture: ① to reduce input costs; ② to Improve the yield per mu. This is true for both smart and circular agriculture models. Therefore, the key to the model of circular agriculture is to establish the industrial value chain, so as to reduce the input cost and improve the yield per mu.

There are many reasons for the failure of Zhaori Farm, but the core reason is the lack of commercial value chain. Therefore, if we want to make use of circular agriculture model, we must make great efforts in commercial value and industrial value.

Section 3 Designing of Phoenix eco-circular agricultural demonstration park

I. The project Overview

In 2015, the "No. 1 document" of the Central Government, "Several Opinions on Comprehensively Deepening Rural Reform and Accelerating Agricultural Modernization", insisted that the basic status of agriculture should not be shaken. At the National Agricultural Work Conference, we deepened rural reform, accelerated the development of modern agriculture, and developed leisure agriculture that integrates agricultural production, agricultural sightseeing leisure vacations, and participation & experience, so as to improve the quality of farmers' life, improve the rural ecological environment, and realize the prosperity of people and the beauty of villages. The 12th Five-Year Plan of National Leisure Agriculture Development by the Ministry of Agriculture, XX District of XX citys' planning on rural construction and so on increased financial support, improved financial service guarantee premise, innovated industrial development model, cultivated new technical personnel, and vigorously developed characteristic and efficient agriculture. Based on the above policy support, the circular ecological model of government leadership + company leadership + farmer participation + mutual assistance and win-win will be realized.

The current agricultural development is in an unprecedented period under the support of the government, and the agricultural market is full of vitality. The development of modern agriculture needs to change the model of agricultural development and actively develop various functions of agriculture. It needs to be driven and guided by projects with characteristics, innovation, ecology and guarantee the vital interests of farmers.

II. The policy analysis

1. Responding to government's policy

Phoenix Eco-Circular Agriculture Demonstration Park tries to become a leading enterprise of characteristic breeding in regional mountainous areas. At the same time, in response to the call of the government, Phoenix Park adheres to the principle of "people run, peopel administrate, and people benefit", makes publicity according to the locality, gives classified guidance, and establishes a closer interest connection relationship with farmers' specialized cooperatives and farmers. We will deepen the model of "leading enterprises + bases + farmers" and accelerate the development of characteristic and efficient agriculture with quick results, high added value and a complete industrial chain. We will promote the development of poverty alleviation projects, speed up, focus on concrete, detailed and effective poverty alleviation work, and help the poor people develop industries, lift themselves out of poverty, and increase their incomes. The Phoenix supports a number of poor people who are able to work and can lift themselves out of poverty through production, and increases support for industrial cultivation. We will promote regional agricultural efficiency improvement, take increasing farmers' income as the goal, and adhere to the principles of positive development, gradual standardization, quality improvement and efficiency improvement. Phoenix is a modern project with advanced production technology, excellent ecological environment and economic development.

2. Specific measures

Phoenix is a comprehensive park, combining ecological breeding demonstration, ecological planting demonstration and agricultural tourism, which has a high level of economic benefits, ecological benefits and social benefits.

Supporting policies for Phoenix Eco-Circular Agriculture Demonstration Park are as follows.

① Give full play to the role of the federation of rural cooperative economic organizations in this district.

② Encourage farmers' specialized cooperatives to actively participate in the exhibition and marketing of agricultural products outside the district and promote local products.

③ Establish a cooperative economic personnel training mechanism.

④ Pay attention to the agriculture and breeding. Hire local farmers, give priority to providing training opportunities for poor households, cultivate high-technology talents, so as to increase the employment rate and economic income of poor households and improve their economic level.

⑤ Develop agritainment sightseeing. The original houses of farmers in Phoenix Eco-Circular Agriculture Demonstration Park are improved, and the innate high-quality environment is also used to build rural leisure tourism, increase income sources and improve the economic income of local farmers.

III. The analysis of location basis and resources

1. Analysis of natural conditions of location

The project is located in a mountainous area of Shandong province, with convenient transportation and pleasant scenery. There is a rural road to the west of the project site, which connects with the national road and leads directly to the city to the north.

The climate here belongs to warm temperate monsoon type. Precipitation is concentrated, rain and heat in the same season, spring and autumn are short, winter and summer are long. The annual average temperature is $11-14\,°C$, and it has sufficient light resources. The annual light hours are 2290 to 2890 hours, and the heat condition can meet the needs of crops for two crops a year. The average annual precipitation is generally between 550 and 950 mm. The seasonal distribution of precipitation is very uneven. 60%–70% of annual precipitation is concentrated in summer, which is prone to waterlogging damage. Drought is prone to occur in winter, spring and late autumn, which has the greatest impact on agricultural production.

2. Site elevation analysis

Phoenix Park has a high altitude, about 49–569 m, and the elevation difference between the highest point and the lowest point is about 75. The east and west sides of the base are high mountains, and there are flat and open terraces facing the gully in the east. The overall topography of the base is high in the south and low in the north, and the terrain is fluctuant, with water and terraces available. The mountain slopes on the east and west sides of the demonstration park. The site was originally a farming field, and the landform has been divided into relatively flat open fields of different sizes. Due to the slopes on the sides of streams & gullies and the mountains on the east and west sides are relatively large, and some of them even have steep slopes, they are suitable for tourism trails and planting fruit trees. Different development and construction methods are adopted for areas with different slope conditions, which is conducive to creating diversity and interest and improving land utilization rate. The terrain of the demonstration park is mainly east, southeast and west, southwest, facing with each other. The stream flow direction of the base is from south to north, the stream on both sides of the terraced fields, the trend of the stream is from north to south, high in the south and low in the north, conducive to the collection of water bodies. The drainage design is guided by the slope direction.

3. Current situation analysis

The terrain of the project is generally high on both sides and low in the middle, with good sight view, more scenic spots, beautiful environment, strong structurability and many advantageous points. Current water sources: there is a natural spring and a natural water source flowing from south to north in this project. The water quality is excellent, and the short-term water flow in the rainy season is large, forming a water peak. The spring water source can basically meet the water consumption in the park, and an irrigation ditch has been distributed in the park from the spring water.

Current transportation: on the west part of the site, there is a country road, about 3.5 meters wide, which is the main traffic for outside vehicles in the park. The country road connects the national road to the city. The interior of the park is mostly rural roads, and internal traffic needs to be newly organized.

According to the analysis of the existing vegetation, the current situation of the vegetation in the area is mainly divided into wheat field, dry field, woodland and shrub. Rice fields are mostly terraced, which occupy a large proportion of the area. The east mountain is mostly shrubbery, the west two mountains have dense woods, arbor trees in the forest grow well, with a good landscape effect.

4. SWOT analysis

Strength: the project area has a good environmental background and can create a relatively rich landscape. The project has a good natural environment and abundant resources; Close to the city, the market potential is large.

Weakness: the existing agricultural industry structure in the project area is single and the proportion of traditional agriculture is heavy; Poor infrastructure conditions, supporting facilities are not perfect.

Opportunity: supported by national policies, the local government attaches great importance to rural development; Guided by the development of modern agriculture. Agricultural development is becoming more diversified; Rural tourism has gradually become a trend to promote agricultural development.

Threat: how to connect all industries in the project to form a complete industrial chain; How to make good use of local resources to form a good ecological cycle; How to combine landscape with economic benefits.

5. Market analysis

(1) Breeding industry market analysis

The breeding industry in the park is mainly divided into rock frogs, yellow mealworms, earthworms, snakes, cattle and chickens. At present, people tend to pursue healthy, high-quality and novel green food. The rock frog, yellow mealworm, cow and chicken bred in the park are all green, ecological and organic breeding varieties, which will bring good economic benefits to meet people's full demands.

① Edible value: the breeding industy provide people with rich foods such as meat, milk, eggs, poultry, which has very good edible value, such as the rock frog meat containing protein, glucose, nitrogen acid, iron, calcium, phosphorus, and a variety of vitamins, chicken raised in the mountain containing 21 kinds of essential amino acids, rich in nutritional value. Mealworms and earthworms are also healthy food for people. They can not only be used as feed, but also have good edible value.

② Medicinal and health care value: the cultivated varieties in the park not only have rich edible value, but also high medicinal and health care value. For example, the chitin and antimicrobial peptide unique to yellow mealworm can reduce blood pressure, improve immunity, prevent cancer, reduce aging and have other effects, which can be used in medical and health care products. The bred cobra is a highly venomous snake. But it is also one of the main medicinal snakes; Earthworms contain earthworm elements (Dilong su), earthworm antipyretic hormone, xanthine, antihistamine, vitamin B and other medicinal ingredients.

③ Other values: the breeding varieties in the park not only have edible value, medicinal value and health value, but also have other values. For example, earthworms and mealworms are the best feed for animals such as chickens and frogs. The use of earthworms can deal with chicken, cow manure. The vermicompost can also be processed into green organic fertilizer, which is high-grade fertilizer for landscaping, flower planting and grass planting. The price of vermicompost in China is about 1000 yuan per ton.

(2) Planting industry market analysis

The planting industry in the park is mainly divided into fruit planting, cash crop planting, and seedling planting. Fruit trees are the main varieties with high economic value and local unique, such as cherry, apple, hawthorn, peach, persimmon, dates, pomegranate, etc. Cash crops are wheat, corn, peanuts, sweet potato. Seedling planting varieties are high-stem heather, osmanthus, Padus virginiana 'Canada Red'. Due to the improvement of people's living standard, the demand for green ecological food is increasing day by day. The demand for organic ecological food in the market is greater than the supply, and the market is good. The green ecological fruits, cash crops and green seedlings planted in the park can supply the market demand and bring high economic value.

① Health value. People begin to care more and more about health, pay attention to food safety and protect the ecological environment because of their instinct and cognition of science. Especially they pay more attention to products with no pollution and no pesticides. The green planting industry in the park produces pollution-free agricultural products, green food and organic agricultural products. Green organic products in the park, such as cherries, peaches, plums and other fruits have maintained a good momentum of development, with high health value.

② Brand value. The existence of geographical indication effect is an objective phenomenon, and the geographical indication effect of green agricultural products is more obvious, which has a great influence on the product quality evaluation and purchasing behavior of agricultural products consumers. The demonstration park will be built into a leading enterprise of green agricultural products in Shandong, which, as a brand effect, improves the popularity of the park and drives the development of planting industry in the park.

③ Other values. The planting industry in the park is the link between breeding industry and agritainment sightseeing. The development of planting industry promotes the publicity of the park, attracts the popularity, and drives the development of breeding industry and agritainment sightseeing.

(3) Agritainment sightseeing market analysis

Phoenix ecological circular agriculture demonstration park plans with "food, housing, travel, play, buy" to drive the development of breeding industry, planting industry and the related industries, and promote the products production and sales of the fruits and grains and others. The special tourism plan with the new development such as breeding produce's processing and production, can bring considerable economic benefits. The park and the surrounding farmers help each other to form the agritainment sightseeing, but also for local farmers to create source and benefits, improve the income of local farmers.

① Agricultural added value. Improve the agricultural added value and drive the development of rural tertiary industry. Rural tourism is rooted in rural areas and closely related to agricultural production. Fruits such as cherries, plums, peaches and agricultural products such as grains in the park directly face consumers, and products can skip circulation links and reach consumers directly, timely solving the difficulties of local agricultural industrialization in the purchase and sales system. Tourism demand also directly increased the demand for farm products in the park, improved agricultural added value, promoted the adjustment of rural industrial structure, and provided a good platform for the development of agricultural industrialization management in the park.

② Other values. Improve the local rural environment and improve the living standards of farmers. Most tourists pay close attention to the sanitary conditions of catering and accommodation in rural tourism destinations, the reception service level and the attitude of residents in tourist destinations, especially the higher requirements for health and safety. This will inevitably promote rural tourism scenic areas at the same time increase the investment in infrastructure, improve the living environment, improve the rural social service system, such as water supply and drainage construction, beautification and cleaning, road improvement, housing renovation, sanitary toilet construction, household garbage disposal and other details of life, so that local residents can objectively enjoy a modern life.

(4) Market analysis

Taking the road of green agricultural industrialization, refinement, brand marketing is the best choice and successful way to promote the take-off and sustainable development of agricultural products industry. It has a strong attraction and driving function to the various links of "production, processing and marketing", and will certainly promote the construction of new countryside and the development pace of building a well-off society in an all-round way. The breeding, planting and sightseeing agritainment in the park is a circular economy industry with a promising prospect and a wide market, with high economic value.

① To base on the local market, and expand the local conventional industry. The site tourism resources are rich, the number of tourists is large, and the market scale is large, which is the basis for the development of breeding, planting and agritainment sightseeing. The road to the market is short, with convenient transportation, low freight cost, and huge demand, which can bring huge economic benefits. In the park grow fruit trees such as pear, peach and plum, and high stalk heather, osmanthus and osmanthus seedling for beautifying, and breed the medium and low-end varieties such as chickens, yellow mealworms and earthworms. The low-end market demand types: chain supermarkets, road greening, breeding bases, food stalls, and agritainment, etc.

② To dig the near-distance market and strengthen the near-distance high-end industry. The site has certain regional advantages, and the surrounding roads are interwoven into a network. The transportation in the nearby market is developed and convenient. Meanwhile, the demand for healthy and green agricultural products is increasing, and the urban people's yearning for rural nature, breeding industry, planting industry and agritainment sightseeing has a broad market potential. The cities in the province are their important markets, and the development of the near high-end market will make the demonstration park industry take off. In the park, fruit trees such as cherry, jujube (dates) and persimmon are planted, as well as seedlings of heather and osmanthus, and high-end varieties such as cattle and snakes are bred. The types of high-end market demand include landscaping, well-known restaurants and star hotels.

③ To vigorously develop Qilu market, and advance to the national market. The provincial market is the extension market of Phoenix eco-circular agriculture demonstration park, and facing the national market is the final goal of the development of eco-agriculture demonstration park. Shandong's convenient transportation is the prerequisite for the development of Shandong and national industries. After being based on the surrounding area, we should jump out of the low consumption area and directly face the key domestic market to carry out pioneering promotion. The park planted cherry, apple and other fruit trees and tall heather, osmanthus green seedlings, breeding rock frog and other boutique varieties, its boutique market demand types: boutique bonsai, theme restaurants, star hotels.

IV. The designing principles

The designing principle of Phoenix Eco-circular Agriculture Demonstration Park is: modern model + farmer mutual assistance + combination of planting and breeding + comprehensive demonstration, according to the overall idea of "moderate scale, standardized production, industrial management, beneficial for the development of circular economy" to plan, design and construct. Build modern agricultural innovation and demonstration bases to carry out breeding, planting new varieties and new technologies, especially the innovation and demonstration of circular agriculture technology. The demonstration park is a comprehensive park that attaches equal importance to economic benefits and social benefits, trying to not only bring economic benefits to enterprises, but also greatly promote the development of local agricultural economy and drive local farmers to get rich together. The demonstration park implements the development policy of "strict protection, unified management, rational development and sustainable utilization", so that the

Chapter 7　Planning & Designing and Case of Modern Ecological Circular Agricultural Parks

ecological park becomes a comprehensive park integrating ecological breeding demonstration, ecological planting demonstration, melon and fruit picking, agricultural tourism combined with leisure vacation, and has high level of economic, ecological and social benefits. At the same time, relying on modern agricultural facilities, modern production process and beautiful ecological environment, it integrates into our long-standing farming culture, provides leisure and experience services for surrounding citizens, and drives the development of ecological agriculture.

1. Ecological priority principle

Eecology is an important premise of Phoenix Eco-circular Agriculture Demonstration Park, which should be combined with the topography, to reflect ecological protection, ecological cultivation, ecological landscape, ecological energy and ecological circulation in the park. The principle of creating harmonious and suitable ecology is the basic principle of natural production and living environment, and the basic foundation of improving the quality of landscape environment in the park.

2. High beneficial principle

From the benefit principle, the planning will consider the best input and the highest comprehensive benefit of the project portfolio. The projects of eco-breeding demonstration, eco-planting demonstration and rural tourism will be organically linked and mutually promoted to create greater economic benefits than the individual management, and obtain ecological and social benefits at the same time.

3. Principle of participation

Direct participation in experience and self-entertainment has become an important aspect of current tourism. The plan will emphasize the close combination of participation, entertainment and knowledge of tourism projects, so that urban tourists can widely participate in all aspects of production and life in the park, get fun and relevant knowledge by doing on themselves, so that they can experience the breeding, planting, picking, processing and harvesting of agricultural products at various levels. The operation of high and new agricultural technology and the taste of rural life, etc. can enable tourists to enjoy the cultural atmosphere originating from the countryside and higher than the countryside.

4. Principle of outstanding characteristics

Characteristics is the life of the development of agricultural demonstration park, the more characteristics it has the stronger its competitiveness and development potential will be, so the planning and designing should be based on the actual situation of the park, be clear aobut the resources, choose the right breakthrough, so that the characteristics of the whole park will become more distinct. In the design, the park should highlight: artificial ecological landscape (especially large area of plant landscape) : ecological technology production (using ecological technology to breed new superior animals&plants, and aquatic products) ; the waterfront landscape (the combination of artificial ecology and natural ecology) makes the park a demonstration park of breeding Rock Frog (Yanwa) in Shandong.

5. Principle of cultural factors

Usually when we talk about agriculture, the first thought is its production function, rarely consider its cultural connotation, as well as its poetic cultural factors. In fact, the development of agriculture is closely linked with the progress of culture, and the planting and appreciating of flowers are influenced by history and culture. Therefore, in the planning and designing of the park, we should dig out the inner cultural resources of the project, and develop them to improve the cultural grade of the park, so as to achieve a high level of utilization of landscape resources.

V. The development direction and strategy

1. Development direction of the Park

The Phoenix Eco-circular Agricultural Demonstration Park has formed a diversified industrial development model of production, processing and tourism by operating projects such as efficient agriculture, facilitied agriculture, three-dimensional agriculture and sightseeing agriculture. Combined with the development model of modern ecological circular agriculture, the park has planned three industrial types: breeding industry, planting industry and rural tourism industry. The characteristic breeding and planting are the key economic projects in the park, while the rural tourism is to let consumers taste the special products in the park on the spot, build up the reputation and enhance the popularity of the demonstration park. The three industries in the park complement each other and grow mutually to form a scientific and ecological circular agriculture.

2. Ecological and circular agriculture

① The completion of ecological circular agriculture project can play a good demonstration role in the radical cure of livestock pollution.

② The completion of the project can be promoted into an industry to attract migrant farmers to return home to start businesses.

③ Become the local demonstration base of scientific circular three-dimensional culture.

3. Positioning of the park

On the basis of circular agriculture + efficient agriculture + facilitied agriculture + three-dimensional agriculture + sightseeing agriculture, "agricultural brand demonstration park" + "agricultural ecological promotion park" + "agricultural leisure and sightseeing park" will be built. The proposed modern ecological agriculture demonstration park will adopt the development model of combining the breeding of rock frog, yellow mealworm and earthworm with the planting of green fruits and fine seedling, so as to realize the circular agriculture of scientific ecology. Base on the synchronous development of industrial cultivation and environmental protection, equal development of economic benefits and ecological benefits. Scientific implement the "biogas ecological project", so that the material and energy can recyle ecologically and virtuously, and become a resource-saving, environmentally friendly circular agriculture demonstration in our province.

(1) Agricultural brand demonstration

Relying on the characteristic breeding of rock frog, yellow mealworm, and supplemented with fruit and vegetable picking, rural tourism, they create green brand agricultural park with leisure, entertainment, and tasting the fresh vegetable and fruits.

(2) Agricultural promotion of ecology

In the park, breeding and planting are scientifically mixed to form three kinds of circular agriculture, which can improve the economic output value and reduce the damage and influence of the agricultural park on the ecological environment.

(3) Agricultural leisure and tourism

The park has a superior natural environment and unique spring water resources. The fruits and vegetables in the park are more sweet under the moisture of the spring. You can also experience Lujia culture, pick fruits and vegetables, and make fruits and vegetables by DIY, which will promote the development of rural tourism and bring popularity to the breeding and sales of the main part of the park.

VI. The key construction projects

1. Ecological recycling breeding industry

(1) Rock frog, yellow mealworm, etc.

Rock frog, because of its delicious meat, fast growth rate and big size, is one of the main edible frogs, Chinese medicine believes that the meat of rock frog tastes sweet, can enter into the lung, stomach and kidney. It has the effect of strengthening the spleen and eliminating accumulation and it is used to treat the symptoms of indigestion, eating less and being weak. Some rock frogs with high nutritional value are bred in the project base, and rock frogs have a high requirement for water, the spring just meets the terms of water to breed rock frogs, and a bait farm of yellow mealworm was built around the rock fog pool, convenient to feed the frogs. At the same time a filtration pool was built beside the feed pool, to create a better breeding environment for rock frogs.

Yellow mealworm: because the yellow mealworm contains higher protein, fat, sugar and other nutrients, with much juice and soft body, strong life, easy to raise. Therefore, it is selected as a good feed by the national zoo and breeding farms all over the country. At the same time, yellow mealworm can also be used as human food, people can use for a long term. It is a good supplement of high protein and other nutritional elements!

Earthworm is also known as Youqing (like centipede), Chinese medicine named Dilong. It is a succulent mollusk with 70% protein content. Earthworms can make use of the waste and residues of the agricultural and sideline products as food, do not compete with other animals for feed, can improve soil, change soil into fertilizer, dispose garbage, purify the urban sourroundings, and improve sanitation. Earthworm is still the best protein feed source for pigs, chickens, fish and various animals and rare aquatic products. Earthworms can be multiplied all the year round under artificial breeding conditions, and its feed is mainly the feces of yellow mealworms. It can be used as protein feed for livestock, poultry, fish and others, can be used to improve the soil for fertility.

As for snake, in the park they mainly breed King snake and cobra. King snake is a kind of snake with delicious meat. Many food lovers in many regions like to eat delicious King snake meat. The King snake is not only eatible, the skin can also be used to make handicrafts and it can have medicinal value. Cobra is a highly venomous snake, which is one of the main edible and medicinal snakes. There are three main purposes of artificial breeding of cobra: one is to collect poison; the second is to earn the seasonal price difference after raising for a period of time; the third is for appreciation. The snake market demand is greater than supply, the price tends to increas year by year.

(2) Cattle grazing in the fields

Cattle farms mainly breed beef cattle and a small part of dairy cows. The sales channel of beef cattle and dairy cows is mainly the market, and the secondary channel is to be processed and eaten in the farmhouse.

Project features: DIY production and processing of inner circulation products + processing knowledge science.

Processing types: DIY handmade milk soap, DIY handmade nougat, DIY milk, DIY yogurt, etc. Next to the cattle farm is a handmade DTY workshop that can be hand-processed into a range of agricultural products and by-products. In addition, visitors can personally participate in the production, especially for children, they can experience process from cognition to learning, learning by lively activities.

2. Ecological circular planting industry

(1) Avenue of fruits, trees and flowers sea

Plants blooming in Spring: cherry, apple, pear, peach. Cherry, apple, pear, peach blossom from March to April, fruit period from June to October. Cherry, apple, pear, peach flowers are colorful and fragrant, important flower trees for the early spring. The blooming flowers are gorgeous, brilliant, full of trees, like rosy clouds, very spectacular. They can be planted in large areas to create a "sea of flowers" landscape, can also be clusters dotted in the green land to form a brocade group, can also be planted alone to form the poetic painting of "a single red in the midst of green foliage". Cherry, apple, pear, peach trees can also be used as road trees or make bonsai.

Plants blooming in Summer: crape myrtle. Crape myrtles are flowering from June to September, fruiting in November. Crape myrtle tree are graceful, with smooth and clean trunk, beautiful color. They are flowering when there are less flowers in the summer and autumn, with long flowering period, so it is named "hundred days of red", complimented as "summer green covers eyes, this red is full of the house". It is the good bonsai wood for appreciating flowers, trunk and root, and its root, bark, leaves, flowers can be used as medicine.

Plants for appreciating leaves in Autumn: red maple and so on. Red maple is purplish red when young, yellowish brown when mature, the fruit core is spherical. The fruit ripening period is October. Red maple is a very beautiful leaf tree species, its leaf shape being beautiful, its red color being bright & lasting, its branches being orderly and hierarchical, its appearance being beautiful, and ornamental value being very high.

Plants blooming in winter: winter sweet (Chimonathus Praecox, Calyx canthus). It blooms from November to March the next year, the fruiting period is from July to August. Winter sweet blossoms proudly open in the cold weather of frost and snow. Its yellow flowers with strong fragrance are the main ornamental flowers and trees in winter. It is not only the ornamental flowers and trees, and its flowers contains many fragrant and aromatic substance, such as linalool, borneol, cineole, alpha-pinene and sesquiterpene, being one of the best tea flowers. The high-grade spices refined from it are expensive, 1000 grams of the spice being eaqual to 5000 grams of gold in the international market price.

(2) Ecological orchard

Plum orchard: plum blooms in March and April, and the fruits are ripe from July to August. They are full and round, exquisite, beautiful in shape and sweet in taste, being one of the favorite fruits. In spring, visitors can enjoy plum flowers, which have a beauty of serenity. In the summer, you can taste the plum fruits, a green fruit hanging in the tree, waiting for you to pick them, taste them, full of sweet fragrance.

Pear orchard: Pear blooms in April, fruits collected from July to September. Cuiguan pear is a kind of sand pear, the fruit is nearly round, Cuiguan fruit shape index is 0.96, yellow green. The pulp is Snow White and tender, crisp & juicy, easily melted in mouth with little residues, with few stone cells, and the taste is thick & sweet. The pulp of Golden pear is tender and juicy, sugar content up to 14.7%. It tastes sweet and smells aromatic. Because of it's fresh tender and juicy, sweet and sour palatability, it is also known as "natural mineral water".

Pomegranate orchard: pomegranate blooms in May to June, fruiting in early to mid-September. Pomegranate flowers are big and gorgeous with long flowering period, from the wheat harvest time to October. And the pomegranate fruit is gorgeous. Because its flowers can be appreciated, and its fruits are edible, they are loved by people, and the pomegranate bonsai is favored. The Chinese regard pomegranate as a symbol of good fortune and many children.

Peach orchard: peach blooms from March to April, fruits from August to September. Peach is known as "longevity peach" and "immortal peach", with the folk symbolic significance of fertility, auspiciousness, and longevity, because of its delicious pulp, is also known as "the world's first fruit".

Cherry orchard: varieties are mainly black pearl cherry, purple cherry. The cherries bloom in the middle and late of February, and fruit in the middle and late April. The appearance of the fruit is bright, crystal, beautiful, red like agate, yellow like fair skin, which is rich in sugar, protein, vitamins and calcium, iron, phosphorus, potassium and other elements.

Jujube orchard: varieties are mainly Zhuyao dates (pig kidney), Luojiang Guifei dates (senior concubine). Jujube blooms in May, and fruits in early and mid August. Dates taste sweet, which contains rich vitamin C, Since ancient times, there has been a saying that "three dates a day, a hundred years old you cannot say".

Persimmon orchard: varieties are mainly Luotian sweet persimmon, chocolate dark persimmon. Flowering in May to June, fruiting in September to October, sweet persimmon is the world's only natural sweet persimmon varieties, mature in autumn, do not need processing, can be eaten directly. Its characteristic is big, gorgeous, sweet, crisp and delicious, with round body and square bottom, thin skin and thick pulp.

(3) Planting and breeding in the forest

Planting in the forest has obvious advantage, for the forest can well control the time and intensity of sunlight, reduce the strong light to harm the flowers and fruits, improve their quality, reduce aging, extend the harvest period, which have the role of natural preservation. Raising chickens after fruits are harvested can prevent from damaging flowers and fruits. While increasing the added value of fruit trees and chickens, a virtuous cycle of mutual promotion and mutual benefit between chickens and fruit groves was also established.

Chapter 7　Planning & Designing and Case of Modern Ecological Circular Agricultural Parks

(4) Varieties of fruits, vegetables and flowers in the corridor

Loofah, white flowered gourd, fruit of fiverleaf Akebia, grape, kiwi, strawberry, wisteria, honeysuckle, etc. The melon and fruit corridor is a corridor full of vegetables, fruits and flowers, trying to create a green, agreeable and interesting corridor, so that tourists can have a good time in the ecological corridor, appreciate some strange fruits and flowers, and pick fruits and fruits. After picking, they can be processed and eaten in the farmhouse, feeling of being surrounded by green nature.

(5) Combination of agriculture and tourism

Farmhouse experience: project characteristics, fresh, natural, simple, cultural experience. The project site cooperated with the surrounding farmers to transform the farmers' houses for agritainment. Visitors can take the children to be here at the weekend to breathe fresh air, taste the farm-style vegetables. Visitors can not only meet all kinds of vegetables in the field, but also experience the fun of picking fruits and vegetables. They can also make soybean milk, feel the farming culture, experience the natural and simple country life, and let the children enjoy their childhood in nature!

Catering and accommodation: project features, leisure, vacation, ecology. When visitors come here, they can temporarily bid goodbye to the congested roads, the noisy urban life, and enjoy the rustic and quiet countryside. Here the fresh air, the lush grasses and trees, the birdsongs, and the green grassland make people integrate into nature and return to the nature at any moment.

The recreation program has four projects: mountaineering trail, magic tree house, recreational fishing and fun stream.

Mountaineering trail. The feature is recreational fitness and jungle adventure. In the area of the Wizard of Oz, a hiking trail is planned, where visitors can enjoy hiking through the forest, listening to birds, watching plants, and enjoying the fun of hiking in the wilderness. The hiking trail also uses wood close to nature.

Magic tree House. The features is magic fun, overlooking the garden scenery, and drinking tea and chatting. In the area of the Wizard of Oz, several magical tree houses were designed along the trail to add a touch of magic to the garden. Visitors can climb into the magical tree houses to get a view of the garden, have a rest, or chat and drink tea.

Leisure fishing. The features is leisure fishing and lotus appreciation. In this area a fishing area is planned to breed grass carp, silver carp and other fish, and in the pond lotuses are planted. Tourists can go fishing and enjoy the lotus at the same time. Lotus flowers are the symbol and messenger of friendship. In ancient China, people have the tradition of folding plum flowers in spring to see off friends and picking up lotus seed to memorizing the beloved in autumn. Guests can bring their own fishing gear or hire fishing gear to catch fish, which belongs to the customers, and can be processed and eaten at the farmhouse or taken away by themselves. Some fishing competitions can be held on weekdays.

Fun stream. The project features is splashing in the water to rest, catching crabs and tadpoles. There is a clear stream running down from the mountain spring. Visitors can play in the stream, play with the water, catch crabs, catch tadpoles and so on. Some photography enthusiasts can also take pictures here.

Bibliography

[1] 2011 – 2015 China agroecological market supply and demand forecast and investment prospect assessment report[R/OL].[2022 – 05 – 01] https://www.docin.com/P-1583929523.html.

[2] CHANG W T, YUAN M, YAN P. Agricultural residue resource utilization technology demonstration and emission reduction effect analysis[M]. Tianjin: Tianjin University Press, 2017.

[3] DENG Y L. Achievements and development of agricultural ecological construction in China[J]. Agricultural environmental protection, 1997 (1) : 32 – 34.

[4] HANG X H, WANG W, ZHANG L, et al. Effect of red clay on the removal of phosphorus from simulated and poultry wastewater [J]. Chinese journal of environmental science, 2012, 32 (6) : 1399 – 1405.

[5] LI C. Agricultural ecology[M]. Beijing: Chemical Industry Press, 2009.

[6] LI W J, LAN M, PENG X J. Removal and control of antibiotics from livestock wastewater by UV/H_2O_2 combined oxidation method [J].Environmental pollution control 2011, 33 (4) : 25 – 28,32.

[7] LIU S H. Green low-carbon circular agriculture[M]. Beijing: China Environment Press, 2016.

[8] LIU T. The connotation and industrial scale of agroecology[J]. Research of agricultural modernization, 2002 (23) : 38 – 40.

[9] LIU W P. Introduction to resource circulation[M]. Beijing: Chemical Industry Press, 2016.

[10] LV W G. Study on the typical model of ecological agriculture in Shanghai[M]. Shanghai: Shanghai Science and Technology Press, 2018.

[11] OUYANG C, SHANG X, WANG X Z, et al. Study on the removal of ammonia nitrogen from pig wastewater by electrochemical oxidation [J]. Water treatment, 2010, 36 (6) : 111 – 115.

[12] YU H P, QIAN Y H, HU T X. Study on the removal of ammonia – nitrogen from aquaculture wastewater by type I attapulgite adsorbent [J]. Guangzhou chemical industry, 2009, 37 (7) : 140 – 141,161.